THE CONNECTIVITY MANAGEMENT HANDBOOK™

A guide to using Cable Management Systems for cable plant and physical network management.

by Walter A. Chiquoine and Elizabeth R. Hist

Transport Management Group, Inc.

Library of Congress Catalog Card Number: 95-61022

ISBN: 0-9647162-0-8

Printed in the United States of America.

Cover design by Marsha Sholl.

IMPORTANT NOTICE:

This book is designed to provide tutorial information and methodology for the physical management of premises communications wiring and equipment. This book is published with the understanding that Transport Management Group, Inc. and its authors are supplying information but are not attempting to render engineering or other professional services. If such services are required, the assistance of an appropriate professional should be sought.

Every effort has been made to make this book as complete and accurate as possible. It is offered on an "as-is" basis, with no warranty or fitness being expressed or implied. Neither Transport Management Group, Inc. nor the authors shall be liable to anyone for special, collateral, incidental, or consequential damages in connection with or arising out of the purchase or use of this book, nor shall be liable for any claim of any kind whatsoever against the user of this book material by any other party.

To our families who have patiently given us love and support, to our friends and associates who have given us guidance and encouragement, to the many pioneers in the field of cable management who have led the way, and to the memory of Carl R. Fyrwald who helped us to believe in ourselves.

ACKNOWLEDGMENTS

EDITORIAL REVIEW BOARD

The authors wish to express their deepest appreciation and gratitude to the members of the Editorial Review Board, who together spent many hours reading, editing, and commenting on the various topics presented in the book. They have provided many different viewpoints from within our industry, and have helped us achieve a thorough and detailed treatment of connectivity management.

Edward C. Barrett
President and CEO
Hi-TECH Connections, Inc.
Provider of *Enterprise Network Systems and Management Services*

Si H. Go
Executive Vice President
Computerized Documentation Services Inc.
Computerized Communications Asset Management (CCAM) CMS Software

Peter Hale
Customer Support Services Supervisor
Accugraph Corporation
MT923 Software

Richard J. Hartland
Manager, Customer Service and Network Support
Zeneca Inc. (USA)

Ray Horak
President
The CONTEXT Corporation
Telecommunications Consulting

Lynne Marie Jackson
Assistant Analyst for Networking Management
Datapro Information Services Group
Analyst for *Cable Management Systems and Software*

James C. Myers
Director, North American Operations
Total Wire Company, Inc.
UltiCAM for Windows CMS Software

Stephen C. Paulov
Publisher and Editor
Cabling Business Magazine

Steven Press
Senior Vice President
Computerized Documentation Services Inc.
Computerized Communications Asset Management (CCAM) CMS Software

Harold E. Sloss
Director, Product Sales
Telco Research Corporation
Tru-Server and Tru-System Telemanagement Software

Eric Teicholz
President
Graphic Systems, Inc.
Author, *Computer-Aided Facility Management* and *CAD/CAM Handbook*

ABOUT THE AUTHORS

Walter A. Chiquoine is a pioneer in the field of cable and connectivity management. He has guided major corporations in developing and implementing effective recordkeeping systems for cable plants in a wide range of commercial and industrial facilities. Chiquoine is founder and president of Transport Management Group, Inc. which provides consulting and services for connectivity recordkeeping. He is an instructor in cabling and wiring, a Contributing Editor to Cabling Business Magazine, and is a frequent speaker at communications industry seminars. His articles on cable and connectivity management appear regularly in leading industry journals. Chiquoine is a member of IEEE and BICSI.

Elizabeth R. Hist, a consultant and project manager for Transport Management Group, Inc., has directed the evaluation, selection, and implementation of cable management systems for companies in the industrial, financial, utility, and commercial fields. She has developed and tested methods for all phases of CMS selection and implementation and is experienced in CMS evaluations, catalog development, physical inventory and data loading. She provides TMG clients with ongoing CMS maintenance when required. Hist has held corporate administrative positions in the fields of wire and cable marketing and business unit relocation.

TABLE OF CONTENTS

Editorial Review Board v

About the Authors vi

Table of Contents vii

List of Tables xiii

List of Figures xviii

Introduction xxi

Part 1: Cable and Connectivity Management - A Tutorial

Chapter 1: The Connectivity Management Process 1
Implementing Connectivity Management

Chapter 2: Cable Plant Architecture 11
The TIA/EIA Standards
Types of Premises Cabling
A Hypothetical Site

Chapter 3: Basic Naming, Labeling, and Recordkeeping 21
Recordkeeping Tools
Key Aspects of Recordkeeping
The Uses of Recordkeeping
Access to Records
Recordkeeping for Quantity and Characteristics
Recordkeeping for Availability and Usage
Recordkeeping for Connections
Naming Conventions
Methods of Labeling

Chapter 4: The TIA/EIA 606 Administration Standard 39
The General Framework of TIA/EIA 606
Required Information and Linkages
Other Aspects of TIA/EIA 606

TABLE OF CONTENTS

Part 1: Cable and Connectivity Management - A Tutorial (Cont'd)

Chapter 5: Benefits, Value, and Costs 53
Benefits
Barriers to Analysis
Estimating Actual Value
Costs of a CMS

Chapter 6: CMS Functionality 69
The Relational Database
Computer-Aided Drafting
Common CMS Functions
Specific CMS Structures

Chapter 7: CMS Administration 81
Putting the CMS to Use
Maintaining the CMS Applications

Chapter 8: Interactions with Other Management Systems 87
PBX Administration
Call Accounting/Directory
Billing (General Ledger)
Network Management
Facility Management

Chapter 9: The Service Order 93
Service Order Functionality
Using a Service Order
Interfacing with an Existing Service Order System

Chapter 10: Trouble Tickets and Problem Management 97
Trouble Ticket Functionality
Using a Trouble Ticket
Interfacing with Existing Problem Management Systems

Chapter 11: Asset Management 101
General Considerations
Asset Information

Part 2: Planning For and Selecting a Cable Management System

Chapter 12: The Project Team 105
The Team Approach
Roles and Responsibilities
Milestones

Part 2: Planning For and Selecting a Cable Management System (Cont'd)

Chapter 13: Defining Project Scope 109
Evaluating the Physical Plant
Evaluating the Cable Plant
Project Scope Document

Chapter 14: Developing a Records Structure 117
Identifying Required Information
Considering the Records Structure
Data Structure and CMS

Chapter 15: CMS Access and Platform 127
Determining Access Requirements
Platform Options

Chapter 16: Using Existing Records 131
Limitations
Evaluating and Using Existing Records

Chapter 17: CAD/Graphics 135
Special Benefits
Determining the Need for CAD/Graphics
Maintaining CAD Records

Chapter 18: Reports 141

Chapter 19: Defining New Workflows 149
Considering Current Workflows
Creating New Workflows
Applying New Workflows

Chapter 20: Planning and Authorization 157
The Authorization Request
Estimating Anticipated Benefits
Project Plan and Timeline
Estimated Costs
Summary of the Authorization Request

Chapter 21: Beginning Your Catalog 169
Creating Support Tables
Desktop User Devices
Network Control Equipment
Cabling and Wiring
Termination Hardware
Pathways

TABLE OF CONTENTS

Part 2: Planning For and Selecting a Cable Management System (Cont'd)

Chapter 22: Preparing a Request for Proposal (RFP) 209
The Value of an RFP
Preparing the RFP
Sample RFP

Chapter 23: Evaluating RFP Responses 261
Preparing an Evaluation Matrix
Vendor Presentations and Demonstrations
CMS/Vendor References
Determining the Final Candidates

Chapter 24: Software Customization 279
General Considerations
Specific Requirements

Chapter 25: Final System Selection 283

Chapter 26: The Purchase Order and Licensing Agreement 285
Preparing Purchase Orders
The Licensing Agreement

Part 3: Implementing a Cable Management System

Chapter 27: Installation and Testing 289

Chapter 28: Setting Up the Catalog and Database 295
Validation Tables and Support Information
Cable and Equipment Types
Termination Hardware
Pathways
Drawing Files
Drawing Symbols/Icons

Chapter 29: Testing the Catalog and Database 301
Creating the Components
Connecting the Components
Creating the Circuits
Tracing the Circuits
Testing Rules and Restrictions
Finishing the Support Tables

Chapter 30: Testing Service Order, Trouble Ticket, and Reports 315
Testing the Service Order
Testing the Trouble Ticket
Testing the Reports

Part 3: Implementing a Cable Management System (Cont'd)

Chapter 31: Working with CAD Drawings 319
Preparing and Importing Drawings
Placing and Moving Icons
Modifying or Changing Plan Drawings
Printing and Plotting the Drawings

Chapter 32: Working with a SQL Report Generator 323
Creating Custom Queries and Reports
Making Global Record Changes

Chapter 33: Documentation 325
Reference Tables and Catalog Diagrams
Procedures

Chapter 34: Training 331
CMS Administrator and Project Manager
Customer Service Center Users
Help Desk Personnel
Voice Technicians
Data Technicians

Part 4: Physical Inventory

Chapter 35: Planning and Scheduling 335
Defining Tasks and Required Information
Estimating the Time to Complete Tasks
Developing a Schedule

Chapter 36: Developing Inventory Procedures and Forms 341
Inventory Procedures for Cables and Equipment
Inventory Procedures for Connections
Inventory Procedures for Office Areas
Developing Inventory Forms

Chapter 37: Labeling 361
Labeling in the Wiring Closets
Labeling Office Outlets
Labeling Desktop Equipment
Selecting the Labels to Use

Chapter 38: Working with Existing Data 367
Organizing the Data
Transferring the Data
Verifying the Data

TABLE OF CONTENTS

Part 4: Physical Inventory (Cont'd)

Chapter 39: Training 371
The Inventory Process
Filling Out the Inventory Forms
Auditing the Inventory Forms

Chapter 40: Data Entry and Auditing 373
The Data Entry Process
Handling Error/Warning Messages
Identifying Rechecks
Auditing the CMS Records

Glossary 377

Index 383

LIST OF TABLES

1.1	The elements of the cable and connectivity management process.	2
1.2	Initial Definition of Requirements list.	6
2.1	Significant industry standards relating to cable and cable management.	11
2.2	Elements of the TIA/EIA cable architecture.	13
2.3	General types of premises cabling.	16
2.4	Significant Canadian standards relating to cable and cable management.	20
3.1	List of possible cable and connectivity recordkeeping tools.	21
3.2	List of jacks in numerical sequence.	23
3.3	List of jacks by office number.	23
3.4	Complete list of jacks with cable pair connections noted.	24
3.5	Complete list of backbone cable pairs with jack connections noted.	24
3.6	"Master" list based on jacks, with cable pair connections shown.	25
3.7	List showing connected jacks and cable pairs only.	25
3.8	Factors to consider for a recordkeeping plan.	26
3.9	Elements of the communications infrastructure or transport system.	28
3.10	Basic tables to track all elements at the circuit level, with usage information maintained.	32
3.11	Characteristics of good naming conventions.	32
3.12	Comparison of physical and logical names.	35
4.1	Summary of record elements in the TIA/EIA 606 administration standard.	40
4.2	Pathway record for a conduit showing required information.	41
4.3	List of some possible space types.	42
4.4	Space record for a Telecommunications Closet showing required information.	42
4.5	Cable record for a copper backbone cable showing required information.	43
4.6	Termination hardware record for an optical fiber patch panel showing required information.	43
4.7	Termination position record for a punch down block showing required information.	44
4.8	Splice records for a splice in separately-administered cables showing required information.	45
4.9	Cable record for a separately-administered spliced cable showing required information.	46
4.10	TMGB record showing required information.	46
4.11	Bonding conductor record showing required information.	47
4.12	TGB record showing required information.	47
4.13	Specific color codes for termination positions.	48

LIST OF TABLES

5.1	Benefits of a good connectivity management plan.	53
5.2	Determining factors for CMS benefits.	54
5.3	Barriers to estimating the value of CMS benefits.	58
5.4	Possible hardware costs for a CMS implementation.	63
5.5	Possible software costs for a CMS implementation.	63
6.1	Major groups for drawing layering.	76
6.2	List of common CMS functionality.	77
7.1	System administration and maintenance tasks for ongoing CMS support.	86
8.1	Examples of other management systems or interfaces.	87
8.2	Correspondence of recordkeeping functions between Computer-Aided Facility Management (CAFM) and Cable Management System (CMS).	92
11.1	List of potential fields in a device record for tracking asset information.	103
12.1	Elements of the team approach.	105
12.2	CMS team member responsibilities.	106
12.3	CMS team Project Leader responsibilities.	107
13.1	Initial data for CMS planning.	109
13.2	List of potential sites for CMS planning.	110
13.3	List of potential buildings for CMS planning.	110
13.4	List of buildings, floors, and outlets for CMS planning.	111
13.5	Primary Planning Table.	111
13.6	Identified backbone cables for CMS planning.	112
13.7	Identified pathways for CMS planning.	114
13.8	Identified horizontal wiring for CMS planning.	115
14.1	Proposed list of data elements that describe cables.	118
14.2	Proposed list of data elements that describe circuits.	119
14.3	Example of a separate horizontal wiring record.	120
14.4	Example of a separate backbone cable record.	120
14.5	Proposed list of data elements that describe pathways.	121
14.6	Proposed list of data elements that describe termination hardware and position.	122
14.7	Proposed list of data elements that describe information outlets or jacks.	122
14.8	Proposed list of data elements that describe controlling equipment.	123
14.9	Proposed list of data elements that describe desktop equipment.	124
14.10	Sample list of final datasets (tables) for proposed CMS recordkeeping.	125
15.1	Typical CMS access levels for various functions.	128
15.2	Sample evaluation of platform elements for CMS planning.	130
16.1	Summary and evaluation of existing recordkeeping systems.	132
16.2	Example of table information for existing asset management records.	133
17.1	Considerations for using CAD records with a CMS.	136
17.2	Some additional requirements for CAD maintenance.	139
18.1	List of minimum recommended report formats for a CMS.	141

18.2	Recommended reports for TIA/EIA 606 recordkeeping.	143
18.3	Sample cable report, listing cables by name.	144
18.4	Sample equipment report, listing devices by category.	144
18.5	Sample (partial) cable connection report by cable pair number.	145
18.6	Sample drawing report by drawing name.	146
18.7	Sample Service Order report by open Service Order number.	146
18.8	Sample Trouble Ticket report by open Ticket number.	146
20.1	Documentation to consider for an Authorization Request.	157
20.2	Benefits of a good connectivity management plan.	159
20.3	Sample project implementation schedule.	163
20.4	Cost ranges for outside project services.	165
21.1	List of supporting tables for a CMS.	169
21.2	Building, floor, and room designation codes for the LRP site.	170
21.3	Initial list of basic desktop equipment and categories.	172
21.4	Detailed list of desktop equipment and categories.	173
21.5	Possible connector configurations for a telephone.	174
21.6	Possible connector configurations for a terminal.	174
21.7	Possible connector configurations for a facsimile machine.	175
21.8	Possible connector configurations for an answering machine.	176
21.9	Possible connector configurations for a printer.	176
21.10	Possible connector configurations for a PC.	177
21.11	Possible connector configurations for a modem.	178
21.12	Sample naming conventions for desktop equipment.	179
21.13	General list of network equipment and categories.	180
21.14	Detailed list of network equipment and categories.	181
21.15	Sample naming conventions for network control equipment.	187
21.16	Sample backbone cable types.	188
21.17	Sample backbone cable types with pair counts and termination hardware.	189
21.18	Sample cable types and configurations.	189
21.19	Sample backbone cables with type and CMS naming.	190
21.20	Sample types of horizontal cables.	191
21.21	Possible horizontal cable connections.	192
21.22	Examples of termination blocks.	201
21.23	Basic information regarding pathways.	207
22.1	Benefits of using a Request For Proposal (RFP).	209
22.2	General sections of the RFP.	210
24.1	Possible areas for CMS customization.	279
24.2	Sample customization request for change in field name.	281
24.3	Sample list of corresponding field names for batch file transfer.	282
25.1	Evaluation summary for final system selection.	283
26.1	Sample statement on placing a copy of CMS software source code in escrow.	286
26.2	Sample set of specifications for inclusion in purchase order.	287
27.1	Elements of initial installation and testing.	290
27.2	Elements of functionality testing.	291

LIST OF TABLES

28.1	Elements of setting up the database.	296
29.1	List of catalog components to test.	301
29.2	Elements of testing the catalog.	302
29.3	Sample of a basic connection record for connecting a horizontal cable to an information outlet.	304
29.4	Pin configurations for 568A and 568B outlets.	305
29.5a	Sample horizontal cable to termination hardware connection record.	306
29.5b	Sample backbone cable to termination hardware connection record.	306
29.6	Sample connection record for a single patch panel connector to the TA ports on the controller.	306
29.7	Sample connection record with 32 patch panel connections to a 3270 controller.	307
29.8	Sample connection record for a card to a network hub.	307
29.9	Sample connection record for a PC to an office outlet.	308
29.10	Sample connection record for a phone to two horizontal cable pairs.	308
29.11	Connection record for horizontal cable pair directly to controlling equipment port.	309
29.12	Connection record for horizontal cable pair to a port on a patch panel.	309
29.13	Sample connection record for a horizontal cable pair block to an Ethernet port block.	310
29.14	Connection record for horizontal cable pair to riser cable pair.	310
29.15	Sample connection record for a horizontal cable station block to riser cable block.	311
30.1	Service order items to be tested.	315
33.1	General list of user-specific CMS documentation.	326
33.2	List showing documentation of possible nomenclature for database components.	329
35.1	Sample list of activities to be included in physical inventory.	336
35.2	Example list of information to be recorded for backbone cables.	337
35.3	Example list of information to be recorded for desktop equipment.	337
35.4	Sample schedule for physical inventory.	339
36.1	Summary of the steps in a complete physical inventory.	342
36.2	Sample procedures for inventory of backbone cables.	343
36.3	Sample procedures for inventory of control equipment.	344
36.4	Sample procedures for inventory of horizontal cables in the Telecommunications Closet.	345
36.5a	Sample procedures for inventory of wiring closet connections - control equipment.	347
36.5b	Sample procedures for inventory of wiring closet connections - backbone cables.	348
36.5c	Sample procedures for inventory of wiring closet connections - horizontal cables and audit.	349
36.6	Sample procedures for inventory of office equipment and connections.	350
36.7	Sample fields from the CMS cable screen.	351
36.8	Sample inventory form for backbone cables.	352
36.9	Sample inventory form for backbone connections with termination hardware tracked separately.	353
36.10	Sample inventory form for backbone cable connections where termination hardware is an attribute of the cable.	354
36.11	Sample inventory form for non-expandable control equipment.	355
36.12	Sample inventory form for expandable control equipment.	356

36.13	Sample inventory form for cards or modules in expandable control devices.	357
36.14	Sample inventory form for recording horizontal cable names by wiring closet.	358
36.15	Sample inventory form for office inventory, arranged by outlet number.	359
36.16	Sample inventory form for office inventory, arranged by office number.	360
37.1	Elements for selecting and using equipment and cable labels.	361
37.2	TIA/EIA 606 Color Codes for termination positions.	363
37.3	Types of labels and label-making devices.	365
38.1	Initial steps required to prepare existing records for transfer to the CMS database.	367
38.2	Individual steps involved in transferring cross connect records to the CMS database.	368
38.3	Elements to consider when auditing the CMS records.	369
39.1	Elements of training for physical inventory.	371
40.1	Elements of training for data entry personnel.	373
40.2	Sample instructions for the order in which inventory records are entered into CMS database.	374
40.3	Sample reports to be used in auditing CMS records.	375
40.4	Elements to consider when auditing the CMS records.	376

LIST OF FIGURES

1.1	The OSI model of network functions, indicating physical media (cabling) as the bottom layer.	1
1.2	Increasing sophistication of a connectivity management plan.	3
1.3	Elements of a circuit description.	4
1.4	Flow chart of the first phase of CMS implementation, from planning for to selecting a CMS.	5
1.5	Flow chart of the second phase of CMS implementation, from system selection to final system setup.	6
1.6	Flow chart of the third phase of CMS implementation, from final system setup to full operation.	7
1.7	Textual representation of a circuit.	8
1.8	Schematic representation of a circuit.	9
1.9	CAD (graphic) representation of a circuit.	10
2.1	Different topologies or layouts of the cable architecture.	12
2.2	TIA/EIA 568A cable plant architecture.	14
2.3	Ground bus system for commercial buildings.	15
2.4	Site plan with first and second level backbone cables.	17
2.5	Site plan with cable pathways.	18
2.6	Site plan with telecommunications grounding system.	19
3.1	Hierarchy of chassis device, cards or modules, and actual ports.	29
3.2	Example of a structured naming system for horizontal cables.	33
3.3a	Circuit and element directionality from the MC to the workstation.	34
3.3b	Circuit and element directionality from the workstation to the MC.	34
4.1	Diagram of singly- and separately-administered spliced cables.	44
4.2	Example of administration-point label color coding.	49
4.3	Telecommunications infrastructure items for TIA/EIA 606 administration-point labeling.	51
4.4	Telecommunications infrastructure items for TIA/EIA 606 administration-point labeling, with labels shown.	52
5.1	Savings formula for Moves and Changes.	59
5.2	Savings formula for trouble calls.	60
5.3	Savings formula for unused active connections.	61
5.4	Savings formula for frame inventory.	62
5.5	Method for estimating inventory costs.	64

5.6	Method for estimating data entry costs.	65
5.7	Method for estimating project management costs.	66
5.8	Summary statement of project implementation costs.	68
6.1	Representation of a database table, showing a record and a field.	70
6.2	Descriptions of the basic building blocks of a database.	71
6.3	Example of connection records tracked in a separate table.	72
6.4	Example of connection records tracked in an item table.	73
6.5	CAD architectural plan drawing.	74
6.6	Schematic representation of a circuit.	75
6.7	CAD drawing of an equipment rack showing patch panels and network hubs.	76
6.8	Sample approaches to device configurations and connection-point naming.	79
7.1	Flow chart of administrative procedures based on key individuals in different groups having a direct interface with the CMS.	83
7.2	Flow chart of administrative procedures based on centralized CMS access through a Service Order.	84
9.1	Sample screen showing administrative information in a Service Order.	94
10.1	Sample screen showing administrative information in a Trouble Ticket.	98
13.1	Sample Project Scope document.	116
17.1	CAD-based floor plan drawing.	136
17.2	CAD-based site plan drawing.	137
17.3	CAD-based site plan drawing showing pathway information.	138
18.1	Graphic report showing all LAN printers on a specific floor.	147
18.2	Graphic report showing the circuit path for troubleshooting a problem on PC512.	148
19.1	Sample workflow diagram for Voice Moves and Changes.	153
19.2	Sample workflow diagram for Data Moves and Changes.	154
19.3	Sample workflow diagram for major Moves and Changes.	155
19.4	Sample workflow diagram of the Trouble Ticket process.	156
20.1	Estimated savings for Moves and Changes.	159
20.2	Estimated savings for trouble calls.	160
20.3	Estimated savings for recovered active but unused circuits.	160
20.4	Estimated savings for avoidance of recurring cable and outlet surveys.	161
20.5	Summary statement of anticipated CMS implementation savings.	162
20.6	Estimate of timing for physical inventory tasks.	164
20.7	Summary of anticipated physical inventory costs.	166
20.8	Summary statement of project implementation costs.	168
21.1	Diagram showing possible connections to a facsimile machine.	175
21.2	Configuration for a PC with an internal modem.	177
21.3	Configuration with connectors on front and back.	178
21.4	Ports located directly on the front of the controller.	181
21.5	Controller ports connected to 3299 device or TMA controller cards.	182
21.6	Controller ports connected to patch panel.	183

LIST OF FIGURES

21.7	Connections between MAUs, PCs and Servers.	184
21.8	LAN hub with expandable chassis.	184
21.9	LAN hub with different cards and connector types.	185
21.10	Sample configurations for network servers.	186
21.11	Horizontal cable connections tracked on a pair-by-pair basis.	193
21.12	Horizontal cable connections tracked on a single connector basis.	193
21.13	Horizontal cable connections tracked on a 3-pair and 1-pair connector basis.	194
21.14	Horizontal cable connections tracked on a two-connector basis.	194
21.15	Office outlets tracked as separate entities.	196
21.16	Office outlets tracked as attributes of the horizontal cable.	197
21.17	Individual office outlets set up for each connector.	198
21.18	Office outlets set up with multiple connectors.	199
21.19	Termination blocks tracked as separate entities.	200
21.20	Termination blocks tracked as attributes of the cable.	201
21.21	Patch panel tracked as a separate entity.	203
21.22	Patch panel tracked as an attribute of the port.	204
21.23	Patch panel with 32 ports for Controller LG1D11.	205
29.1a	Hybrid 8-pair horizontal cable consisting of four Category 3 pairs and four Category 5 pairs.	304
29.1b	Example of hybrid horizontal cable consisting of one Category 3 connector and one Category 5 connector.	305
29.2	Circuit components with termination hardware tracked as separate entities.	312
29.3	Circuit components with termination hardware as attributes of the cable and control equipment.	313
33.1	Key cable configuration diagram.	327
33.2	Key device configuration diagram.	328
33.3	Possible configuration diagram for an expandable Ethernet hub.	329
33.4	Diagram showing possible nomenclature for database components.	330
35.1	Flow chart of the third phase of CMS implementation, including physical inventory activities.	335
36.1	Diagram of the three types of connections to record during frame inventory.	346
37.1	Recommended labeling for a patch panel where network ports are presented.	362

INTRODUCTION

Outside cable plant has been tracked and managed to some degree for many years by local telephone companies, but the recordkeeping of premises cabling by individual user organizations did not begin in earnest until the mid 1980's. This followed the breakup of AT&T and the divestiture of the local Bell operating companies, when the Federal Communications Commission (FCC) began the process of deregulating inside wiring. For the first time, user organizations now had responsibility for installing, maintaining, and troubleshooting much or all of their cable plant. Many users were unprepared, with little experience and even less cable documentation to address the challenges that arose. Add to that the incredible growth in LAN cabling over the last decade, and it becomes easy to see how many organizations have struggled to bring their wiring and connections under control.

Since then, there has been much discussion and many attempts to improve the management of premises wiring, particularly through the development of wiring standards and the use of computerized recordkeeping tools to track the location and availability of cable. But while today's communications hardware continues to get more intelligent and self-aware, cable still remains a dumb and passive component of our networks, requiring a very labor-intensive approach to make or change all the physical connections necessary to create individual circuits. Wiring standards and recordkeeping tools are helpful, but they are only part of the solution. The user must recognize that cable and connectivity management is a *process*, one that includes individual and organizational behaviors, practices, and procedures as well.

Our view of connectivity management in this book is that of the process - directing and controlling the usage of manpower, cable capacity, and equipment connections in the most efficient manner. Such a manner is usually unique to each group's environment and budget, so there are many approaches to consider. What works best in one environment may be inappropriate in another. Generally speaking, a good connectivity management plan must include the following five elements:

- o Well-planned Cable Plant Architecture
- o Naming/Labeling of Cables and Devices
- o Recordkeeping
- o Control of Access
- o Administrative Practices and Workflows

Disregard for any of these elements will likely compromise the effectiveness of your cable management implementation, even increasing your work load without providing significant benefit.

When we speak of cable management today, most people think of a software-based recordkeeping system commonly called a Cable Management System (CMS). Such a system could be a spreadsheet, a database, a CAD (Computer-Aided Design) system, or some combination of database and CAD. There are many CMS packages available today with a broad variety of functions, features, and platforms. Many telemanagement and LAN management systems include some features of a CMS. How do you select the right one for your organization? How do you implement it so it becomes an effective tool? How do you integrate it alongside of other management tools and procedures, or across organizational boundaries? Our intent of this volume is to help you understand the context of cable and connectivity management in your organization - to help you design the processes that will fit in and work for you.

Many users have approached cable management with two false impressions which have led to ineffective or abandoned efforts. The first impression is that cable management recordkeeping applies only to cable. In reality, and to be most useful, most users expect their cable documentation to describe complete circuits and all their components. Additional information about network and desktop equipment must now be included, suggesting that "connectivity management" may be a better name for your efforts. Recordkeeping of communications hardware is a more complicated problem that may now involve other groups and individuals who must be part of the overall cable and connectivity management process. In other words, the scope of the process is much bigger than was first imagined, and the project must be approached accordingly. This book addresses the issues of complete connectivity recordkeeping and will help you build the processes and consensus necessary for success.

The second false impression is that recordkeeping, in and of itself, will automatically bring benefits to the organization, as if choosing a slick and sophisticated recordkeeping system is the only strategic decision that matters. Quite the contrary, experience suggests that recordkeeping, in and of itself, is nothing more than another chore and burden to the organization no matter how slick and sophisticated it is. Records must be *used* and *maintained* to be of any value. It is the organizational procedures and behaviors of people that dictate how effective your cable and connectivity information will be. Again, this volume will help you explore the procedural side of connectivity management as you make decisions about a recordkeeping strategy.

It is our firm belief that, given the specific circumstances within each organization, there can never be just one "right" approach to connectivity management. Similarly, there is no one CMS that will work perfectly for everybody. Therefore, we cannot offer a simple recipe or a standard cookbook approach, and we choose not to build an approach around a particular CMS. Instead, this volume tries to frame the questions that must be asked, and ultimately answered, when considering a management system implementation. To the best that our experience and insight allow, we discuss the implications of different answers - what will happen if, etc. Sometimes an answer has a good side and a bad side, and you must live with the consequences of both sides. Things that are not convenient can usually be addressed, and you certainly don't want unnecessary surprises along the way. The economics of each decision must be considered as well. The ultimate decisions are yours to make, reflecting the preferences and consensus that you can achieve within your organization. Your approach will be unique in many ways.

As to the structure and format of connectivity records themselves, there are also many different approaches, some paper-based and some computerized. Paper records, while unglamorous and arguably inefficient, are still a viable option in some situations. Basic paper records (drawings and/or lists) which are adequately maintained are certainly better than a CMS that is awkward to use or poorly maintained. While we will describe several approaches to paper recordkeeping, keep in mind that you may find other ways to do it. The true measure of a beneficial approach is that which is accessible, useful, and maintained. And only you can be the judge.

If your interest is in electronic records, there are at least a dozen commercial software packages that focus on cable and physical network management in a comprehensive fashion. There are others that are less comprehensive, offering a simpler set of descriptions of our cable plant. There are yet other systems that focus more on the configuration of network devices, with some ability to describe the cable and connections in between. Add to these the myriad of telecommunications management software (basically call accounting and PBX management systems) that include some form of cable recordkeeping, and you have a total of nearly 100 software packages that claim to function as a CMS. While many of these systems share common features and similarities in data architecture, it is fascinating to note how different they are! It's like buying a car: they all have four wheels, seats, and an engine, but how they look and how they drive can be quite different.

This volume does not attempt to describe any particular system in detail, since such an evaluation requires some context of goals and requirements. Systems are also subject to change and ongoing development. Instead, our objective is for you to understand the context of your requirements and the general functionality which best meets these requirements, for only then can you compare these to the records structure and functionality of specific systems. We strongly recommend that you use a Request for Proposal (RFP) to evaluate CMS products. This approach allows you to understand what you are getting, what you are not getting, and where you must adapt. Equally important, it can help you hold the CMS vendor accountable to delivering what is promised with respect to functionality, compatibility, and performance. You will find examples of specific CMS functionality throughout this volume where it is appropriate to illustrate certain points.

As part of your records structure, you may choose to follow the TIA/EIA 606 format, "Administration Standard for the Telecommunications Infrastructure of Commercial Buildings." We will be offering many examples of data formats and cataloging that are compliant with the TIA/EIA 606 standard, others that are modified or somewhat simpler, and details of the standard itself. But there are many issues of connectivity management that are naturally not addressed by TIA/EIA 606, as you shall see. And just like any recordkeeping system, it too must be evaluated as to how useful it might be and how difficult to maintain - how well it will fit into your organization. As we explore the connectivity management process, TIA/EIA 606 will serve as a valuable reference point for comprehensive recordkeeping of cable and other elements to which it applies.

One final editorial point concerns our interest in developing the most general case of cable and connectivity records, for two important reasons. (By general case, we mean that it can usefully handle all the specific types of "things" you want to track, rather than trying to include every conceivable "thing".) First, in many organizations, there is an overlap of responsibilities with regard to the cable plant. Different network support groups, the traditional separation of voice and data services, and the possible involvement of Facilities or Building Services groups sometimes create a fragmented and disjointed plan for cable management. Our approach here is to treat the cable plant as a single, unified system where rules and responsibilities are established and enforced for each group involved. When it comes to defining a records structure for enterprise connectivity information, such a structure must span *all* services and not be limited to simply voice or data. To achieve this, common descriptions of cable and components must be developed that each group can live with.

Secondly, a recordkeeping shortcut today can be a painful limitation later. Certain narrow methods of recordkeeping, and even certain CMS software, may work very well under a particular set of conditions. If the conditions change, or new types of components or records must be added, it can be difficult or impossible to adapt the system. For example, a spreadsheet to track horizontal wiring and its port connection may not be adaptable to tracking multiple backbone cable segments and longer circuits. It is possible, even desirable with certain cable architectures, to segment each network service and build a cable management plan around that service and cable plant only, as the example above might be used for some LAN wiring. Such cases are usually the exception.

Using all of these points of experience to guide our discussion, the book is divided into four parts. Part 1 is a tutorial on cable and connectivity management, and we strongly suggest taking the time to read it thoroughly. It explains the cable and connectivity management process in general terms, framing and addressing the issues that should be considered during an implementation project. Part 2 continues with the actual planning of a recordkeeping process, and the evaluation and selecting of a CMS. Each task is broken down in detail, with planning forms and tables provided. The advantages and disadvantages of different approaches are given, illustrated by numerous examples. Part 2 also explains the process of cataloging, by which you develop names and types for the cable and equipment that you wish to track, and ultimately define the connection point configurations for each. Part 3 discusses the various steps necessary for implementing your system, including installation, testing, training, and documentation. Part 4 covers the processes of physical inventory and data entry. Taken all together, these sections will give you what you need to implement, use, and maintain your connectivity records to your benefit and satisfaction.

PART 1:
CABLE AND CONNECTIVITY MANAGEMENT - A TUTORIAL

THE CONNECTIVITY MANAGEMENT PROCESS

1

Cables, terminations, active devices, and interconnections are the transport system for voice and data network signals, providing the highways that move information from user to user within the organization. It is often regarded as the bottom layer of a stack of electronic functions that allow networks to operate (shown in Figure 1.1 below). Each layer must operate properly to support the layers above. If the cable plant functions poorly, then the network functions poorly.

The network exists solely to support business or organizational needs. The network provides valuable capabilities and benefits, but at a cost to build and maintain. So a poorly designed or poorly managed cable plant can limit the benefits a network can provide and will increase the cost of maintaining it. Your cable plant directly affects the bottom line. Therefore, good management of your cable plant will provide the capabilities that the network requires while reducing costs for building and maintaining the network. These are your goals.

But what do we mean when we speak of cable and connectivity management? Depending upon whom you ask, it could mean a computerized recordkeeping system, pathway enclosures for running cable in a building, panels that keep patch cords neat and tidy, a service order system, or a host of other definitions. Traditionally the answer has depended on your point of view, often leading to some confusion and misunderstanding. None of these answers, by itself, is adequate. The ideal case would be a broad and general definition that can encompass all these things, since all of these ideas tend to help us better manage our cable, connections, and equipment.

To reach a clear understanding of cable management, let's turn the question around somewhat and ask, what does it mean to manage

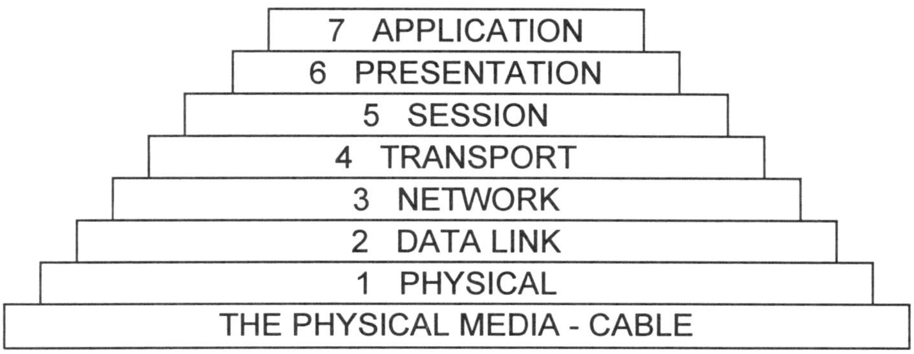

Figure 1.1: The OSI model of network functions, indicating physical media (cabling) as the bottom layer.

your cable and connections? Restating the goals for cable management differently, we can say that managing cable and connections means maximizing your capability while minimizing costs. And since this management involves both cable and equipment, we will choose to call it *connectivity management*, although at times we will refer to it by the more traditional name of cable management. The next step must be to consider the specific issues that affect capability and cost, which will lead you to the elements given in Table 1.1. For example, there are three main issues that affect the capability of the transport system:

- o Capacity: Having sufficient cable pairs or fiber strands available.
- o Performance: Having capacity that will perform as required.
- o Configuration: Having suitable architecture to deliver service to any location in a direct and timely fashion.

Good design practices are obviously a major element of cable and connectivity management. Being able to create a new circuit or establish a new service quickly has an advantage for the business. Selecting pathway and termination hardware that help you identify, organize, or administer your cable has benefits, and so these components can be considered a part of cable management. But while cable plant design has a strong impact on overall capability, what about after installation? You must also administer this capacity effectively on an ongoing basis.

Good connectivity management can help

- SOUND CABLE PLANT DESIGN
- NAMING AND LABELING
- RECORDKEEPING
- CONTROL OF ACCESS
- ADMINISTRATIVE PRACTICE

Table 1.1: The elements of the cable and connectivity management process.

avoid installing cable you don't necessarily need. But the overwhelming cost for managing cable is people - people who add new wiring, people who perform Moves and Changes (MAC), and people who do the testing and troubleshooting when things go wrong. Therefore any management plan must consider the activities of people and how their time is spent. Connectivity management implies people management as well.

This is especially true in larger organizations, where the cable and connectivity management process involves a number of groups with different responsibilities and likely spans a number of different facilities. This situation requires us to take an *enterprise-based approach* that provides for coordination between groups and facilities. Throughout this book, we will address the full set of issues that arise when introducing connectivity management on an enterprise basis.

To be truly cost-effective, cable and connectivity management must help people do certain tasks more quickly and efficiently. For people who work directly with wiring systems, the traditional problem has been the amount of time spent simply trying to identify and locate physical circuit components (which includes cable, interconnections, and controlling equipment). Most cable and equipment has been installed over the years with little thought as to its identification. Basic methods for naming and labeling cable and devices are necessary to help with navigating around a particular equipment room, as well as around your facility as a whole. So good connectivity management includes a plan for naming and labeling your cable and equipment, and the discipline to execute it consistently.

To improve efficiency even further, it would be useful to know ahead of time what you need to work on and where it is. From the troubleshooter diagnosing a network problem to the planner arranging for a 200-person Move, physical information about the facility has been in short supply and must usually be determined by exhaustive on-site inspection. Even when some information is maintained, it is often not generally available to all parties, or is incomplete and inconsistent. To go directly where you need to go, and to know exactly what you need to work on, requires that you maintain appropriate information about your cabling and make this information generally available to all interested parties.

This process of recordkeeping therefore becomes the next element in your understanding of cable and connectivity management. You may choose a records format for cable alone such as TIA/EIA 606, "Administration Standard for the Telecommunications Infrastructure of Commercial Buildings", or go with a custom approach of greater or lesser extent. Your documentation could be simple paper records, or a computerized hierarchical or relational database that we will refer to as a Cable Management System (CMS).

The tasks of naming, labeling, and recordkeeping are not one-time jobs that you can do once and then forget about them. Improving efficiency and raising your level of service require you to put those records to use and, of course, to maintain them adequately. Records by themselves do nothing for us. It's like having a big library in your house, but not knowing how or what to read - of what value is your library? So in order to make the records useful, you must address two more elements of cable management. The first is controlling the access to your facilities, and the second is creating the workflows or procedures that make information available, put it to use, and assure that it is maintained.

Controlling access to wiring closets and equipment rooms is a critical aspect of cable management. As soon as the element of recordkeeping is added to a cable management plan, you must be concerned with the accuracy and completeness of the records. Any action that affects an item in the records must be reported so the records can be updated. By controlling access to the wiring closet, you have some assurance that all such actions are reported, and that actions are not being taken "outside" of the management process. If access cannot be reasonably controlled, then recordkeeping must be limited to items over which you have full control. You can compare it to a business accounting system that controls access to funds through purchase orders, invoices, expense reports, and the like. Without controls, the business could not account for its money.

The final aspect of managing cable and connections is that of administration, or the processes and procedures that are followed to make it all work. It is reasonable to say that the traditional approaches to MACs and problem resolution have been rather haphazard, with little in the way of defined procedures between the original request for service and its final installation. On an enterprise basis, there are often several individuals involved in making this happen. Once again, as soon as you add the element of recordkeeping to your cable management plan, you must be concerned that these records are used by all persons who need them and that they are appropriately maintained. This often requires a more formal approach as to how things get done, a transition that gets more important as your recordkeeping gets more complex, as more people get involved, or as it becomes tied in with other administrative recordkeeping systems. These procedures must be integrated into the overall flow of administrative tasks which include network management, call accounting, PBX management, asset tracking, problem management, and so forth.

By looking at those areas where you can

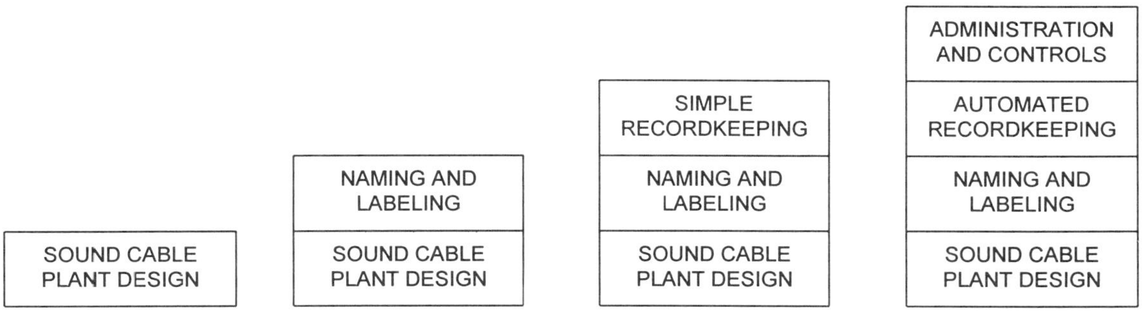

Figure 1.2: Increasing sophistication of a connectivity management plan.

improve your capability and reduce your costs, you can see that good management of cable and connections is the process of directing and controlling the usage of both manpower and cable capacity in the most efficient manner. The first element of your management strategy is sound cable plant design, including the use of specific hardware and devices that help you to identify, organize, and administer these facilities. Second, you can then adopt naming conventions and labeling techniques that help you navigate around your facilities. Third, you can use recordkeeping tools, paper or computerized, that tell you what things are where and what is connected to what, so you can plan and execute your actions more efficiently. Having made a decision to use recordkeeping tools, you must then determine how to control access to your facilities and plan a set of workflows or procedures that will use and maintain the records effectively. Taken together, these five elements are what we mean by the terms cable management or connectivity management. Each element can be added until you attain the degree of control that you desire, as shown in Figure 1.2. Throughout the rest of this book we will fully explore these ideas so you can plan a management strategy suited to your organization. While we have geared this book towards implementation of fully functional, automated recordkeeping systems, you will find lots of guidance on simpler, less intensive approaches that may be suitable for some environments.

Like any good endeavor, implementing a cable management strategy starts with some planning. Understanding how these elements interact is important. For example, your recordkeeping should only include those pieces of information that you know will be used routinely and that you know you can maintain adequately. Naming conventions must be suited to the recordkeeping format and how you will query the system. Other recordkeeping systems must be considered. Responsibilities for updating must be assigned as part of your procedures. So it pays to understand the big picture before you try to implement any of the pieces, otherwise your results will be less than satisfactory. Buying a fancy CMS before you're prepared to use it can waste a lot of money.

As for the recordkeeping itself, there are a number of approaches to choose from, and they are described in detail in Chapter 3. At the simplest level, you can keep a list of cables and pair counts. As things get more involved, there is a strong need to track connections as well as cables, since this allows you to define a sequence of components we could call a circuit. But to truly describe a complete circuit will also require that you know its equipment connections at each end. Cables, connections, port, and user device are the complete information required to define a circuit, as shown in Figure 1.3. Managing this information is much more complex than a simple list of cables and pair counts. Exactly how you choose to describe all these "things" will have a major impact on how friendly and useful your records will be.

Even more significantly, the extent of circuit information may affect who has to be involved in maintaining the records. If there are additional people or groups who modify network or desktop equipment, these persons must be included in the management process. This underscores the earlier recognition that controlling access and developing standard procedures are necessary components of the management plan.

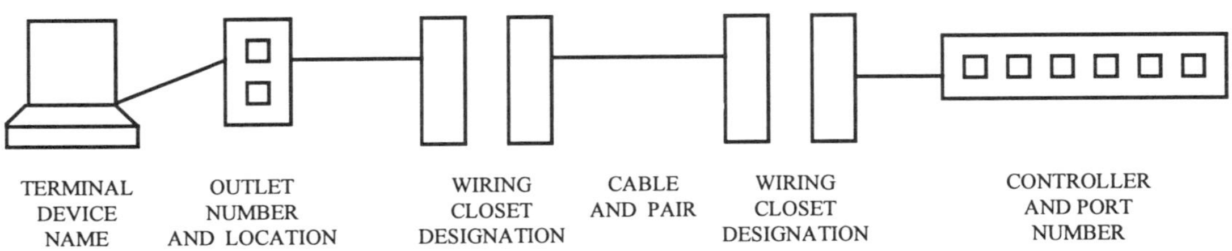

Figure 1.3: Elements of a circuit description.

Your need for information may overlap with other recordkeeping, and duplications must be resolved.

Implementing Connectivity Management

An enterprise CMS implementation project can generally be broken down into three phases. The first is the initial definition of your requirements leading to evaluation and selection of a particular CMS suited to your needs. This phase is described in the project flow chart shown in Figure 1.4 and is covered in detail in Part 2 of this book. The second phase includes the actual installation and testing of the CMS based on the proposed catalog and procedures which you have previously developed, and as modified by the specific format and functionality of the CMS you have chosen. These tasks are shown in Figure 1.5, and are covered in detail in Part 3. The third phase includes the gathering of information, loading data into the CMS, documentation and training, and any final refinements of functionality or procedures which may be required. This phase is described in Figure 1.6 and is covered in detail in Part 4.

Any well-run project will require a good sense of the entire set of tasks which must be accomplished in order to appreciate the costs and timing associated with each. Of course, the actual project flow may vary from organization to organization, depending on specific circumstances and requirements. If you familiarize yourself with these flow charts early on, you will avoid surprises and be able to see how all the pieces fit together. Whatever you can do to anticipate later project requirements will save you time later. Adapted from the project flow charts, the discussion that follows will give you a sense of how a project will unfold and the issues that must be considered.

Implementing a management strategy begins with some evaluation of how things currently operate and what you think are the key areas or tasks that will benefit the most. You will likely need to discuss with all affected individuals a number of issues such as who has control over

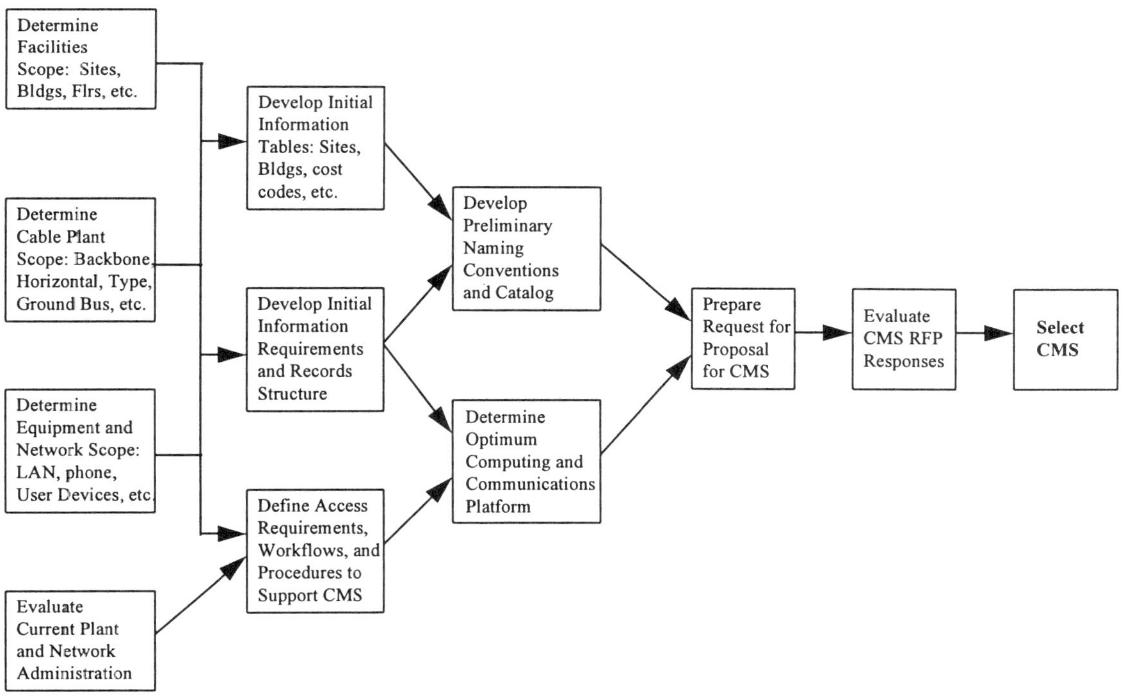

Figure 1.4: Flow chart of the first phase of CMS implementation, from planning for to selecting a CMS.

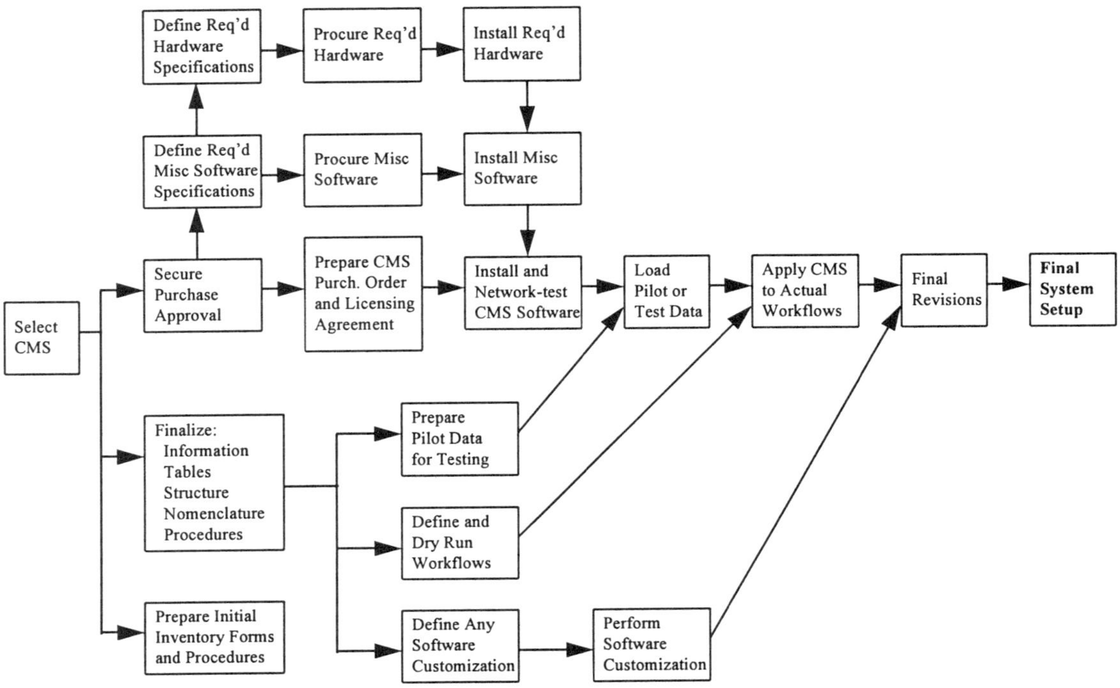

Figure 1.5: Flow chart of the second phase of CMS implementation, from system selection to final system setup.

what equipment and connections, what other recordkeeping currently exists or is required ongoing, and how selected information will be maintained. A typical planning list, called a Definition of Requirements, is given in Table 1.2.

Your first goal must be to define specific recordkeeping requirements. This starts with determining what items or things you want to manage, such as 4-pair horizontal wiring, termination hardware devices, or desktop terminals and PCs. Then you must identify the specific information that you want to track about each type of thing. This could be a description, model number, serial number, trouble history, or whatever. But for every thing you want to track, and for every piece of information you want to track, you must determine:

- SCOPE OF SITES, BUILDINGS, CABLE, AND EQUIPMENT
- PROPOSED RECORDS STRUCTURE
- NAMING CONVENTIONS
- ACCESS AND PLATFORM
- ADMINISTRATION PLAN

Table 1.2: Initial Definition of Requirements list.

- o How you will obtain it;
- o How you will use it;
- o How you will maintain it.

This test must be absolute - if you can't answer each question in a clear and positive fashion for each piece of information, then scratch that piece of information off your list. If you need to develop new procedures, then set them down on paper. You will need to get these changes approved and accepted by management and staff, and be ready to provide comprehensive training as part of your implementation. If you have to scale back or simplify your recordkeeping, that's okay too. It's better to start with a modest, manageable system that can grow than to pursue something too big and unmanageable at the outset.

The initial planning process must reconcile recordkeeping with administration and procedures. Simply introducing an automated recordkeeping tool with little other planning will have minimal impact on your operations. Recordkeeping will require some adaptation, some new responsibilities, and some accountability for their proper use and maintenance. So the initial definition of requirements must include an administration plan.

The administration plan will identify how people will access information in the system. Here, there are several choices. You can allow each individual to access the system, obtain information, and update records according to their actions. You could provide this access at each individual's PC or terminal, or from a central location. Otherwise you could have a single point of records administration, with one or more individuals who interface with the system and issue paper instructions to guide the action of others. Changes and updates must then feed back through the central group for entry into the system. Such an approach will usually require a much more precise and disciplined workflow built around a service order document, since all actions must be accounted for within this process. Yet another approach might be to issue instructions from a central group, but require each individual to perform the records updates according to the actions they take. Each approach here has advantages and disadvantages that will be discussed later, but whatever approach you choose must be clearly defined and documented so that each individual knows what is expected of them. Determining the need for access to the system in this fashion will usually identify the proper platform for keeping the records, such as a LAN

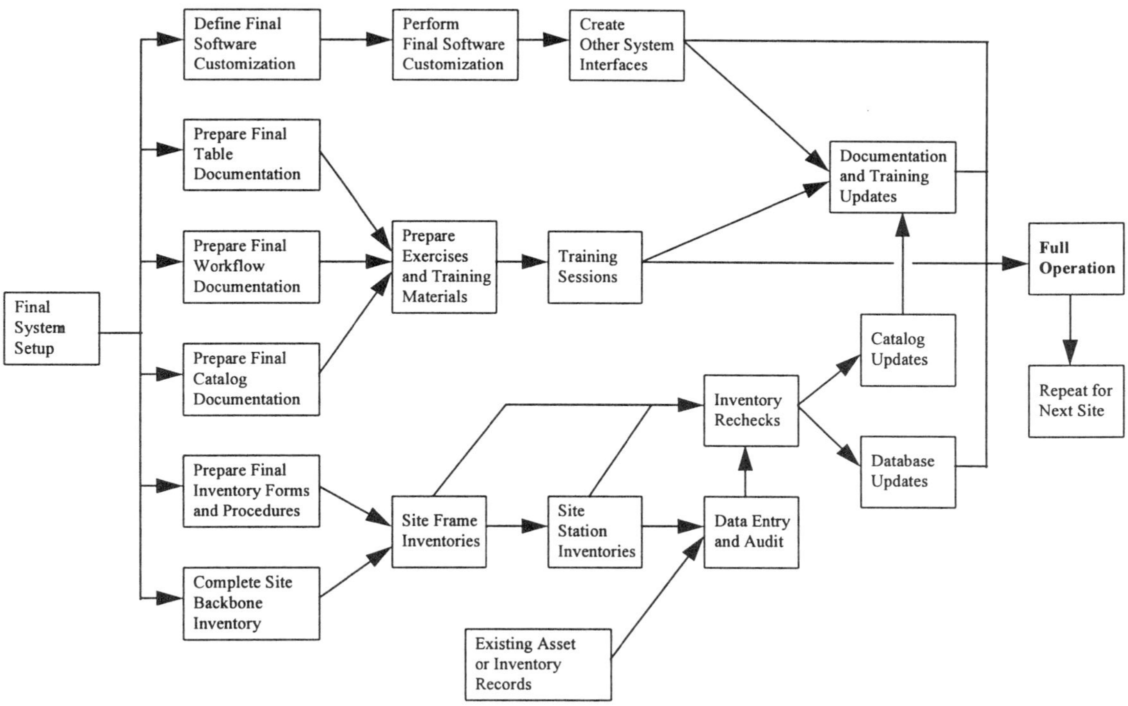

Figure 1.6: Flow chart of the third phase of CMS implementation, from final system setup to full operation.

server, stand-alone PC or workstation, or mainframe-based system.

Reconciling recordkeeping requirements with the administrative procedures will provide the list of "things" and information that belong in the recordkeeping system. The final part of your initial definition of requirements is to develop naming conventions for this information and determine how it ties together. In other words, you can develop a tentative structure for this information that will present it in the most useful fashion. For example, what common pieces of information describe a circuit for each different service? How do you want this circuit information organized so a single records structure adequately represents all circuit types? What is the desired presentation of this information, in both screen and paper formats, so that it best serves the purposes you envision? Simply comparing textual, schematic, and CAD presentations of the same information gives an illustration of this point as shown in Figures 1.7, 1.8, and 1.9.

To summarize, your initial planning will identify the items and information that you wish to track, naming conventions (nomenclature) for this information, who affects this information and how individuals will access it, the platform required to achieve this access, and the detailed workflows and procedures that will be followed in order to use and maintain this information. Consensus and approval must be obtained to then implement this plan.

The next implementation step is to evaluate different CMS software packages to see which may be the best fit for your requirements. This can be done to varying degrees, from very informal to a comprehensive Request for Proposal (RFP). In either case, no software will be a perfect match for your requirements. You will likely have to tailor your recordkeeping somewhat, and make any necessary adjustments to your nomenclature and procedures. Some customization of the software may be considered. Your implementation planning cannot be completed until you have actually selected a system, at which time you can establish all final requirements for nomenclature, procedures, and platform.

It is strongly recommended that you conduct a formal evaluation of candidate CMS software through an RFP, especially if you are considering an enterprise-wide plan that covers a number of different groups and tasks. There are several good reasons for this. First, it forces you to be thorough in your planning, looking at your requirements from all angles. Second, it gets you a much more precise response from the vendor,

CIRCUIT MANAGER - Circuit Number 312-322-0304

Component Count 6

Seq	Mult	Pr	Component ID	Pair/Port/Jack	Pr	Component Type
01	01	01	PBX1	1-1111	01	PBX
01	01	02	PBX1	1-1111	02	PBX
02	01	01	CA1	01		CABLE
02	01	02	CA1	02		CABLE
03	01	01	J1.01.1304	A	01	JACK
03	01	02	J1.01.1304	A	02	JACK

Figure 1.7: Textual representation of a circuit (*Courtesy of CRMS from IntegraTRAK, Inc.*).

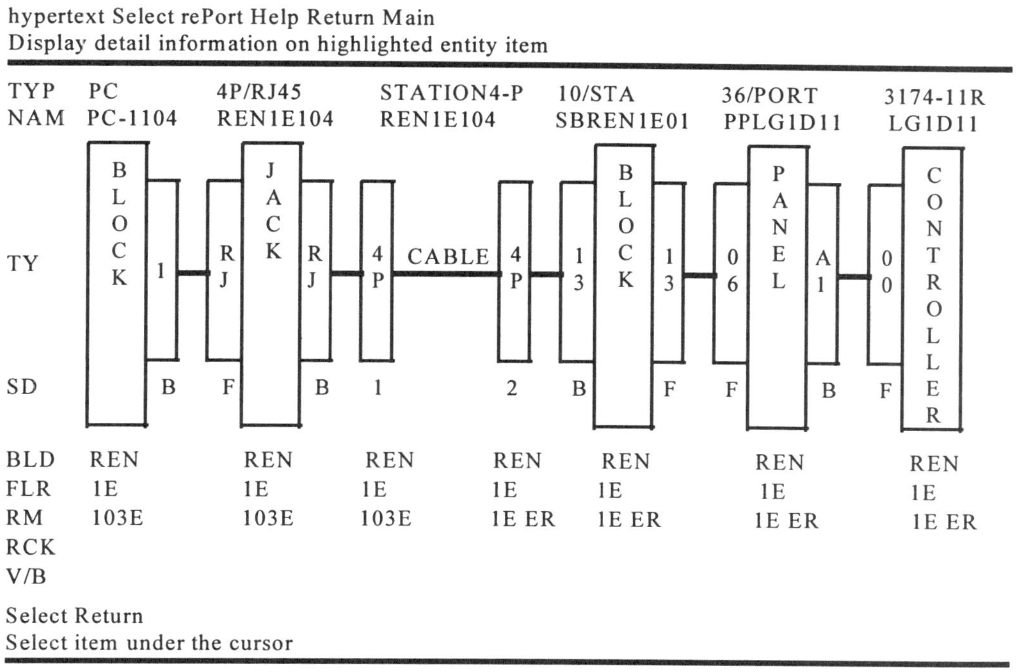

Figure 1.8: Schematic representation of a circuit (*Courtesy of CCAM from CDS, Inc.*).

usually from a qualified technical person. When a vendor indicates compliance or non-compliance with the items in an RFP, it is easier to hold the vendor accountable and you have more assurance of getting what you ask for. Finally, the RFP response will give a better indication of the areas where a product is weak versus your requirements. In these weak areas, you may have to adapt your requirements or simply live with functionality that may be awkward or undesirable. It is much better to know this and accept it up front, anticipating and responding to these areas, rather than being surprised by them later on.

Once you have selected a CMS and made any necessary adaptations to your requirements, you are ready to begin the real implementation phase of the project. The software must be loaded and tested to assure proper operation and access, especially when located on a Local Area Network. Final naming conventions must be developed and entered in to the master catalog of the system, after which you can run a live test to see that it all comes together in proper fashion. For example, if you define certain components and then put them together into a circuit, does the right information appear at the right time and in the right place? If you are using a service order, does it convey all required connectivity information for the technician? Do you have to refer back to other screens or reports to fill in missing information? It is easier to modify the catalog at this point rather than after a significant amount of live information has been loaded.

When you are satisfied with your catalog structure and the CMS functionality, the next step is to develop documentation that can be adapted for training of the various individuals who will become part of the process. Training for different individuals must include a general sense of the information and records structure, how information will be provided to them, and what their responsibilities are with respect to reporting back their actions in the proper format. At the same time, you can begin planning the inventory and data entry processes required to populate the system. Backbone, frame, and station inventories

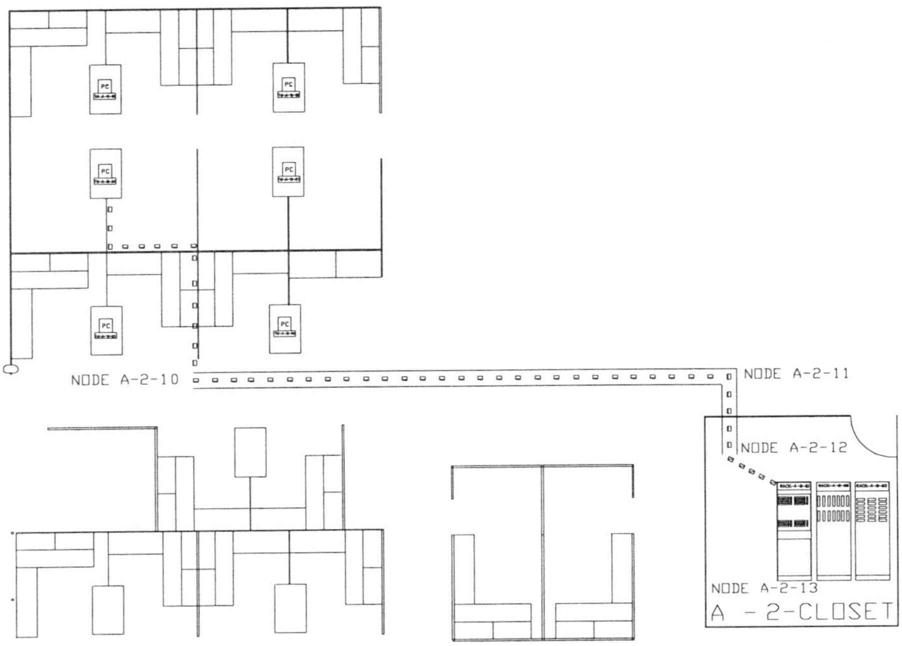

Figure 1.9: CAD (graphic) representation of a circuit (*Courtesy of COMMAND from ISICAD.*).

must be coordinated, audited at every step as well as during data entry. Updates to the catalog, documentation, and training material may be required during inventory. Any required software customization should be completed by this point.

As you complete the inventory of a particular area, the staff must be ready to use that information immediately, according to the procedures which have been developed. Break the project down into areas that can be completed within a few weeks, then work each area thoroughly. When you have completed a particular area or site, be ready to move on to the next area and repeat the inventory process until all areas are complete.

At all times, it is imperative that individuals be held accountable to these procedures. Some ongoing auditing of the database is recommended.

So, what will life be like after you have implemented your connectivity management plan? You will need to sustain all of the elements of good cable and connectivity management discussed earlier in the chapter. Additions and renovations to the cable plant should be suitably designed, with all cable and hardware installed and labeled according to your standard methods. Information should be captured and entered into the CMS at the earliest possible time. Since you have made its use routine, the CMS should make many of these tasks faster and easier.

Routine operations will also be closely tied to use of the CMS, providing for frequent reference and frequent updating. A well-defined service order process may be a critical component. Customer Service, Help Desk, and technical support groups should all have adequate training and access. Remember, if you don't use the information, it won't provide any benefits!

CABLE PLANT ARCHITECTURE

2

In order to make good decisions about recordkeeping and administration, it helps to have a solid understanding of what your cable and communications architecture looks like. The job of defining your recordkeeping requirements and setting up the records requires a working knowledge of the different components, how they interrelate to make circuits, and how they function as a network. In this chapter we will review the basic elements of the cable plant, its evolution over the last several decades, and the relatively recent trend towards standardization around TIA/EIA 568A, "Commercial Building Telecommunications Wiring Standard".

Historically, there are a variety of cable types and cable architectures used to support different communications services. Twisted pair wiring for telephony is probably the oldest and most ubiquitous type of wiring, and has become the popular choice for many data and Local Area Network (LAN) applications. Coaxial cables have been used primarily for video and older data networking, including T1 and T3 digital voice networking. Since the mid-1980's, optical fiber has become popular for high-speed digital backbones and is growing in popularity as the medium of choice to the desktop.

Unfortunately, earlier wiring systems tended to follow the architecture of the network which they supported, be it voice, data, or video. Up until 1984, all telephone wiring was managed by the local phone company while coaxial data networks were managed by the old Computer

TIA/EIA 568A	Commercial Building Telecommunications Wiring Standard
TIA/EIA 569	Commercial Building Standard for Telecommunications Pathways and Spaces
TIA/EIA 606	Administration Standard for the Telecommunications Infrastructure of Commercial Buildings
TIA/EIA 607	Grounding and Bonding Requirements for Telecommunications in Commercial Buildings

Table 2.1: Significant industry standards relating to cable and cable management.

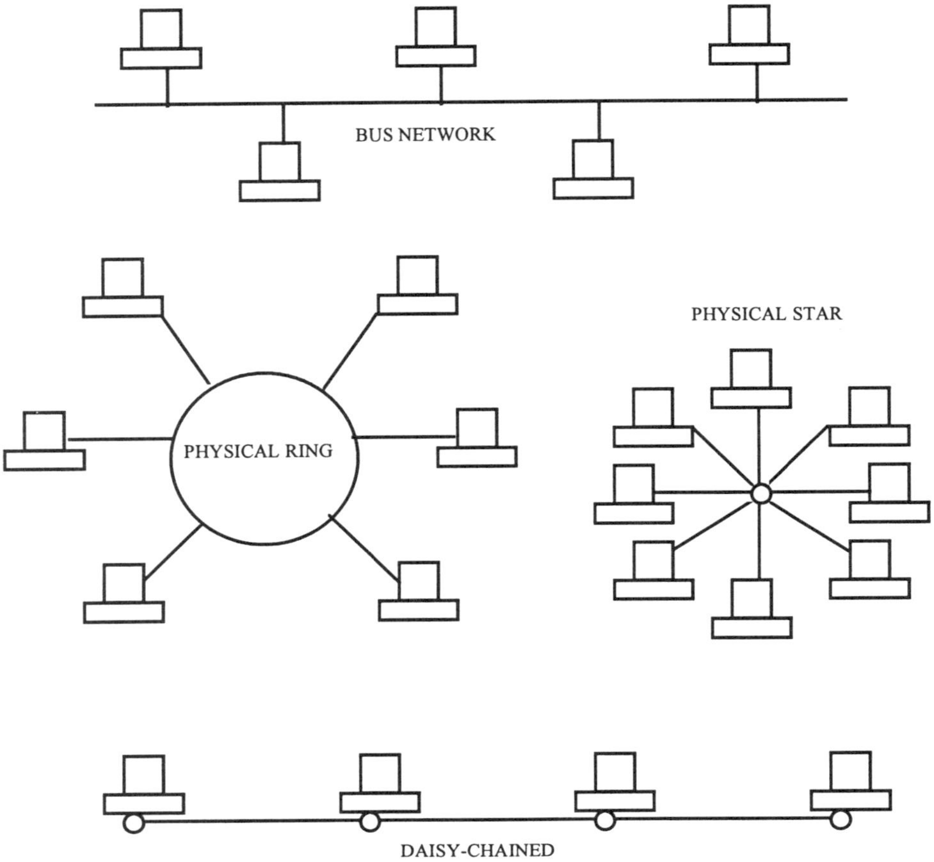

Figure 2.1: Different topologies or layouts of the cable architecture.

Department. Many users likely have a combination of cable plants among their buildings, and ideally your CMS implementation should be able to handle them all. And even today, there are different terms used to describe the cable plant, and different names used for very similar things. For example, cables running between wiring closets are stilled called Backbone, House, Riser, and Gray cables. Developing a consensus around naming and description, a common "jargon" if you will, is an important part of beginning a cable management project.

Many earlier network systems used specific wiring configurations, or topologies, as part of their architecture. Differences in topology led to incompatibilities between networks, which was a major complaint from users trying to integrate or migrate between systems. When trying to implement cable and connectivity recordkeeping, it is useful to understand the nature of these topologies indicated in Figure 2.1.

The first topology is referred to as a bus network, where a long backbone cable is run throughout a facility with drops or taps used to tie individual devices into the bus cable. The bus, in effect, is the central portion of the network rather than some active device. The second topology is a physical ring, where cable is run from device to device in a way that loops back on itself so the cable becomes continuous. There is no natural beginning or end of the cables in the ring, and one may have to be assigned arbitrarily.

The third topology is a physical star, where individual cables all run back to a central point. It assumes there is an active device (often called a hub) at this central point that is managing control of the network. This is similar to traditional telephone wiring, and has become the preferred method of cabling. Sometimes this architecture is referred to as a tree topology, hierarchical star wiring, or a tree of stars whereby backbone cable is star-wired to a central location and horizontal wiring is star-wired to the end point of each backbone cable. This is the nature of TIA/EIA 568A wiring described in the next section.

Finally, there is a topology referred to as daisy-chained, where cable runs from device to device in the network without looping back on itself as a ring would do. There may or may not be some type of control device at one end of the daisy-chain.

The TIA/EIA Standards

Fortunately, the industry has recently developed a set of standards covering the description, design, and administration of cable facilities. Over the last several years more and more new cable installations have been designed with reference to the cornerstone of these standards, now known as 568A. It is a comprehensive and detailed document, specifying the layout and characteristics of premises wiring systems. Formally, it is known as TIA/EIA 568A, and it is the first of four significant standards listed in Table 2.1.

The second standard listed is TIA/EIA 569, which covers the design and layout of pathways (conduit, duct, raceway, etc.) and spaces (wiring closet, equipment room, work area, etc.). Next is TIA/EIA 606 covering recordkeeping, labeling, and administration. It will be covered in detail in Chapter 4. Finally there is TIA/EIA 607, which covers telecommunications grounding and bonding systems. Together, these standards form a coherent framework for understanding the cable plant architecture. Our discussion here will be a very basic one, simply to familiarize yourself with the basic components of a cable plant if you are not already. The corresponding Canadian standards, CSA T527 through T530, are listed in Table 2.4 at the end of this chapter. These standards are very similar, although not necessarily identical.

We strongly recommend that you purchase a copy of the 568A and 606 standards, and that you consider obtaining the others as well. If you desire even more information on cable design practices, then you might want to get The Telecommunications Distribution Methods Manual, a three-volume set prepared by Building Industry Consulting Service International (BICSI). If parts of your cable plant are very vendor-specific, then you might want to acquire any information available from the vendor as well.

Collectively, the TIA/EIA standards define spaces, pathways, cable, terminations, and the ground bus system, as listed in Table 2.2. They do not cover user devices or network equipment, components which do tend to be part of your enterprise recordkeeping. As we discuss each of these items, please refer to Figure 2.2.

These standards describe two types of cable: Backbone and Horizontal. The horizontal wiring runs from the outlet at each workstation or office back to the first wiring closet. Backbone cable runs from a central point of distribution out to each wiring closet, where it makes connections to the horizontal wiring. In fact, there can be up to two runs of backbone cable out to each wiring closet, with the backbone cables interconnected at an intermediate point. The length of horizontal wiring is never to exceed 90 meters (295 feet), while the maximum total length of backbone cable depends on the type of cable itself. Certain types of cable are allowed as horizontal or backbone cable. For the specific types and maximum lengths, consult the 568A standard.

Pathways are enclosures of various types in which cable is pulled or otherwise installed. Horizontal pathways could include metallic

- CABLES AND TERMINATIONS
- CABLE PATHWAYS
- ADMINISTRATIVE SPACES
- GROUND BUS SYSTEM

Table 2.2: Elements of the TIA/EIA cable architecture.

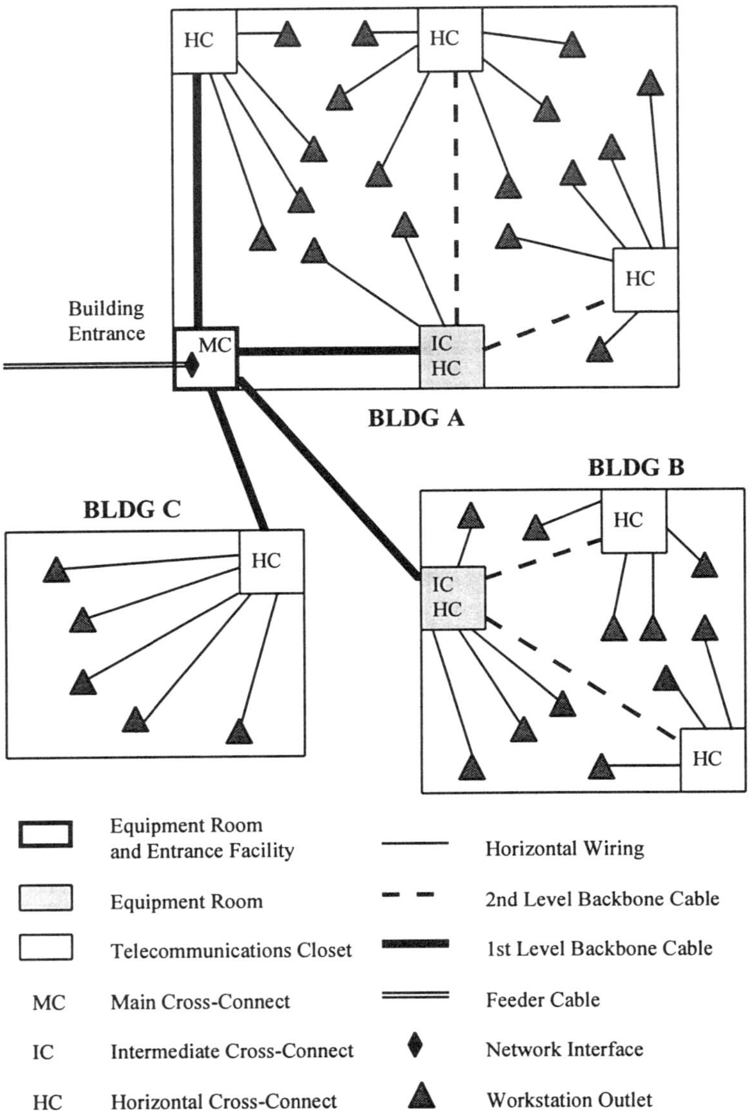

Figure 2.2: TIA/EIA 568A cable plant architecture.

conduit, underfloor and overhead raceways and cable trays, access floors, and perimeter raceway systems. Backbone pathways could be in similar facilities, or may be as simple as a metallic sleeve penetrating between floors. Interbuilding backbone pathways would likely be tunnels, or underground metallic or plastic conduit systems. Cable is not necessarily required to be in an enclosed pathway, such as direct-buried interbuilding cable or horizontal wiring which is run directly through a drop ceiling space.

Where pathways are used, the user must be concerned with both the total capacity and the remaining available capacity of the pathway. For suspended systems such as overhead raceways, the total weight or loading of installed cable must also be tracked since there is a total maximum weight allowed. While the TIA/EIA 569 Standard discusses different pathway types and locations in detail, they are all essentially the same for recordkeeping purposes. In other words, the same set of information would be adequate to describe the different types of pathways encountered.

A space is a point or location where access and administration of the cables and pathways is required. This includes the Equipment Room (ER), defined as a centralized space for telecommunications equipment. The ER can also be a location where different segments of backbone cable are terminated. Next is the Telecommunications Closet (TC), where backbone cable and horizontal wiring are terminated. Other spaces include manholes, Entrance Facilities, and the Work Area itself.

In each space, there is typically some type of interconnection between different cables, or between cables and equipment. These points can be considered hierarchical, according to the types of cables being interconnected. For example, connections between horizontal wiring and first or second level backbone cable are known as Horizontal Cross-Connects (HC). The connections between first and second level backbone cables are known as Intermediate Cross-Connects (IC). The connections between first level backbone cable and communications equipment or the network interface itself are known as Main Cross-Connects (MC). Note that a given space, such as an Equipment Room, may have more than one type of cross-connection present, as shown in two of the Equipment Rooms of Figure 2.2.

Inspection of Figure 2.2 shows that circuits from Building C start at the workstation outlet, go through a Horizontal Cross-Connect in the Telecommunications Closet, and then over a single run of backbone cable to the Main Cross-Connect in the Equipment Room. Either horizontal or backbone cable could be in an enclosed pathway. Circuits from Building A could be one of two arrangements. Some circuits go through the Horizontal Cross-Connect in the Equipment Room onto backbone cable, while other circuits go through the Horizontal Cross-Connect at a Telecommunications Closet onto second-level backbone cable, through an Intermediate Cross-Connect in an Equipment Room onto first level backbone cable, and finally to the Main Cross-Connect. Circuits in Building B will be one of these two arrangements as well.

The TIA/EIA standards place particular emphasis on the installation of a sound, dedicated grounding system for telecommunications use as spelled out in the 607 standard. When installed, it too must be considered for tracking in a recordkeeping system, since such records will have value for planning and for troubleshooting. Such a system is shown in Figure 2.3.

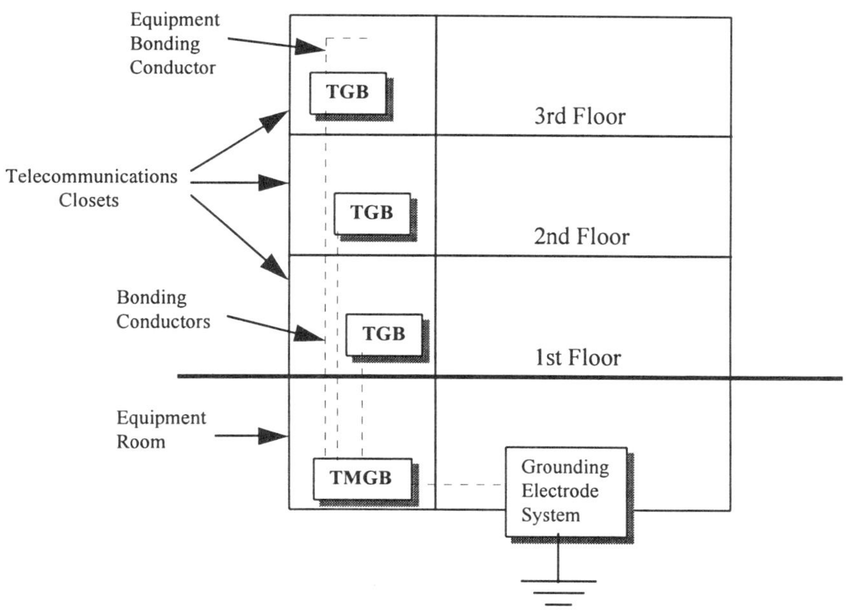

Figure 2.3: Ground bus system for commercial buildings.

The ground system begins at the Grounding Electrode System (GES), a set of buried electrodes designed to achieve good ground potential and continuity in the local environment. Under certain conditions, the incoming Power Service Ground may be used. From the GES, a bonding conductor runs to the Telecommunications Main Grounding Busbar (TMGB) usually located near the Main Cross-Connect. From the TMGB, individual bonding conductors run to the Telecommunications Grounding Busbar (TGB) in each Equipment Room and Telecommunications Closet. Equipment bonding conductors are run from the TGB to individual devices and ground terminations in each closet. For additional grounding system requirements, refer to the 607 standard itself.

Types of Premises Cabling

As for types of cabling, there are a number of different yet generalized constructions to which we will be referring, listed in Table 2.3. The first is what has become known as UTP, for Unshielded Twisted Pair cable. UTP is what you have always thought of as the traditional distribution system for voice services, with insulated conductors that are paired and then twisted together They are terminated using modular jacks and plugs or insulation-displacement connectors such as 66 and 110 wiring blocks. Horizontal UTP includes four pairs and comes in different performance categories, while backbone UTP is available in multiples of 25 pairs from 25-pair up through many hundreds of pairs in a single cable.

Another traditional style of cable is called coaxial, with an insulated center conductor wrapped with a foil and/or a braid that makes the outer conductor. While there are many different styles of coaxial cable, for purposes of recordkeeping they can be treated in a similar fashion. A coaxial cable typically uses a special, individual connector (such as BNC, TNC, N, F, etc.) which is terminated on patch panels in the wiring closet. Because of differences in balance and impedance, a balun is required to convert from coaxial to twisted pair and vice versa.

More recently, other forms of twisted pair wiring have been popular for high-speed data networking. STP, or Shielded Twisted Pair, contains one or more twisted pairs that are individually enclosed in a metallic foil shield for better noise immunity. STP often requires special connectors that provide for grounding and/or shield continuity, and it is not amenable to the traditional 66 and 110 wiring block terminations. In common usage, STP is used to refer specifically to the 150 ohm IBM Type 1 or 1A style of cable.

Another form of twisted pair is now being referred to as ScTP, for Screened Twisted Pair. This type of cable has two or more pairs wrapped with an overall metallic foil shield, again for higher noise immunity. While it is possible to use some twisted pair termination methods with this cable, it does require some method for independent grounding of the shield itself.

Fiber optic cable includes one or more individual strands of optical fiber. Tight-buffered cables have strands with a thick plastic coating directly over the fiber, while loose-tube designs allow the fiber to move around within the buffer.

- UNSHIELDED TWISTED PAIR (UTP)
- COAXIAL
- SHIELDED TWISTED PAIR (STP)
- SCREENED TWISTED PAIR (ScTP)
- FIBER OPTIC CABLE
- COMPOSITE OR HYBRID CABLE

Table 2.3: General types of premises cabling.

Each strand requires a separate connector, most commonly the ST, SMA, or SC connectors being used on premises. As with coaxial cables, a patch panel or termination enclosure is required.

The final type of cable to note is not really a unique cable, but a combination of any of the previous types. This style of cable is often referred to as hybrid or composite cable, and most frequently would include UTP and STP, UTP and optical fiber, or STP and optical fiber. It may require some special termination panels suited to the particular combination of cables included. Use of composite cable can present a challenge for CMS recordkeeping, since the cable is made of subunits with different characteristics. The records structure must have some provision for composite cables, or else each subunit must be tracked as an individual cable and then related through their naming conventions.

A Hypothetical Site

To give a real understanding to the TIA/EIA architecture, and to offer later examples of planning and implementing connectivity records, we can describe a typical site laid out according to 568A. The example site shown in Figure 2.4 is ABC Company's Landisville Research Park consisting of three buildings. The Administration Building has a basement and floors 1 and 2 where the Information Technologies Group and administrative personnel are located. The Research Engineering Building has a basement and floors 1, 2 and 3, which are divided into the East and West Wings. This building houses the research and engineering personnel, and includes some lab areas. The third building is the Facilities Operations and Maintenance Building which is responsible for operations and maintenance of the site facilities and provides some operations assistance for the headquarters building located downtown.

There are approximately 500 employees located at the Landisville Research Park. These employees utilize a number of different services including voice, an IBM Token Ring LAN and 3270 Mainframe service. The majority of voice

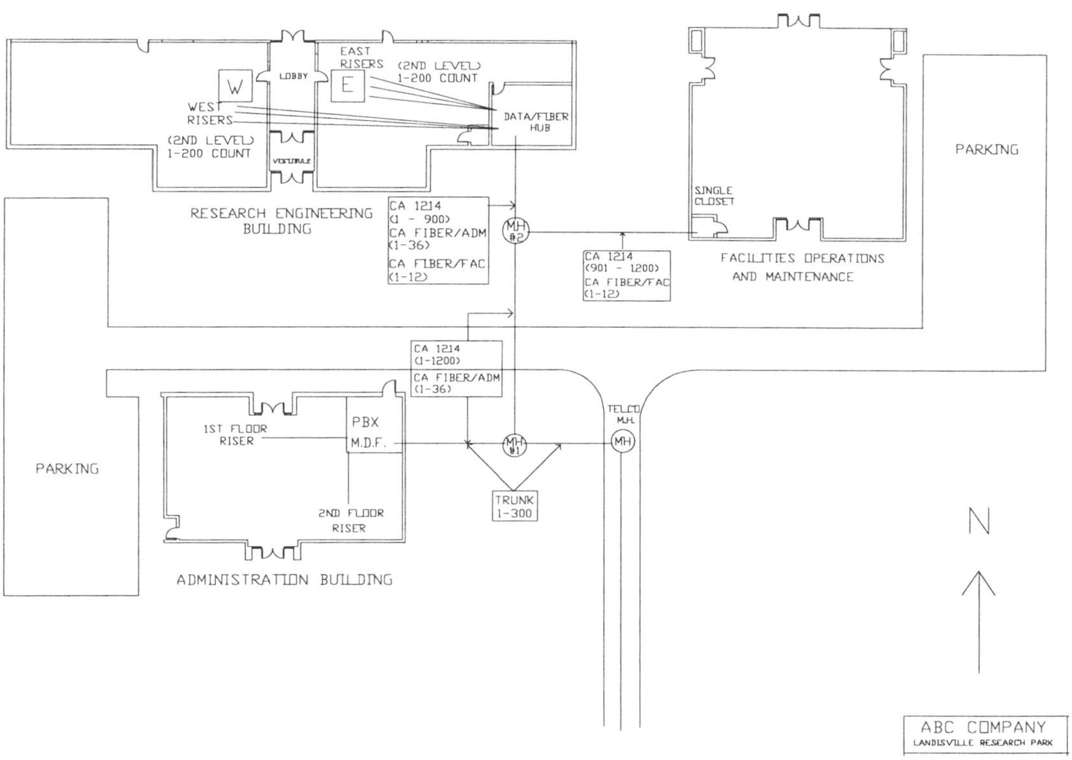

Figure 2.4: Site plan with first and second level backbone cables.

lines are handled through the local company PBX, which is located in the basement of the Administration Building. There are also some additional voice lines which are leased from the local telephone company. All servers for the Token Ring LAN are located in the Data Center in the basement of the Research Engineering Building, with horizontal cable connections made to LAN hubs in the local telecommunications closet. Individual 32-port mainframe controllers are also distributed in the local telecommunications closets on each floor.

Each user typically has one voice connection and one data connection per office, however, some workstations may have two data connections or an additional voice line which may be used for a dial-up modem line or answering machine. Voice lines are basically digital for office phones with analog lines used primarily for modems and answering machines.

The number and types of telecommunications closets vary from building to building. The Main Cross-Connect point is located in the basement of the Administration Building along with the local voice PBX. The Data Communications Hub is located in the basement of the Research Engineering Building, which is also the main distribution point for the site fiber cable.

Horizontal cables consist of 4-pair UTP (Category 3 and Category 4), RG-62 coaxial cables, and IBM Type 1 cables. The number and type of cables run to each office varies from building to building. The Administration Building has 50 offices per floor with two 4-pair UTP Category 3 and two 4-pair UTP Category 4 run to each office. In the Research Engineering Building there is only one 4-pair UTP Category 3 and one 4-pair UTP Category 4 run to each of the 100 offices per floor. There is also an IBM Type 1 cable run to each office on the 2nd floor of this building. The Facilities Operations and Maintenance Building has three 2-pair cables and one RG-62 coax cable run to each of the 40 offices on the first floor, with two 2-pair cables and one RG-62 coax cable run to each of the 20 offices on the second floor. The personnel in the Facilities building are primarily 3270 Mainframe users.

The main feeder cable from the local telephone company is a 300-pair cable which

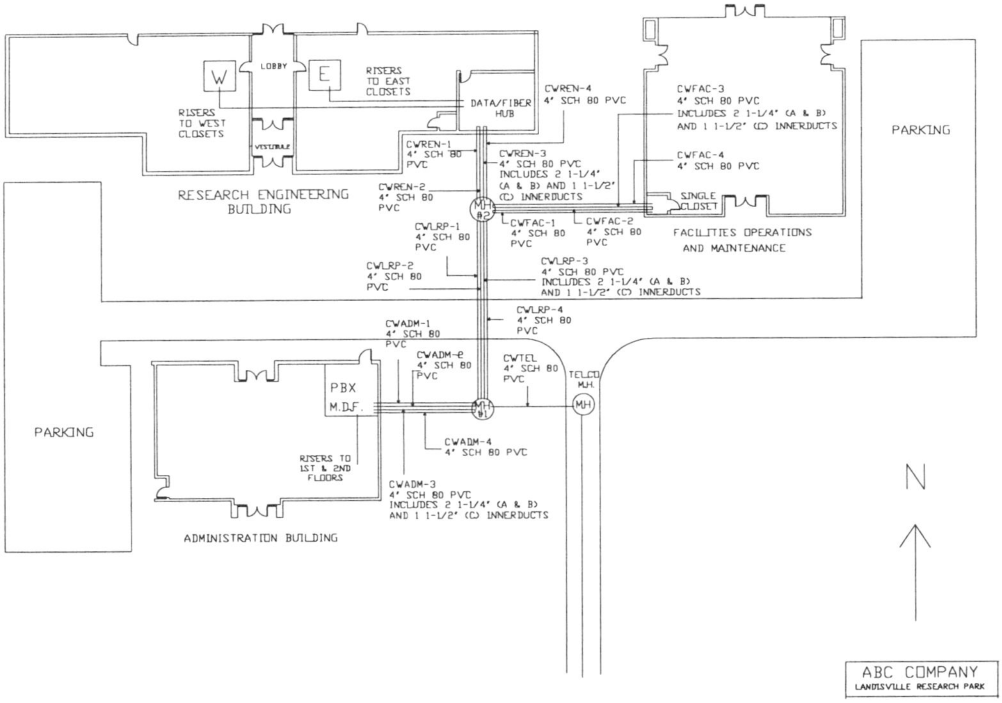

Figure 2.5: Site plan with cable pathways.

terminates in the basement of the Administration building, where the local voice PBX is located.

The first level copper backbone cable (Cable 1214) is shown in Figure 2.4, and runs between the Main Equipment Room in the Administration Building and the Intermediate Cross-Connect points in the Research Engineering and Facilities Buildings. Cable 1214 is a 1200-pair copper cable with pairs 1 to 900 terminating in the Research Engineering Building and pairs 901 to 1200 terminating in the Facilities Building.

The first level fiber backbone cables originate in the Intermediate Cross-Connect point in the basement of the Research Engineering Building and terminate in the Administration and Facilities Buildings. The fiber backbone consists of two separate cables, a 36-strand cable (labeled Fiber/ADM) which terminates in the basement of the Administration Building, and a 12-strand cable (labeled Fiber/FAC) which terminates in the telecommunications closet on the first floor of the Facilities Building.

The second level backbone consists of both copper and fiber cables and are located in the Research Engineering and Facilities Buildings. The Facilities Building has only a single Telecommunications Closet on the first floor, where all first and second floor horizontal cable is terminated. There is no second level backbone cable in that building.

The second level copper backbone cable in the Administration Building consists of two individual cables, one run to the Telecommunications Closet on the first floor and one run to the Telecommunications Closet on the second floor. Each cable is separately identified and labeled, and has a count of 1-200. The second level backbone cables in the Research Engineering Building are similar in that there are six 200-pair cables, one run to each of the telecommunications closets in the East and West Wings on floors 1, 2 and 3. These cables are also separately identified with individual counts of 1-200. There is no backbone cable run to the West Wing basement since there is only the single closet in the Research Engineering Building basement, located with the Data Communications Hub.

Second level fiber backbone is similar to the copper backbone, in that a 12-strand fiber cable is run to each telecommunications closet on

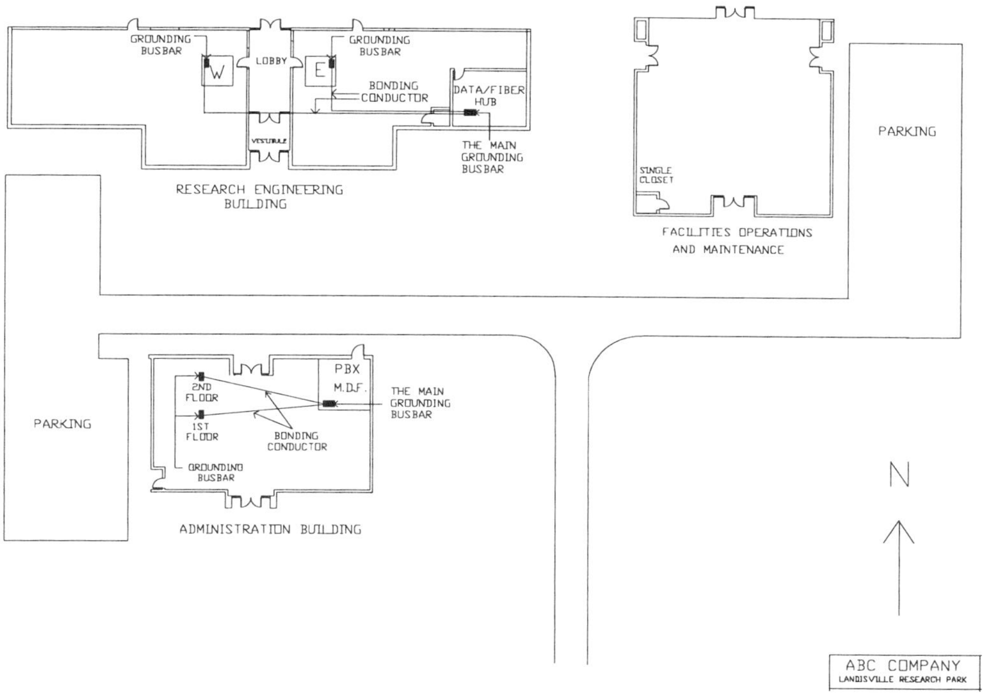

Figure 2.6: Site plan with telecommunications grounding system.

CSA STANDARD NUMBER	CORRESPONDING TIA/EIA STANDARD	STANDARD NAME
T527	607	GROUNDING AND BONDING, DESIGN GUIDELINES FOR TELECOMMUNICATIONS IN COMMERCIAL BUILDINGS
T528	606	DESIGN GUIDELINES FOR ADMINISTRATION OF TELECOMMUNICATIONS INFRASTRUCTURE IN COMMERCIAL BUILDINGS
T529	568	DESIGN GUIDELINES FOR TELECOMMUNICATIONS WIRING SYSTEMS IN COMMERCIAL BUILDINGS
T530	569	BUILDING FACILITIES, DESIGN GUIDELINES FOR TELECOMMUNICATIONS

Table 2.4: Significant Canadian standards relating to cable and cable management.

floors 1 and 2 in the Administration Building and to the East and West Wing telecommunications closets on floors 1, 2 and 3 of the Research Engineering Building. Each cable is separately identified and labeled.

The pathways which contain the first level backbone cables are shown in Figure 2.5. There are four individual pathways running between each of the site buildings and manholes. All are 4" Schedule 80 PVC conduits, each labeled with a unique identifier. The third conduit between each building and manhole is used for the first level fiber backbone cables. Each of these conduits contains three innerducts. Each of the innerducts have the individual unique identifiers A, B and C to identify them as separate pathways.

Innerducts A and B are used for the Fiber/ADM and Fiber/FAC cables, while C is currently spare. The fourth conduit between each of the buildings and manholes is also spare and can be used for expansion and growth.

The Telecommunications Main Grounding Busbars are located in the Main Equipment Rooms of the Administration and Research Engineering Buildings, as shown in Figure 2.6. Bonding conductors are run to the Telecommunications Ground Busbar (TGB) in each of the local Telecommunications Closets. Equipment bonding conductors are then run from each TGB to the individual equipment and grounding terminations in each closet.

The Landisville Research Park site will be used throughout the book as a reference to help illustrate different points of labeling, naming and facilities documentation.

BASIC NAMING, LABELING, AND RECORDKEEPING

3

The use of basic naming and labeling techniques can be a great benefit for navigating around wiring closets and the cable plant, independent of any recordkeeping that you may choose to do. Labeling of cables and devices speeds up the ability to identify and create circuits and connections, with fewer errors. It becomes more straightforward to assess current capacity and availability simply by visual inspection. This holds true for active hardware and devices as well as cabling. Of course, termination hardware which allows for easy labeling will help you in this effort and is recommended.

It makes sense to name and label some things automatically, such as cable pairs on a wiring block. If you desire additional information for recordkeeping then other things will need to be named, and sometimes labeled. Since recordkeeping will impact naming requirements, let's look first at the nature of recordkeeping, some specific approaches you can take, and the benefits and limitations that each naturally has. We will return to the matter of what must be considered when developing specific naming conventions and the methods for labeling cable and equipment accordingly.

Recordkeeping Tools

The most basic forms of recordkeeping will be based on pencil and paper. Lists, drawings, and scribbled notes already make up the documentation of the cable plant. Updates should be made by erasure and correction rather than crossing out or writing over, so only lead pencils and heavy stock paper should be used. They may be kept loosely, in a notebook, or tacked up on a wall somewhere useful.

Unfortunately, all paper records have at least two serious drawbacks. First, the master copy can exist in only one location at a time, meaning you will always have to go to the records, they will not come to you. Sometimes this means only one person can work with the records at a time. Copies of the master can be made and distributed, but all updates must be done on the master itself and copies should be replaced frequently. If several masters are distributed at different locations such as each wiring closet, they can be convenient to use while at that location but it becomes very inconvenient to reference information from one list to another. Second, paper records work well for lists but don't do a very good job of relating items on different lists. For circuit information, for example, we must relate several different items from different lists in a specific sequence, a job that can be challenging on paper.

- PAPER LISTS
- PAPER DRAWINGS
- COMPUTERIZED SPREADSHEET
- COMPUTERIZED DATABASE
- COMPUTER-ASSISTED DRAFTING (CAD)

Table 3.1: List of possible cable and connectivity recordkeeping tools.

Computerized records could take the form of a spreadsheet, a pure CAD (Computer-Aided Drafting) system, an in-house database, or a commercial CMS database (with or without CAD capabilities). A spreadsheet is like an electronic list, with the ability to automatically sort and query - a significant benefit over paper records. But while it may be more convenient for an individual to use and maintain, it shares the same limitations as paper records: it is list-based, and may be on a stand-alone PC. CAD systems, like some spreadsheets, can be single- or multi-user, and can be accessible at more than one location. As a graphical format, they can capture textual information in the form of labels or notes, but they cannot do anything else with it. The graphical format does convey significant information about the location or routing of cable and devices, which can be of value during certain tasks.

Probably the most functional type of recordkeeping system is the database, particularly those that have been developed as commercial CMS applications. The database has an inherent structure that can be made to describe cable and connectivity information quite well. The description of each item becomes a record, the list of items becomes a table of records, and records from different tables can be related in new formats or views. In other words, the database allows for lists of "things" as well as the association of "things" on different lists, just like we need to associate the items of a circuit. Tables can be sorted and queried, while reports can be written which extract information from different tables. Interestingly there are many ways that these records, tables, and relationships can be structured, and we see these different structures reflected in the different software available. It can take a pretty thorough evaluation of the CMS to determine if it functions the way you anticipate.

Certain CMS software includes some degree of CAD functionality, either as a native, integral part of the system or as a supplementary system which interfaces with the actual CMS software. These systems, in theory, offer the benefits of both CAD and database systems, but they must be evaluated carefully. A CMS that is built into a CAD system will require you to have or create plan drawings, and will expect you to use the CAD interface as the principal means of accessing your records. Making and breaking connections may not be as simple as picking up an icon and dropping it somewhere else. This may require some additional training not commonly given to a clerical administrator. Also, the database side may not be as robust and functional as you would otherwise find, and CAD drawings are more difficult to manage on a network. However, they do offer the graphical plan and arrangement drawings that can be very useful for facilities management, where routing of cables and pathways is valuable information. The system can be more easily shared between facilities and communications groups.

Other CMS systems may include some limited CAD functionality, such as moving walls and placing text on the drawing, while allowing CMS database information to be overlaid on the drawing as icons and symbols. These systems are not dependent on having drawings to build the database. Other systems have a display-only capability, allowing drawings to be viewed but not edited. For either of these cases, you will likely have to purchase a standard CAD system to maintain the drawings themselves. We will discuss both database and CAD functionality in Chapter 6.

Key Aspects of Recordkeeping

Recordkeeping for your cable plant and communications hardware can encompass a broad range of items and information, as well as a set of relationships or associations among these items. For example, you can track both backbone cable and horizontal wiring and the pair-to-pair relationships between these cables that would describe a circuit connection. The more things you want to track, the more extensive the lists of things become, and the relationships between things will get more complicated. Your goal should be to choose a recordkeeping approach that only includes the things and relationships that will be useful and of value in your particular environment. Given the almost countless ways to choose and organize information into a records structure, how do you determine what your approach should be?

An obvious response to this question is to follow TIA/EIA 606, "Administration Standard for the Telecommunications Infrastructure of

Commercial Buildings," or to simply follow the format of a particular CMS product. Blindly embracing either approach will likely produce two results. First, you may not understand how to get maximum value out of your records, and secondly you will run into some problem areas leading to breakdowns and frustration. Our discussion of recordkeeping is intended to be much more fundamental so that you can determine what aspects of TIA/EIA 606 will provide benefits and be manageable, and what particular features and functionality you desire when evaluating a CMS. The TIA/EIA 606 Standard is covered in detail in the next chapter, and it is recommended that you come back and review the current chapter after reading about the 606 standard.

A second point worth noting is that people tend to be parochial about recordkeeping. If you do certain tasks, you expect records to be organized and available in a fashion most beneficial for those tasks. Given the potential extent of cable and connectivity recordkeeping, it is likely that people with different needs and points of view will be required to develop a single recordkeeping structure. This effort will take some accommodation and adaptation, leading to a common description of your environment that is useful to everyone involved. It is only when you look at recordkeeping in a very fundamental way that you can anticipate individual needs and develop a global, enterprise-based system.

It helps to take this rather complex problem of recordkeeping and break it down into a number of simpler aspects or factors that can guide your thinking and allow you to see how well a particular approach can meet your objectives in a particular environment. We will discuss these aspects more in Chapter 5, but first we must consider the most basic building block of the records themselves - the list.

A list of things or items is such a simple concept that it is taken for granted. How you organize and use a list is not always as simple as it seems. Take for example a list of your favorite possessions - say a teddy bear, a guitar, a pocketwatch, and a hat. You could arrange these items alphabetically, randomly, by age, by value, by importance, or any number of other methods. Once the list is organized by some criteria, it is easiest to use (to find something on the list) when you apply that particular criteria. If your list of favorite possessions is arranged alphabetically you would expect to find "teddy bear" at the end, since "t" comes last in alphabetical sequence. If you wanted to find all musical instruments, you would have to scan the entire list since such an item could fall anywhere on the list. The list is not as convenient to use when you apply a criterion other than that around which the list was organized.

Imagine now that you add the dollar value of each item to the list, and arrange it alphabetically. Finding a particular item on the list is straightforward. But suppose you wanted to find the item that has a particular dollar value? Again, you would have to carefully scan the entire list until you found that particular value, and then you would be able to see which item that was.

Applying this to your cable plant, imagine a list of jacks arranged in some alphabetical or numerical sequence, as in Table 3.2. For each jack, the office location is given. Finding the location of any particular jack is easy

JACK NUMBER	OFFICE NUMBER
101	1110
102	1112
103	1114
104	1116
105	1118
106	1120
107	1111
108	1113
109	1115
110	1117

Table 3.2: List of jacks in numerical sequence.

OFFICE NUMBER	JACK NUMBER
1110	101
1111	106
1112	102
1113	107
1114	103
1115	108
1116	104
1117	109
1118	105
1119	111

Table 3.3: List of jacks by office number.

since the jacks are listed in sequence. But if you wish to find the jack at a given location, you have to scan the entire list until you find that location, thus identifying the jack. For a list of hundreds of jacks, this may take some time and effort. On the other hand if the list is sequenced by office number, as in Table 3.3, finding the jack in a given office is easy while finding the office of a given jack is more involved.

These are simple examples of a built-in dilemma with recordkeeping and the use of lists, particularly lists on paper. "Look-ups" like the previous examples involve using something you know to find out something you wish to know on a particular list or table. This process is distinct from a database "query", where records are selected from a table based on a match with a specified attribute value. For our purposes, the look-up is like asking "What color is this cable?" while the query asks, "Which cables are blue?" As soon as you settle on a method for organizing the list, it will be convenient for look-ups based on that criterion, and less convenient for look-ups based on other criteria. This will always be true for a paper list. Using a computerized list, such as a spreadsheet or relational database, can avoid this problem as long as you can sort and order the list according to the criterion of your look-up. Thus you see one of the significant benefits of computerized records over paper records.

Lists also present a second problem, perhaps more challenging than the first. There needs to be a way to cross-reference the information on one list with information on another list. In a database, this is the same as relating a record from one table to a record from another table. To represent a circuit, for example, you must associate a number of items in sequence - the user device, the outlet, cable pairs, and perhaps a port. But for any set of lists, including both paper and computerized lists, there are only three basic ways to relate information together.

The first method is to maintain separate and independent lists of each item of interest, and make duplicate notes of related or cross-referenced information on two lists. Imagine a local taxi cab company that maintains a complete list of taxi cabs and a complete list of cab drivers. On any given shift, the dispatcher must know which driver is in which cab - a record on the driver list must be related to a record on the cab list. In order to do easy look-ups by cab or driver, the supervisor notes the driver of each cab on the cab list, and the cab of each driver on the driver list. You see that two entries are required.

Separate lists of jacks and backbone cable pairs, as shown in Table 3.4 and 3.5 for example, would require you to do the same thing. You can note the jack connected to each backbone cable pair on the backbone list and then note the backbone cable pair on the jack list. This method requires double entry, but capacity and availability for each list are easily obtained. For assignment and troubleshooting, the connection information is easy to obtain from either jack number or cable pair. Doing this for several lists representing circuit items may mean making several entries twice to capture circuit information, and this could be a very time-consuming task.

A second approach is to make one type of component a "master" list, noting the associated information against items on this list. The master

JACK NUMBER	CABLE PAIR
101	2
102	4
103	9
104	1
105	
106	7
107	5
108	
109	3
110	10

Table 3.4: Complete list of jacks with cable pair connections noted.

CABLE PAIR	JACK NUMBER
1	104
2	101
3	109
4	102
5	107
6	
7	106
8	
9	103
10	110

Table 3.5: Complete list of backbone cable pairs with jack connections noted.

JACK NUMBER	CABLE PAIR
101	2
102	4
103	9
104	1
105	
106	7
107	5
108	
109	3
110	10

Table 3.6: "Master" list based on jacks, with cable pair connections shown.

list represents a complete list of that particular item, but the list of other items goes away. The dilemma is, which list becomes the master and which is eliminated? If the dispatcher simply notes the assigned car on the driver list, he loses information about available cars. If the dispatcher notes the driver on a list of available cars, he loses information about available drivers. It's a simpler approach, but not as useful.

For circuits, you could make your list of jacks the "master" list as shown in Table 3.6 and note each backbone cable pair connected to each jack on this list. You reduce the connection information to a single entry, but your list of backbone cable pairs is essentially eliminated. Jack usage and availability are easy to see, but much more analysis is required to determine backbone cable usage. If you know the jack, the connected cable pairs are easily identified; however it may require searching the entire cable pair column to find the jack if only the backbone cable pair is known. For a large number of records, this can be very tedious. There would be no means of tracking bad cable pairs since bad pairs would not be listed by any particular jack. Reducing connection records to a single entry system reduces the upkeep but may result in giving up a lot of capability.

The third approach involves creating a list of related information for only those records which require it, instead of two separate lists or a master list. For the cab dispatcher, this would mean simply writing down each pair of assigned cars and drivers, without a complete list of either. While this may be the easiest method, it is the least useful in an overall sense.

Our list would simply track the cable pairs and jacks of connected circuits, as shown in Table 3.7. Both spare cable pairs and jacks would have to be inferred, since they are not listed anywhere. You would have to estimate spare capacity, and tracking of bad cable pairs again is not accomplished. If put in a spreadsheet, a circuit list could be sorted by jack number or cable pair so that information is easy to find and could be routinely helpful. This method works best where circuit information is valuable, but where backbone cable or station wiring capacity is not a concern or is not under your control.

These problems with lists are reduced but not altogether avoided in a relational database. Any primary table (list of records) in the database can usually be sorted and ordered according to your look-up criteria, as you would expect. However, the database also creates new tables, or relations, between records from the primary tables. For example, backbone cable pairs are listed in a primary table. To create a connection or association between different cable pairs, such as would be done to describe a circuit, the database typically puts this relationship as a record into a new table. The way these relationships are defined and organized is very structured and specific, and in some cases you can only look-up into the relational table by the criteria around which it is organized. If a particular database relates all circuit items such as cable pairs, terminations, ports, etc. to a Circuit ID, for example, then we

JACK NUMBER	CABLE PAIR
101	2
102	4
103	9
104	1
106	7
107	5
109	3
110	10
112	15
115	17

Table 3.7: List showing connected jacks and cable pairs only.

• USEFULNESS OF SPECIFIC INFORMATION
• ACCESS TO THE RECORDS
• SCOPE OF ITEMS OR ELEMENTS TO TRACK
• ANTICIPATED BENEFITS

Table 3.8: Factors to consider for a recordkeeping plan.

can only look up a particular circuit by Circuit ID. Knowing the jack ID of the circuit may not be enough to get you the information that you want. In a CMS, the more ways you can look up relations the better off you will be, and this becomes one measure by which you can evaluate CMS software. On the other hand, there will almost always be some look-ups that are more cumbersome than others depending on the structure of that particular database.

Now that you understand more about the behavior of lists, we can return to the factors that must be considered when developing a records structure, given in Table 3.8. The most fundamental aspect is the usefulness or purpose of keeping particular information about various things. What is specific information used for, and what can you do with it? The second aspect is that of access to the records. Where will they reside? Will they be centralized or distributed? Will the information be available at the right time and place? Thirdly, what is the scope or types of items for which you would keep such records? What would a record look like for each item? And finally, you must consider the environment - how will knowing specific information be beneficial for a particular cable plant architecture?

The Uses of Recordkeeping

Of course, one of the basic reasons for recordkeeping is to know quantity - how many do you have? The simple act of making a list, in any format, will meet this goal. But beyond a simple indication of quantity, tracking additional information can serve other purposes. The first type of information you can track is that of the characteristics, attributes, or qualities of the item in question. Location is probably the most significant attribute. Using the previous example, the list of jacks included the office number or location of the jack. In this case, the office number can be considered an attribute of the jack. If the list was kept by office number, then the jack ID would be an attribute of the office.

Shape, size, color, flavor, or even astrological sign are all examples of characteristics or attributes. For horizontal wiring, the fire safety rating, performance category, and length are all meaningful characteristics. Attributes of a pathway might include diameter, material, size, and length. Knowing the location of each jack is a significant benefit. In a similar fashion, you can consider attributes or characteristics of any type of thing and make them part of the record or list of that particular thing. While it is tempting to think of all the possible characteristics of an item and make them part of that item's record, you must choose carefully. Only those attributes which truly need to be known and which will be used in some fashion are worth tracking. For example, if all of your horizontal wiring is plenum cable, then there is little value in tracking this as part of a record since it can be taken for granted.

The second purpose of records is to indicate both the availability and usage of the item. For example, knowing the pair count of a particular backbone cable gives an indication of it's capacity, and can be regarded as an attribute of the cable. In order to know usage, you need to set up a table that lists each pair and whether or not it is in use. This provides a whole different level of information that can be used for planning and assignment as well, while simply listing the pair count could not. Communications hardware could be tracked in a similar fashion, listing the availability and usage of the individual ports on each device. Note that tracking information on availability and usage of a cable or device requires a completely different records structure than would be used for tracking attributes alone. Rather than a single record, there must be a set of records for each cable and device, one record for each usable entity (pair or port) contained within that cable or device. This is arguably a more complex records structure, and will likely take more time to set up and maintain.

The third purpose for our recordkeeping is to identify the connections or associations between various things in a way that represents the components of a circuit. There are many ways to do this, with different benefits and limitations in different situations, as you shall see. And tracking connections, of course, requires that you be keeping records of the things that will be connected. In other words, you must be maintaining records of individual cable pairs rather than a simple list of cables and pair counts. It is only when you get to this level of recordkeeping that your connectivity information becomes useful for assistance during MACs and troubleshooting. If you intend to track termination hardware, ports, or user devices as circuit components then you may have to maintain records of these as well, down to the individual port or connection point.

Indicating a connection can be as simple as noting it on an existing item record (list), as sophisticated as building it into a relational table in a database, or just about anything in between. As a general rule, the more sophisticated this representation becomes the more useful it likely will be, and the more effort it will likely take to maintain. However, the usefulness and ease of maintenance will depend on additional factors, such as whether your recordkeeping will be centralized or distributed.

Access to Records

Whether paper or computerized, records that are kept in a single location are centralized. When records are centralized, it becomes easier to cross-reference information from one list to another. The entire collection of records is available to you to fill in whatever information you need. For example, suppose you keep four lists. The first list is a list of PCs by office number, the second is a list of jacks by office number, the third is a list of backbone cable pairs showing the jack they are connected to, and the fourth is a list of IBM 3270 ports and the backbone cable pairs that each is connected to. If you now wanted to identify the port that a particular user and PC is attached to, you would obtain the office number from the first list, the jack from the second list, the backbone cable pair from the third list, and finally the port number from the fourth list. There would be no need to keep a separate list of circuit and connection information since it can be derived from your original lists. Circuit information can be assembled, given to a technician, and used for a new assignment or for troubleshooting.

Centralized records can be convenient for planning major moves, checking the capacity of a particular cable, disaster recovery, and simply knowing whether an office location has a spare jack. They can help you avoid a long walk to the other side of campus. But if you work routinely out in the wiring closets, the only information available to you is that which you copy and take with you. Additional information is unavailable and hence of no value in this situation. If, for example, you needed to identify a different cross-connection, you would have to spend the time to trace the physical connection itself. You may even have to test it to see if it is active, since many dead connections are often left on the frame. And you may be left guessing what happens at the other end of each cable.

Distributed records, on the other hand, could be located out in the wiring closets and equipment rooms. Here they can be referenced by the technician working in the wiring closet, and all the information about usage and connection for that closet is available. Connections do not have to be traced or tested. Unfortunately, you may still be left guessing what happens at the other end of each cable, but the records should at least indicate where each end is. But more importantly, distributed records are not as convenient to use for planning, pre-assignment of cable pairs or ports, and in some cases troubleshooting. Generally, centralized recordkeeping favors planning and administration, while distributed records tend to favor the technician doing specific tasks. This suggests, again very generally speaking, that characteristics and usage information are more appropriate for centralized than distributed records, while distributed records require connection information to be most useful. A list showing local connections would be easier to use than a list of complete circuit information. Here you see an example of how the method of access influences the preferred records format.

Paper lists will be either centrally located or distributed. Computerized records, either database or spreadsheet, can be a combination of both. Many CMS installations are on a stand-alone PC, and hence are centralized. Others can be on a network, and available at any PC with network access. Still other networked systems may allow access only from a PC with pre-loaded software, as with client-server applications. In the latter cases, you enjoy the benefits of centralized records while allowing access at multiple locations. But to get maximum use of the records in all situations, then the CMS should be accessible in or near each wiring closet, or at least each major equipment room.

Basic Recordkeeping for Quantity and Characteristics

Having considered the various uses of different information and the method of accessing it, we can now look at your selection of the types of items you wish to track. It helps to consider how one would structure information about different types of things to meet each of the three possible objectives of recordkeeping: Characteristics, usage, and connections. Different items use somewhat different information to meet these objectives. For example, the descriptions of a 400 pair riser cable, a bank of conduit, and a network wiring hub are all quite different. Each requires its own records format yet at the level of connections or associations they must be defined in compatible ways. A cable pair does not simply connect to a wiring block, it connects to a pair of pins or contacts. Another way of saying this is that the cable pair occupies some connection position on the block, and must somehow be associated to the record of that position. On the other hand, it is the cable and not a cable pair that is associated with a pathway, and there must be a cable record to associate with a pathway record if required. So representing a cable may require two types of records, one for cables and one for cable pairs.

- PATHWAYS
- BACKBONE CABLE
- HORIZONTAL CABLE
- TERMINATION HARDWARE
- USER DEVICES
- CONTROL DEVICES

Table 3.9: Elements of the communications infrastructure or transport system.

If you consider the variety of things that make up our communications infrastructure, you will find six basic types of components, listed in Table 3.9. Just about anything you would want to track will fall into these categories. You do not have to track every category with a separate set of records, and some categories you may ignore completely. But the categories that you do choose, and the purpose of the records for each category you choose, will create certain options as to the overall organization of the records, particularly with respect to connection information. Let's now consider how recordkeeping objectives are met for different categories of cable and equipment.

For tracking the characteristics of various components, simple tables can be developed for each type of component. Each table has an entry or record for each item of that type, followed by the attributes you choose to track for that component. The minimum attributes given here are consistent with the TIA/EIA 606 standard, discussed in a later chapter, but are not the minimum requirements set in the standard.

For backbone and horizontal cable, you should include the cable name, type, the pair or strand count, and the location of each end as a minimum. Each pair count going to a different location must have its own record. Additional information, such as details of the cable construction, its length, the data of installation, and the termination method could be included. For hybrid or composite cables, each subunit should be identified and its characteristics noted.

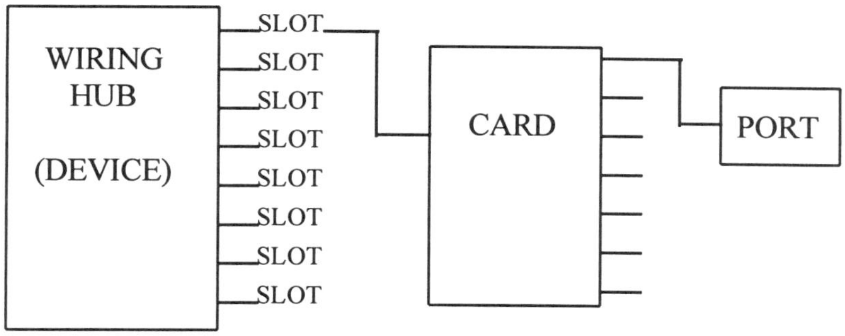

Figure 3.1: Hierarchy of chassis device, cards or modules, and actual ports.

Each pathway record should include the pathway name, type, usage (power, telecom, CATV, etc.), percent fill, and percent loading. To associate a cable to a pathway, the easiest method is to indicate the cables in each pathway as attributes in the pathway record. Conversely, the pathways through which a cable runs could be shown as attributes on each cable record, but they should then be listed in the sequence they occur. For the most convenient look-ups both methods can be used, relating cables to a pathway and pathways to a cable. As an alternative, pathways can be indicated on a plan drawing rather than a list of pathway records, with the included cables marked in text. Tracking pathways should be regarded as optional, and should only be done to meet an established need.

Keeping records of termination hardware such as punch-down blocks and patch panels is usually not practical for the sole purpose of tracking their quantity, characteristics, or attributes. Information about these devices may be meaningful when we discuss availability and/or connection recordkeeping, but even then must be evaluated carefully.

Tracking only the type and characteristics of controlling equipment as a part of your cable management records is also optional. In many cases, these records are available in other formats, such as in network or PBX management systems. Such records should be kept somewhere in some fashion. It can be helpful to track the location of active network devices on a plan drawing or CAD system, and such tracking must ultimately become part of an enterprise physical network management plan. But for purely cable management purposes, interest in such devices only occurs when you consider connection recordkeeping.

The same can be said of user or desktop devices such as terminals, PCs, printers, etc. Basic lists of both controlling and user equipment showing quantities, types, and other characteristics can be broadly defined as the function of Asset Management. This topic and its relationship to cable management will be discussed in a later chapter. If you are not concerned with connection recordkeeping, then it is recommended that asset management of hardware and devices be maintained separately and independently of cable records. On the other hand, if you desire connection information, then they will have to be integrated in some fashion.

An additional consideration arises with modern control equipment. An expandable 10BaseT wiring hub is a good example. More often than not, these devices accept cards or modules which go into slots in a chassis of some sort. Individual ports are a characteristic of the card, not the chassis itself as shown in Figure 3.1. Since the cards or modules can be swapped in and out, the configurations of ports on the overall device can change. Further, you may wish to have an inventory of the cards and available positions in each chassis in order to track capacity. You will have to decide if you can simply track the number of slots and the total number of each port type as an attribute of the chassis device, or else keep a record of each card and the number of each port type on the card.

Recordkeeping for Availability and Usage

From the previous examples, it should be evident that simple lists of cable and devices must be expanded in order to track the availability and usage of their components. In fact, you need a record or entry, in some way, shape, or form, for every available item. For the backbone cable, that is a conductor pair or fiber strand. So every pair or strand gets its own record where you can indicate its status. You could track cable at the individual conductor level, but this would rarely provide any additional benefit. You could also indicate the color code, type of connector or termination at each end, or testing data as attributes of the pair.

You can expand the lists of horizontal cable in a similar fashion to track it at the pair level. But you might want to consider whether this is necessary. In many facilities, a 4-pair horizontal cable is never used for more than one service. Use one pair and you've used them all, so to speak. If the technicians are very familiar with each circuit type, the actual pairs in use are known.. In this case, there is no benefit to tracking the status of individual pairs, and we can simply indicate the status of each cable on the original list of cables (or list of jacks, since they are really one and the same). Be sure to specify the wiring closet and jack location at opposite ends, and consider noting the wiring configuration (568A or 568B) of the jack.

Coaxial cable can also be treated differently, since it contains only a single pair of conductors. The list of coaxial cables themselves will suffice. Composite cables, with different cable types inside, can be treated as a list of pairs provided that the nature or type of each pair is now indicated, such as UTP, STP, coax, or fiber strand. The status of each can also be noted. As an alternative, you can create a name and record for each cable sub-unit and list the pairs or strands of each sub-unit cable along with its status. The type of cable should be listed for each sub-unit.

The availability and usage of a pathway such as conduit or cable tray is typically given in percent fill and/or percent loading, and can be tracked simply as an attribute of each pathway on your drawing or list of pathways. "Percent Fill" is the percentage of usable area that is already used within a tray or duct, while "Percent Loading" is the weight of the included cable as a percentage of the total weight allowed in a cable tray.

As noted in the previous section, keeping records of termination hardware becomes meaningful when tracking connection information. As long as you are tracking cable pair usage, then tracking the availability and usage of terminations may offer little additional benefit. The exception is when the termination itself is an active device, such as a smart jack or intelligent patch panel. Knowing the availability of controlling equipment and user devices may be very beneficial but still remains in the domain of asset management and can be maintained independently of the cable records. Including such equipment in your cable management records is reasonable as long as it can be accessed and maintained by the proper personnel. It is when you get to connection records that active devices can play a big, if not required, role.

Referring back to Figure 3.1, knowing which ports on controlling equipment are in-use and which are available requires you to track down at the port level. This can be done with a single table for each device listing the installed ports and their respective status. If you desire to know slot availability, then you must track all three levels. You can keep a master table for each device, with general characteristics and the total number of slots available. For each installed card, create a separate table giving it's chassis name, chassis slot number, other characteristics, and a list of port numbers for that card. The status of each port can then be indicated.

Recordkeeping for Connections

The tracking of connections or associations between the different components of a circuit requires some method of relating two or more pieces of information that may or may not exist as items (records) on different lists. Tracking connections automatically indicates availability and usage, and builds upon the ideas of the previous section. The items for which you desire availability information (pairs, jacks, ports, etc.)

are the items that must be related as a circuit, in some fashion. There are many ways of doing this, following the three general forms of records discussed in an earlier section: duplicate lists, a master list, or a single list of connected cables and devices. The intent, of course, is to ultimately show the sequence of cables and other elements that define a circuit between two active devices, such as between the PBX and a telephone or between a wiring hub and a terminal. There is no right or wrong approach, only a good approach - one which captures only the information you will use and maintain in a friendly format.

At a minimum, you must track a user device, a description of the horizontal cable, any backbone cable pairs, and some representation of the port and control device. At a greater level of detail, you can track the user device, a unique connection point on the user device, a mounting cord, the outlet connector, each horizontal cable pair or conductor, the horizontal cable termination, the patch cord or jumper at a pair or conductor level, the backbone cable termination, the backbone cable at a pair or conductor level, the other end backbone cable termination, the patch cord or jumper at a pair or conductor level, the port termination, wiring from the port termination to the port, the port itself, and the port control device. You can go into even further detail if the circuit includes additional segments of backbone cable, additional intermediate terminations and patch cords/jumpers at each wiring closet, and any boosters, regenerators, protection devices, splices, conversion devices, etc. Can you see how the apparently simple concept of cable management could become a very complex issue, with a CMS database containing potentially hundreds of thousands of records? Certainly it is in your interest to keep it simple, keep it useful, and to have it maintainable.

Whatever your final definition of a circuit, certain things will likely be true. First, the elements that you can tie together into a circuit will depend upon the original tables or lists of things that you create in the first place. You can make these tables or lists as simple or complex as you wish. Secondly, there must be a consistency from element to element as to what is being connected. In other words, a telephone circuit uses backbone cable pairs rather than an entire backbone cable. Thirdly, simply connecting cable pairs does not define a circuit - it must be referenced to either a desktop device, a port ID, or an independent circuit ID (or any combination thereof) in order to be useful. And fourth, circuit descriptions can go well beyond the realm of what you need to track availability and usage of cables and devices, and any circuit descriptions beyond this level are of questionable value.

For paper recordkeeping, circuit information must be approached in a simple fashion, since almost all methods involve multiple entry. Lists of ports, backbone cable pairs, horizontal cable pairs are recommended. Each list can have an entry to cross-reference each record to the next element in the circuit, i.e. the port list would indicate a connected cable pair, etc. Further, the horizontal cable list should indicate a user device ID or circuit ID to be complete. Availability and usage for each list is maintained, and a circuit can be derived by cross-referencing from table to table. A master table of horizontal pairs with backbone cable pair and user device ID indicated would work, but backbone cable pair availability is harder to discern.

To maintain cable availability information, at a minimum there must be lists of both backbone and horizontal cable pairs, with port and horizontal pair indicated on the backbone list and backbone pair and user device ID indicated on the horizontal list. Another minimum alternative is to place a list of both local backbone and horizontal cable pairs in each wiring closet, indicating any other pair to which it is connected. Such lists avoid the tracing of jumpers and do provide availability, but the lists are distributed and cannot be accessed at a central location.

The only paper approach that is always single-entry is the use of a circuit table, where the port, backbone pairs, horizontal pairs, and user device ID are all listed in a single table. This can work well for troubleshooting, but usage and availability information is difficult to determine.

If you consider keeping lists of all six elements in Table 3.9 and want to track circuit connections, the number of tables required expands to eleven, listed in Table 3.10. There are six to eight or possibly more entries (representing connections or associations) per circuit, not counting fixed connections. For a cable plant of 2,000 backbone pairs and 4,000 horizontal pairs, this translates into possibly 25,000 or more

- PATHWAYS
- BACKBONE CABLES
  - BACKBONE CABLE PAIRS
- HORIZONTAL CABLES
  - HORIZ. CABLE PAIRS
- TERMINATION HARDWARE
  - TERM. HDWE. POSITION
- USER DEVICE
- CONTROL DEVICES
  - CARDS OR MODULES
    - PORTS

Table 3.10: Basic tables to track all elements at the circuit level, with usage information maintained.

individual records that must be maintained. This is a typical site of less than 500 users. Applying the TIA/EIA 606 standard, without a user device table or control device information, only reduces that number to between 10,000 and 20,000 records. That's several hundred pages of records.

The alternative, of course, is some computerized method of recordkeeping. While tracking 20,000 records in several spreadsheets is possible, it is unlikely that all the proper associations of different records can be made - for this, a relational database is the tool of choice. But when it comes to designing a database, many different approaches can be used for a table structure, provided that what you ultimately see on the screen relates the right things in the right fashion. It is the process of "cataloging", or developing the categories of elements and their descriptions in the database, which is covered in Chapters 14, 21, and 28. Further information is also provided in Chapter 6 on "CMS Functionality". The intent of the current chapter is to give an overview of the general possibilities and their implications.

Naming Conventions

The naming conventions or nomenclature for cables, pathways, and devices must have the characteristics listed in Table 3.11, especially when being developed for a CMS database. This will be true for naming types of things as well as the individual things themselves. For example, the database may require names for different types of PCs (PS/70, PS/80, 386, 486, etc.) as well as a name for each individual PC. In addition, we will discuss the directionality of cables and the difference between logical and physical descriptions.

The first characteristic of nomenclature is that each type name and item identifier must be unique. Type and item names are often used as key fields for each record, and the database simply won't work properly if duplicates are present. For your own purposes, names must be unique to avoid duplicates, which would cause improper accounting, incorrect entries, and other types of confusion. It's not unusual to have jacks numbered 001 to whatever in each wiring closet, with the jack numbers duplicated from floor to floor. Some means must be adopted to make them unique, either by renumbering or adding additional characters to the jack ID. Note that upper and lower case in most databases represent unique characters.

Second, item names should be structured, that is, have some form, purpose and consistency from name to name. For example, individual horizontal cables (jacks) might use a ten-character name where the first three characters represent the building, the next three characters represent the floor or Telecommunications Closet, and the next four characters represent an alphanumeric jack designation as shown in Figure 3.2. This addresses the previous problem by making existing jack numbers unique. The structure must apply to, and be applied to, all items of that type. The process of developing a detailed naming

- UNIQUENESS
- STRUCTURED
- QUERIABLE
- COMMON ACROSS GROUPS

Table 3.11: Characteristics of good naming conventions.

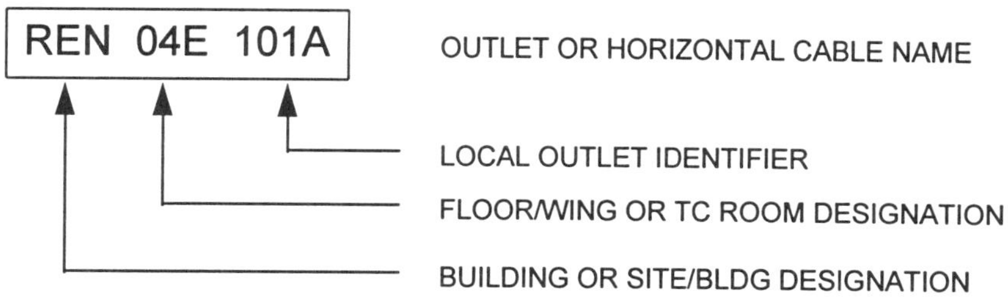

Figure 3.2: Example of a structured naming system for horizontal cables.

structure for cable and devices is explored in Chapter 21.

The structure used to name different cable and equipment can vary according to need. Actual room numbers or locations can be embedded in names for those things that will never move, such as a cable or termination device. A type of service or particular system designation can be used such as network type, or a department code, manufacturer's ID, or other method can be applied. For the names of cable and equipment types, a structured method can be useful when differentiating between closely related types. For example, it's easy enough to define devices called "Phone" and "PC", if that is as far as you need to go. But you may wish to differentiate between "4PR-CAT3" and "4PR-CAT5" cable types, in which case the structured naming is appropriate. Structured naming of type or item should only go far enough to achieve uniqueness by making differentiations that matter and are useful in your work.

Certain items can be numbered sequentially without regard to naming structure, and some CMS databases do this automatically. The advantages here are uniqueness is always maintained, and you never have to remember the naming structure or the next name in sequence. The disadvantages are that you cannot create item names in the field during inventory, items following a pattern can have quite different numbers (such as termination devices), and you will have to use different fields in the item record to look-up or query for that item type.

The third consideration for naming is closely related to the idea of structure - the naming structure should be friendly and useful for performing queries against those records. In a prior example, the use of Building and Floor designations for jack naming allows you to display or report on all jack records for a selected Building and/or Floor, since most databases allow the use of wildcard characters. Final adaptation to a CMS must consider this point. In this example, if a particular table can be queried through any field in the record, and Building and Floor fields are included, then this may not be a concern.

If a field is not available for certain information, then encoding it in the name can be beneficial. For example, you may wish to indicate whether a particular backbone cable is first or second level. The cable type field, meanwhile, only indicates the physical type of the cable, i.e. 300-pair ARMM. Therefore the first or second level designation can be built in to the cable name, provided it is done consistently for all backbone cables. Now you can query for all first or second level backbone cables. More generally, you can often include logical information (the cable use or purpose) in your cable naming where the table fields provide for physical information (the cable construction). This is not generally the case for devices, for which we will discuss physical and logical names in a moment.

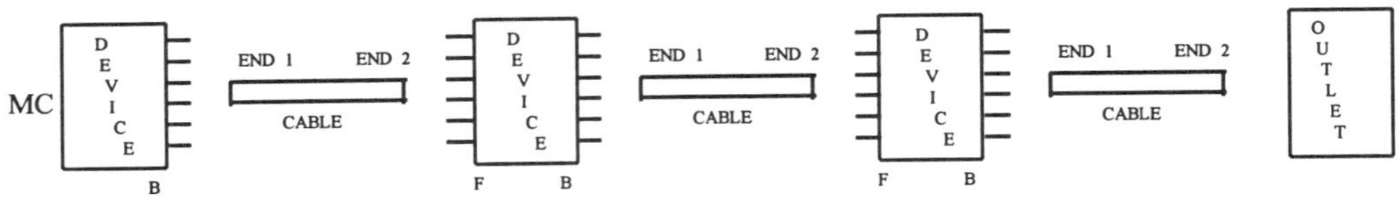

Figure 3.3a: Circuit and element directionality from the Main Cross-Connect to the workstation.

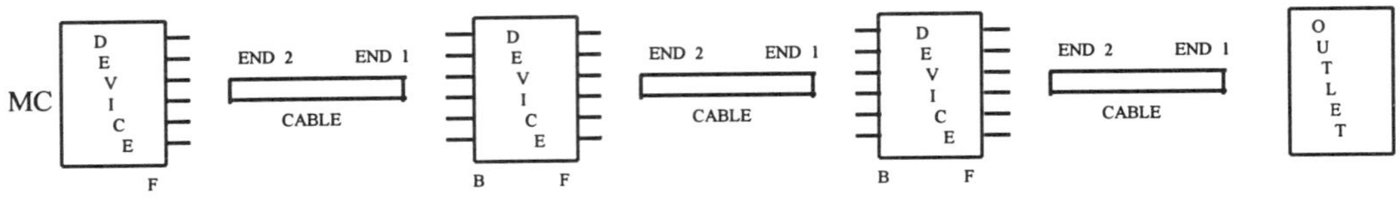

Figure 3.3b: Circuit and element directionality from the workstation to the Main Cross-Connect.

Queriable naming implies structure that is easy and friendly to use, not obscure and hard to remember. You will benefit if you can use encoding which already exists and with which people are already familiar. If you make up new codes in your nomenclature, be sure they are part of your final documentation and training and that they are readily available to each CMS user.

Finally, nomenclature must be made common to all groups and across all systems, to the greatest extent possible. There should be only one name for each cable type, and one name for each type of device. You will need to review current labeling and recordkeeping systems to see what is in use, where overlaps occur, and where adaptations must be made. Adaptation should be done to achieve the greatest functional benefit from two or more systems, even if that means changing something already in place. By using common names where possible, you will be able to cross-reference one system of records to another, such as from the CMS to the PBX administration software. Without such naming, remembering two or more schemes is a burden and users will tend to favor one system over another.

If you decide to adopt a new naming convention rather than using names currently referenced in other systems, make sure that the names can be modified in the existing system. You will then need to develop an action plan to assure the other recordkeeping systems are updated for all affected items.

Another issue you will face in naming is that of cable directionality - a cable has to start at one end and stop at the other end. In other words, it is important to understand that a cable runs from one location to another, and the database functionality will operate on it accordingly. The display, querying, and reporting of cable records may be based on End 1 (From) and End 2 (To) being specifically identified. This convention may need to be set according to the functionality of a specific CMS; otherwise it is totally arbitrary and up to you. It needs to be applied as consistently as possible.

Having said that, there may be a general, intuitive sense among the technical staff of which end is which. The traditional convention from telephony is that cables start at the Main Cross-Connect (End 1) and run outwards from there, as indicated in Figure 3.3a. For second level backbone, the cable starts at the Intermediate Cross-Connect (End 1) and runs to the Telecommunications Closet. Horizontal wiring then runs from the Telecommunications Closet (End 1) to the workstation outlet. Tie cables between other Equipment Rooms and Telecommunications Closets are less clear, although larger closets where cabling tends to converge and be centralized would be favored as End 1.

Cabling that is daisy-chained from device to device should have End 1 at the connection to the hub or control device. Each cable in the chain should have End 1 nearest the control device. For bus networks, either end of the bus cable can be designated End 1.

Given that CMS-based administration no longer requires building circuits from the MC out, and that different, often less-centralized communications services must be considered, the opposite approach to cable directionality is appealing. With a CMS, the prime point of consideration for service is the user, specifically the workstation outlet. By having the user as a reference, the building or tracing of the circuit starts at the outlet and proceeds in a direction towards the appropriate control equipment. In this case, the outlet becomes End 1 of the horizontal cable, and End 1 of the backbone cable is in the Telecommunications Closet as shown in Figure 3.3b. The MC is then End 2 of all first level backbone cable. You should consider the nature of your CMS administration and the CMS functionality to see how circuits will be assigned, and then consider each of these approaches.

A similar situation may arise for devices, where the logical "front" and "back" of a device must be determined. Again, CMS functionality may dictate this convention or it may be arbitrary. It's possible that you can take any set of connections or ports and call them front or back, no matter where they are physically located. Diagram the devices in circuit sequence, and see how the individual connections/ports line up. Starting at whichever end of the circuit you think of as the beginning, go along the circuit diagram noting the first side of each device as the front, the second side as the back. The respective sets of connections/ports would be front or back, respectively. Similar to the cable end example of Figure 3.3b, we recommend that the "front" of a device point towards the user, while the "back" of a device point towards the network. As an alternative, the ports or connections can be configured with respect to their physical location (front or back) on the device, such as might be done with an outlet or patch panel. This approach may or may not support the circuit tracing functionality in a given CMS.

We have acknowledged that cable and connectivity management can mean different things to different groups that support voice and data communications. PBX managers, LAN administrators, Facilities staff, and wiring technicians all have different expectations, since they usually rely on different descriptions of the world around them. Sometimes these descriptions differ with respect to the logical and physical information which is significant to each group, as suggested in Table 3.12.

For example, network recordkeeping is more dependent on logical descriptions and addresses rather than the physical descriptions required by a connectivity database. In an IBM SNA network, administrators are used to thinking of system connections in terms of a Logical Unit (LU) number. A PBX uses an extension and a physical port address such as Equipment Line Location (ELL). A Token Ring network uses the ID of the network board in each device and another logical ID based on the network operating

DEVICE	PHYSICAL NAME	LOGICAL NAME
CONTROLLER	SERIAL NUMBER OR ASSIGNED NUMBER	NETWORK ADDRESS
PORT	ASSIGNED NAME OR CONNECTOR NUMBER	NETWORK ADDRESS
TERMINAL DEVICE	SERIAL NUMBER OR ASSIGNED NAME	NETWORK ADDRESS
CABLE	SHEATH	CABLE AND PAIR COUNT

Table 3.12: Comparison of physical and logical names.

system. Ethernet networks require a unique logical ID for each connection (which may be a physical board ID) and then overlay other names from higher level protocols such as DECNet.

Let's look for a moment at control equipment - devices that support what you would consider to be the physical ports of a system. This equipment must usually be included in the CMS in order to depict a circuit. As a generic device in the database, it may have its ports or connectors identified only by number, such as 1 to 8, 1 to 32, etc. If logical addresses are used to identify physical ports, two situations in particular can cause a problem. First, consider the case where an individual controller is assigned a new network address. It has the same physical connections, but now you may have to remove the original device, cards, ports, and their connections from the database and then reenter them.

In the second case, imagine that you are tracking individual assets (such as a controller) by serial number. You may do this for billing of depreciation and maintenance, so accuracy is essential. What happens when it fails and must be replaced? If entered as a logical address, you don't have to change much - the replacement unit is likely to have the same address assigned. The serial number, tracked as an attribute of the logically-named device, would have to be changed to match the serial number of the new equipment. But then you would *lose track of the first unit*, whose maintenance, repair, and depreciation information is important! Therefore, if a logical address is used to name control equipment and/or its ports, it might have to be changed when no connections are actually broken, and it won't work for tracking equipment for asset management.

What happens if you use a physical name for each device? Initially, you have to develop a system for naming these devices on a physical basis, such as serial number. Because the serial number itself tells you nothing about the device type or configuration, you could consider a structured name based on equipment type or location, as discussed in a previous section. In either case, individual devices can be tracked whether they are installed or not. If you replace a device, you will probably have to change its connection records as well. The real risk in using physical names is that they don't necessarily relate to an administrator's understanding of a network.

While neither a physical or logical naming system is perfect, physical naming is preferred in a CMS. The value of asset tracking is too great to be designed out of a system. The ideal system would track devices and ports by physical names, but would also be able to assign other physical or logical attributes to the device or port itself - a fair compromise. Perhaps charge codes could be assigned to each port. Use of physical nomenclature allows the logical names of network connections to be assigned independently, and according to each network's needs. It makes it easier for network information from different networks to exist side-by-side.

A similar situation arises when you consider physical and logical descriptions of terminal (desktop) equipment. In the case of voice equipment the telephone extension number is often used as a name, an obvious logical description that works well. For data systems, your choice of description may depend on how many different network connections the device supports. For example, dumb terminals serving a single network can be assigned logical network names, but again asset tracking is lost. On the other hand, PCs can support a variety of network connections via add-on boards. This situation almost forces you to name the PC on a physical basis, independently of any network. And of course, you then get the advantage of asset tracking.

Without realizing it, much of your cable will be named on a logical basis rather than physical. Remember that many multipair backbone cable installations are given pair counts on a logical basis, and the same logical cable runs to many locations in different cable sheaths. For that matter, you could have two or more logical cables in the same physical sheath.

In a CMS, unique cables must be defined between each starting and ending location which likely correspond to a particular pair count of a major backbone cable. For example, Cable 1202, pairs 301 to 500 run from the MC to a particular Telecommunications closet. This cable and pair count must have its own record as a unique cable in the CMS. In this case, you can force a modified name that indicates both cable and count, or use a name that indicates the termination points and the count. While these remain logical names, they more closely resemble the physical cable sheaths themselves. If the cable records are kept in a single table, it is useful to indicate the starting and ending locations and the sheath ID of each separate pair count.

Methods of Labeling

As mentioned earlier, naming and labeling of cables and devices provide benefits whether or not actual records are kept. Naming would apply to many of the items previously discussed such as cables, cable pairs, outlets, work areas, user devices, control devices and ports, pathways and splices, and possibly terminations. If connection recordkeeping is desired, then items should be labeled at the level where connections are made. For example, if your records show a cable pair connected to a port, then the cable pair and port terminations (the two points where the physical connection is made) should be labeled accordingly. The TIA/EIA 606 Standard has specific requirements for labeling which are discussed in Chapter 4.

Labeling must be approached with precision. Handwritten labels must be legible and in the proper case. A method should be adopted to distinguish the number "1" from the letter "l", the number "0" from the letter "o", the number "2" from the letter "z", and the number "5" from the letter "s". All labels should be in a highly-visible location.

Backbone cables should be labeled at each end, with individual cable pairs labeled at their points of termination. Horizontal cables should be labeled at each end and at their point of termination. Each work area outlet should be labeled with the horizontal cable name. If labeled on a removable cover, each outlet should be labeled internally in some permanent fashion in case the cover is removed.

Often overlooked is the need for specific work area names and labels. Each office or cubicle should be uniquely labeled near its entrance. User devices such as PCs, terminals, and printers should be labeled with an ID or tag number using a permanent marker, an engraver, or a permanent adhesive label visible from the front of the device.

Control devices should be labeled with a logical network ID or a physical asset ID. For connection tracking, the port termination should be labeled on the wiring block or patch panel, and on the control device itself if each port is a separate connector on the device. For some devices, this can be met if the device itself is labeled and the individual port connectors are identified with a sequence number such as 01, 02, 03, etc.

Each pathway should be labeled at each end, at each space where pathways come together, and at intermediate points if the pathway is accessible. Splice enclosures should be labeled on the cover, and it is helpful to label the cable ends within the splice enclosure. For labeling of passive termination devices, refer to Chapter 4. If labeling of termination hardware is desired, you may wish to consider additional labeling of the hardware with the terminated cable pair or port information as well.

There are different label types used for the different administration elements. For cables, there are self-laminating plastic labels which have a space for writing the element name and a clear tail which, when wrapped around the cable, overlaps and covers the written label and protects it from wear. Larger cables can have labeling written directly on the jacket or be tagged in some permanent fashion.

Adhesive labels for outlets and devices should be a permanent style suitable for indoor or outdoor use as required. Labels tend to stick on metal surfaces better than plastic, and will frequently peel off quickly from a textured surface. Active devices tend to generate heat and operate at elevated temperatures, which tends to weaken the adhesive over time. Labels can be made up on the sheet ahead of time and attached to equipment during inventory to save some time and avoid duplications.

Machine-printed labels can be used if the adhesive is adequate, and may be easily printed on a computer printer or using a hand-held device. The W. H. Brady Company makes the most comprehensive line of labels and label printers suited for labeling according to TIA/EIA 606. The Brother Company and Dymo also make hand-held labelmakers suitable for indoor labeling.

All handwritten labels should be done with a permanent marker, such as the Sharpie Fine Point or Extra Fine Point marking pen. Permanent marking can usually be removed with acetone or nail polish remover. Some devices such as a 66 block have little space for applying an adhesive label, and should be marked directly with a marker. Label covers for the 66 block fanning strip are not recommended, since they can be removed and reinstalled out of sequence.

Other termination hardware such as the 110, BIX, and Krone blocks have designation strips and label holders as part of the device. These should be used whenever possible. Different color strips can be used to indicate different types of connections, either according to TIA/EIA 606 or a color code of your own choosing.

You may wish to consider bar-coding the labels on certain cables and devices for easier inventory and tracking of the device location. Specific requirements for using bar-codes are given in TIA/EIA 606. In such cases the bar-code reader must be tied in to the inventory recordkeeping system. As with all aspects of your recordkeeping, bar-coding will only be useful if it can be applied consistently and used on a routine basis.

THE TIA/EIA 606 ADMINISTRATION STANDARD

4

As an outgrowth of the TIA/EIA 568 standard for commercial building wiring, the TIA/EIA Working Group TR-41.8.3 has developed and issued a standard for the administration of cable facilities. This document, entitled "Administration Standard for the Telecommunications Infrastructure of Commercial Buildings" (TIA/EIA 606 Standard) provides a general model for structuring and recording information about the physical transport system.

The purpose of the standard is to provide a uniform administrative scheme that is independent of the specific communications systems operating over the facilities. The standard states that it intends to reduce the number of "incompatible and incomplete" administrative schemes and to allow for the development of new products and systems which can be administered within its requirements. However, the standard itself leaves much latitude in how the required information and relationships are actually structured in an administrative system. "Compliance" can take many forms, particularly with respect to CMS software. Many existing systems are either already compliant, close to compliance, or can be made compliant through an acceptable implementation approach.

The General Framework of TIA/EIA 606

The 606 Standard encompasses telecommunications spaces, cable pathways, grounding systems, wiring, and termination hardware. The telecommunications spaces could be anything from TC rooms to manholes to work areas (offices). Cable pathways include raceways, conduit, ducts, sleeves, etc. The 606 Standard *does not* include administration of user equipment in work areas, nor does it include controlling equipment such as hubs, controllers, multiplexers and the like. The administrative scheme defined by the standard basically involves documentation aspects such as labels, records, drawings, reports, and work orders.

As you might expect, the 606 Standard conforms to the TIA/EIA 568A standard for commercial building wiring, the TIA/EIA 569 standard for pathways and spaces, and the TIA/EIA 607 standard for telecommunications grounding and bonding systems. In fact, the 606 standard is intended for facilities that are designed and maintained according to these standards, may not be readily applied to existing building wiring systems, and recognizes that modernization of older systems may be a significant expense. But even existing facilities can use the Standard as a reference point and adapt it to meet specific requirements, just as you might apply TIA/EIA 568A on a selective basis.

TIA/EIA 606 distinguishes between mandatory and advisory portions of the standard. Mandatory elements are those that specify the minimum acceptable requirements, while advisory portions are optional elements which can enhance performance and provide additional benefits. Mandatory elements in 606 are very basic. Most administrative systems will have to deal with many of the advisory elements, and how they do this will affect the usefulness of the system. For example, the termination of a 4-pair horizontal cable could be treated as four individual pairs or as a single four-pair entity. The first case is most flexible in showing various connections, and is also most complex. The second case is simpler but more limited. In the standard, defining a unique termination position for this cable is mandatory, but the particular description is not. Similar situations are found throughout the standard.

Also according to TIA/EIA 606, information is either an "identifier" (the unique name of a particular item) or information about an item which we refer to as attributes or characteristics. An example would be a cable record consisting of a unique cable name (the identifier) and the cable type (an attribute). The standard also requires a "linkage" between certain

	RECORD	REQUIRED INFORMATION	REQUIRED LINKAGES
PATHWAYS AND SPACES	PATHWAY	Identifier Type Fill Loading	Cable Records Space Records Pathway Records Grounding Records
	SPACE	Identifier Type	Pathway Records Cable Records Grounding Records
WIRING	CABLE	Identifier Type Unterm. Pair/Cond. Nos. Damaged Pair/Cond. Nos. Available Pair/Cond. Nos.	Term. Position Records Splice Records Pathway Records Grounding Records
	TERMINATION HARDWARE	Identifier Type Damaged Position Numbers	Term. Position Records Space Records Grounding Records
	TERMINATION POSITION	Identifier Type User Code Cable Pair/Cond. Numbers	Cable Records Other Term. Pos. Records Term. Hardware Records Space Records
	SPLICE	Identifier Type	Cable Records Space Records
GROUNDING AND BONDING	TMGB	Identifier Busbar Type Grounding Cond. Identifier Resistance to Earth Date Measurement Taken	Bonding Cond. Records Space Records
	BONDING CONDUCTOR	Identifier Type Busbar Identifier	Grounding Busbar Records Pathway Records
	TGB	Identifier Type	Bonding Cond. Records Space Records

Original Note: All records may also contain variable optional information and linkages to other records both within and outside the scope of this standard.

Table 4.1: Summary of record elements in the TIA/EIA 606 administration standard.

items or records, something we have called a connection or association, which represents some relationship between two things. Relating a cable pair to a specific termination position on a wiring block is an example of a linkage. For each infrastructure element, the standard specifies the required identifiers, attributes, and linkages, as shown in Table 4.1. Diagrams of administration elements and administration point labeling are given in Figures 4.3 and 4.4 at the end of this chapter. Optional attributes and linkages are also suggested, but not on a complete or all-inclusive basis. Most significantly, the length and format of these fields is not specified. The user is on his or her own to develop nomenclature to fit the required information and linkages of the standard.

The content and format of specific reports are offered on an advisory basis only. A circuit report showing the end-to-end connectivity of the circuit is only recommended, not required. The standard suggests this report should include a user code, cables, and termination positions from end-to-end. A number of different circuit formats could meet this recommendation.

Required Information and Linkages

Having covered the background, we can now consider the highlights of the TIA/EIA 606 standard in more detail. As shown in Table 4.1, records are broken down into general categories of pathways and spaces, wiring, and grounding and bonding. Further breakdown is given in each category, and the standard discusses mandatory and advisory elements for each. While the following discussion covers all significant aspects, consult the standard itself for complete details.

Starting at the top of Table 4.1, each pathway (conduit, tray, raceway, etc.) that conveys telecommunications media from space to space must have a unique identifier and be so labeled at each end point in a TC room, Equipment Room, or Entrance Facility. Intermediate points or spaces where three or more pathways converge, such as a manhole or pull box, shall have each endpoint labeled at that space. Partitioned pathways such as a duct bank or conduit with an innerduct must have unique identifiers for each partition and have each partition so labeled. Additional labeling may be desired. A complete pathway that joins two or more different pathway types or sizes must have each segment treated as a separate pathway. Partitioned systems such as cellular floors, which are impractical to label, must be documented on drawings.

Information required for each pathway include its type, its capacity (how full it currently is), and its loading (how heavy it currently is). Each pathway identifier must be linked to the cables within it, the telecommunications space at each end, any access spaces along its length, any

PATHWAY RECORD	SAMPLE DATA	COMMENTS
REQUIRED INFORMATION:		
PATHWAY IDENTIFIER	CD34	
PATHWAY TYPE	50MM (2") EMT	TRADE #2 CONDUIT
PATHWAY FILL	20%	PRESENT FILL
PATHWAY LOADING	N/A	NO SPEC FOR CONDUIT
REQUIRED LINKAGES:		
CABLE RECORDS	C0001,C0002	CABLE IDENTIFIERS
SPACE RECORD END 1	D306	IDENTIFIER FOR OFFICE
SPACE RECORD END 2	3A	IDENTIFIER FOR TC ROOM
SPACE RECORDS (ACCESS)	N/A	INTERMEDIATE POINTS
OTHER PATHWAY RECORDS	N/A	CONNECTING PATHWAYS
GROUNDING RECORD	N/A	

Table 4.2: Pathway record for a conduit showing required information.

connecting pathways, and the method of grounding. Examples of records for conduit and cable trays are given, with an example for a conduit shown in Table 4.2. Optional information such as length, maximum fill, maximum load, condition, usage, number of bends, drawing number, and product code is listed.

Each telecommunications space (types are listed in Table 4.3) must also be uniquely identified and labeled, preferably at the entrance to the space. The identifier and the type of space must be recorded, and the identifier must be linked to the pathways, cables, and ground buses that terminate in this space. Examples of records for a Telecommunications Closet, a work area, and a manhole are given, with an example for a TC room shown in Table 4.4. Optional information such as space size, space location, area served, and door lock key number is listed.

The standard recommends a pathway summary report that lists all pathways, their types and present fill and load. Similarly, a space summary report is recommended that lists type and location for each space. Record drawings must be maintained which indicate the name, location, and size of each pathway and space.

The Wiring System is broken down into the separate administration of cable, termination hardware, termination positions, and splices. Each cable must be uniquely identified and labeled at each end, with labeling at intermediate locations recommended. Spliced cables of the same type and count are administered as one cable, while spliced cables of different type or pair count may be administered separately. It is recommended that the individual components of a hybrid cable be named and labeled separately.

- EQUIPMENT ROOM
- TELECOMMUNICATIONS CLOSET (TC ROOM)
- WORK AREA
- ENTRANCE FACILITY
- MANHOLE
- HAND HOLE
- PULL BOX
- SPLICE BOX

Table 4.3: List of some possible space types.

The individual cable record must indicate the cable type by manufacturer and manufacturer's designation, and then include every pair/conductor in the cable. Unterminated, damaged, and available pairs must be designated. The cable identifier must be linked to all pathways in which it runs, any splice information, and the grounding record. Individual pairs/conductors must be linked to a termination position record for each end. The Standard gives examples for a fiber backbone cable, a copper backbone cable, and a horizontal UTP cable, with an example for a copper

SPACE RECORD	SAMPLE DATA	COMMENTS
REQUIRED INFORMATION:		
SPACE IDENTIFIER	3A	CLOSET 3A
SPACE TYPE	TC	TELECOMMUNICATIONS CL.
REQUIRED LINKAGES:		
PATHWAY RECORDS	CD34, CT64, SL02-5, CD02	PATHWAYS THAT TERMINATE IN THIS SPACE
CABLE RECORDS	C0001,C0011,CB02	CABLES TERMINATING IN 3A
GROUNDING RECORD	TGB35	GROUNDING BUS BAR

Table 4.4: Space record for a Telecommunications Closet showing required information.

CABLE RECORD	SAMPLE DATA		COMMENTS
REQUIRED INFORMATION:			
CABLE IDENTIFIER	CB02		BACKBONE CABLE CB02
CABLE TYPE	100-PAIR CMR		PHYSICAL DESC.
UNTERM. PR./COND. NOS.	0		LIST OF UNTERMIN'D PRS
DAMAGED PR./COND. NOS.	0		LIST OF DAMAGED PRS
AVAILABLE PR./COND. NOS.	0		LIST OF PRS NOT IN USE
REQUIRED LINKAGES:	END 1	END 2	
PR 1 TERM. POS. RECORD	C4R6-001	3A-A17-001	PAIRS TERMINATED ON
PR 2 TERM. POS. RECORD	C4R6-002	3A-A17-002	PUNCH-DOWN BLOCK
PR 3 TERM. POS. RECORD	C4R6-003	3A-A17-003	
. . .	. . .		PAIRS 4-96 NOT SHOWN
PR 99 TERM. POS. RECORD	C4R6-099	3A-A17-099	
PR 100 TERM. POS. RECORD	C4R6-100	3A-A17-100	
SPLICE RECORDS	N/A		
PATHWAY RECORDS	SL02-05		SLEEVE SL02-05
GROUNDING RECORD	N/A		

Table 4.5: Cable record for a copper backbone cable showing required information.

backbone cable shown in Table 4.5. Optional information such as length, product code, and ownership is listed.

Each piece of termination hardware such as a patch panel or wiring block must be uniquely named and labeled. Termination hardware labeling in the work area is not required if the termination position label is used. The hardware type and any damaged positions must be indicated. It must be linked to a space (location), a grounding record, and each termination position record that applies. Examples for an optical fiber patch panel and a punch-down block are given, with an example of the optical fiber patch panel shown in

TERM. HDWE. RECORD	SAMPLE DATA	COMMENTS
REQUIRED INFORMATION:		
TERM. HDWE. IDENTIFIER	B101-01	FIBER PANEL B101-01
TERM. HDWE. TYPE	F900-12	12-POSITION PANEL
DAMAGED POSITION NOS.	NONE	
REQUIRED LINKAGES:		
TERM. POS. RECORD 1	B101-01-A1	FIBERS TERMINATED WITH
TERM. POS. RECORD 2	B101-01-A2	DUPLEX CONNECTORS
TERM. POS. RECORD 3	B101-01-A3	
TERM. POS. RECORD 4	B101-01-B1	
TERM. POS. RECORD 5	B101-01-B2	
TERM. POS. RECORD 6	B101-01-B3	
SPACE RECORD	B101	LOCATION OF TERM HDWE
GROUNDING RECORD	N/A	

Table 4.6: Termination hardware record for an optical fiber patch panel showing required information.

TERM. POS. RECORD	SAMPLE DATA	COMMENTS
REQUIRED INFORMATION:		
TERM. POS. IDENTIFIER	B101-01-A1	SPACE B101,PNL 01, POS A1
TERM. POS. TYPE	DUPLEX OPTICAL FIBER PORT	PHYSICAL DESCR. OF TERM POS., PART NO., ETC.
USER CODE	N/A	USED ONLY AT WORK AREA
CABLE PR./COND. NOS.	1	ONE 2-STRAND PAIR
REQUIRED LINKAGES:		
CABLE RECORD	F16	CABLE IDENTIFIER
OTHER TERM. POS. REC. 1	4A-01-A1	SPACE 4A, PNL 01, POS A1 POSITION A1
OTHER TERM. POS. REC. 2	N/A	FOR CROSS CONNECT
TERM. HDWE. RECORD	B101-01	
SPACE RECORD	B101	EQUIPMENT ROOM B101

Table 4.7: Termination position record for a punch down block showing required information.

Table 4.6. Optional information such as installed protection devices is listed.

Each termination position must therefore also be named, and labeled if possible. In cases of high termination density, the termination hardware label is sufficient and the termination positions can be determined by convention for that device. The termination position must be identified as to type, the pair/conductor terminated there, and a user code (for a telecommunications outlet connector only). These positions can be defined individually or in sets, and each must be linked to a space (location), the cable name of the pair/conductor terminated there, and the termination hardware record. The termination position at the other end of the pair/conductor and the termination position of a cross-connection are also linked to this record. An optical fiber panel connector, a punch-down

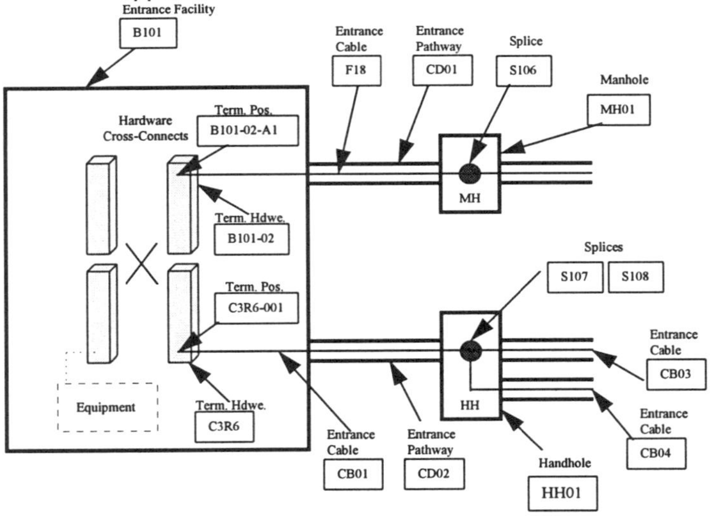

Figure 4.1: Diagram of singly- and separately-administered spliced cables.

SPLICE RECORD	SAMPLE DATA	COMMENTS
REQUIRED INFORMATION:		
SPLICE IDENTIFIER	S107	SPLICE S107
SPLICE TYPE	CRIMPED	METHOD OF SPLICING
REQUIRED LINKAGES:		
CABLE RECORDS	CB01, CB03	CABLES SPLICED
SPACE RECORD	HH01	LOCATION OF SPLICE

SPLICE RECORD	SAMPLE DATA	COMMENTS
REQUIRED INFORMATION:		
SPLICE IDENTIFIER	S108	SPLICE S108
SPLICE TYPE	CRIMPED	METHOD OF SPLICING
REQUIRED LINKAGES:		
CABLE RECORDS	CB01, CB04	CABLES SPLICED
SPACE RECORD	HH01	LOCATION OF SPLICE

Table 4.8: Splice records for a splice in separately-administered cables showing required information.

block position, and a telecommunications outlet connector are used as examples, with an example for a punch-down block shown in Table 4.7. Optional information such as signal carried and cross-connect type is listed.

Documentation of splices depends on whether the spliced cables are treated as a single cable or as more than one cable, as diagrammed in Figure 4.1. In either case, each splice must be uniquely named and labeled on the splice closure, and its type and space (location) must be identified and recorded. For a splice in a singly administered cable, the splice name can simply be linked to the cable identifier.

For splices joining two or more separately administered cables, it gets a little more complicated. A separate splice identifier, type, and location is required for each splice between any two cables, and the splice identifier must be used as the termination position for each individual pair/conductor in each cable.

For example, suppose Pairs 1-50 of Cable A are spliced to Pairs 1-50 of Cable B, and Pairs 51-100 of Cable A are spliced to Pairs 1-50 of Cable C. This arrangement requires you to name two separate splices, one between Cables A and B and one between Cables A and C. The termination position linkage for Cable A, Pairs 1-50 shows each pair/conductor terminated at Splice 1, while Pairs 51-100 are shown terminated at Splice 2. Pairs 1-50 of Cable B are terminated at Splice 1, while Pairs 1-50 of Cable C are terminated at Splice 2. Examples of the two different types of splice records and related cable records are given in the Standard. Examples of splice records for a separately administered spliced cable are given in Table 4.8, while an example of one of the spliced cable records is shown in Table 4.9.

TIA/EIA 606 recommends a cable summary report that lists all cables, their type, and the termination hardware identifier for each end. An end-to-end circuit report is also recommended which would include the user code, cables, and termination positions for each circuit. Note the use of termination position rather than the cable pair/conductor number to describe the circuit. Each telecommunications space should have a report listing the cross-connects by termination position in that space.

Certain information about wiring is required in drawing format. This includes the location and identifier of all cable terminations and

CABLE RECORD	SAMPLE DATA		COMMENTS
REQUIRED INFORMATION:			
CABLE IDENTIFIER	CB01		COPPER BACKBONE CB01
CABLE TYPE	100-PR OUTDOOR		PHYSICAL DESC., PART NO.
UNTERM. PR./COND. NOS.	0		LIST OF UNTERM'D PAIRS
DAMAGED PR./COND. NOS.	0		LIST OF DAMAGED PAIRS
AVAILABLE PR./COND. NOS.	0		LIST OF PAIRS NOT IN USE
	END 1	END 2	
REQUIRED LINKAGES:			
PR 1 TERM. POS. RECORD	C3R6-001	S107	PAIRS 1-50 SPLICED TO CB03
. . .	. . .	. . .	
PR 50 TERM. POS. RECORD	C3R6-050	S107	
PR 51 TERM. POS. RECORD	C3R6-051	S108	PAIRS 51-100 SPLICED TO CB04
. . .	. . .	. . .	
PR 100 TERM. POS. RECORD	C3R6-100	S108	
SPLICE RECORDS	S107, S108		
PATHWAY RECORDS	CD02		CONDUIT CD02
GROUNDING RECORD	N/A		

Table 4.9: Cable record for a separately-administered spliced cable showing required information.

backbone cables. Floor plans are required to show the location of all telecommunications outlets. Drawings showing the routing of cables and the location of splices are recommended.

The TMGB or Telecommunications Main Grounding Busbar for each building is tied into the building's Grounding Electrode System or the main power distribution ground. The TMGB for each building must be labeled as "TMGB" and given a unique identifier. The conductor between the TMGB and building ground must have a unique identifier and carry a warning label on each end to notify the Telecomm Manager if the conductor is loose or removed.

From the TMGB, a backbone bonding conductor runs to a Telecommunications Grounding Busbar (TGB) in each equipment room and wiring closet. Each TGB must be assigned a

TMGB RECORD	SAMPLE DATA	COMMENTS
REQUIRED INFORMATION:		
TMGB IDENTIFIER	TMGB	MARKING ON TMGB
BUSBAR TYPE	6 X 25 X 300 MM	COPPER BUSBAR
GRNDING COND. IDENTIFIER	SPECIAL LABEL	LABELED PER FIG. 4.4
RESISTANCE TO EARTH	10 OHMS	
DATE MEASUREMENT TAKEN	2/5/92	
REQUIRED LINKAGES:		
BONDING COND. RECORD	BC01	
SPACE RECORD	BA7	BASEMENT GRID A7

Table 4.10: TMGB record showing required information.

BONDING COND. RECORD	SAMPLE DATA	COMMENTS
REQUIRED INFORMATION:		
BONDING COND. IDENTIFIER	BC01	BONDING CONDUCT'R BC01
CONDUCTOR TYPE	#2 CU	NO. 2 COPPER CONDUCTOR
BUSBAR IDENTIFIER	TMGB	ATTACHED TO TMGB BUS
REQUIRED LINKAGES:		
BUSBAR RECORDS	GB35	ATTACHED TO GB35
PATHWAY RECORDS	CD02	BC01 IN CONDUIT CD02

Table 4.11: Bonding conductor record showing required information.

unique identifier beginning with the prefix "TGB" and be so labeled. Each backbone bonding conductor must be uniquely named and labeled at the TMGB and the respective TGB. It is recommended that any grounding conductors from the TGB to local equipment be uniquely named and labeled near the TGB.

The TMGB record must include the TMGB identifier, the busbar type, the grounding conductor identifier, the measured resistance to earth (in ohms), and the date of the measurement. The TMGB identifier must be linked to the attached bonding conductor records and the space identifier. An example is shown in Table 4.10.

The backbone bonding conductor record must include the conductor identifier, the conductor type, and the TMGB identifier. It must be linked to each attached TGB record and any pathway records in which it runs. Note that grounding conductor pathways are administered in the same fashion as normal cable pathways. Optional information such as length is listed. An example of a bonding conductor record is shown in Table 4.11.

For each TGB record, the busbar identifier and the busbar type must be maintained, with linkages to the bonding conductor identifier and the space record. An example of a TGB record is shown in Table 4.12.

A grounding/bonding summary report is recommended, which would include each TMGB and showing all TGBs and bonding conductor identifiers attached to that TMGB. Record drawings for the grounding/bonding infrastructure must show the location of the grounding electrode, the routing of the grounding electrode conductor

TGB RECORD	SAMPLE DATA	COMMENTS
REQUIRED INFORMATION:		
BUSBAR IDENTIFIER	TGB35	GROUNDING BUSBAR TGB35
BUSBAR TYPE	6 X 25 X 300 MM	COPPER BUSBAR
REQUIRED LINKAGES:		
BONDING COND. RECORD	BC01	BONDING CONDUCT'R BC01
SPACE RECORD	3A	BUSBAR IN ROOM 3A

Table 4.12: TGB record showing required information.

TERMINATION POINT	COLOR	PANTONE	COMMENTS
DEMARCATION POINT	ORANGE	150C	CENTRAL OFFICE TERM'S
NETWORK CONNECTIONS	GREEN	353C	NETWORK CONNECTIONS OR AUXILIARY CIRCUIT TERM'S
COMMON EQUIPMENT PBX, HOST, LAN, MUX	PURPLE	264C	MAJOR SWITCHING AND DATA EQUIPMENT TERM'S
FIRST LEVEL BACKBONE	WHITE		MC-IC CABLE TERM'S
SECOND LEVEL BACKBONE	GRAY	422C	IC-TC CABLE TERM'S
STATION	BLUE	291C	HORIZONTAL CABLE TERM'S
INTERBUILDING BACKBONE	BROWN	465C	CAMPUS CABLE TERM'S
MISCELLANEOUS	YELLOW	101C	AUXILIARY, MAINTENANCE, ALARMS, SECURITY, ETC.
KEY TELEPHONE SYSTEMS	RED	184C	

Table 4.13: Specific color codes for termination positions.

from grounding electrode to TMGB, and the location of all grounding busbars. The routing of each backbone bonding conductor may be given.

Other Aspects of TIA/EIA 606

The rules for labeling under TIA/EIA 606 are based on requirements established in the companion TIA/EIA Standards 568A (Wiring), 569 (Pathways and Spaces), and 607 (Grounding and Bonding). Labels themselves are classified as either adhesive, insert, or other such as tie-on labels. All labels must meet the legibility, defacement, and adhesion requirements specified in UL 969, "Standard for Marking and Labeling Systems". Further, labels must meet the appropriate exposure requirements for indoor or outdoor use as listed in UL 969.

Self-laminating adhesive cable labels are recommended, with a white printing area and a clear "tail" that wraps around the cable at least one and one-half times and seals the printed area. Label materials and substrates should be suitable for the environment and the surface to which they are applied. In harsh environments, sleeving or tags may be most suitable.

Labeling practices for infrastructure elements should be consistent throughout a site. The label should be readily visible. Sometimes it is appropriate to label the cover of a device, in which case you must make sure the cover can be associated with its respective device, especially if the cover is removable.

Bar coding on labels must be either Code 39, conforming to USS-39, or Code 128, conforming to USS-128. The USS codes are from the Uniform Symbology Specification, developed and maintained by Automated Identification Manufacturers, Inc. All bar code labels must also have the identifier written in human-readable form as well.

The use of label color coding is required for all cable terminations. The colors used and their applications are given in Table 4.13. You can see that the color coding helps in two ways: it differentiates the types of cable runs (backbone, station, interbuilding, etc.) as well as the types of service connections (Central Office, network, PBX, key telephones, etc.). The 606 Standard requires that the termination labels at each end of a cable be the same color, and that you use the color code provided. The exceptions are at the telecommunications outlet, where no color code is required, and that in buildings not containing the

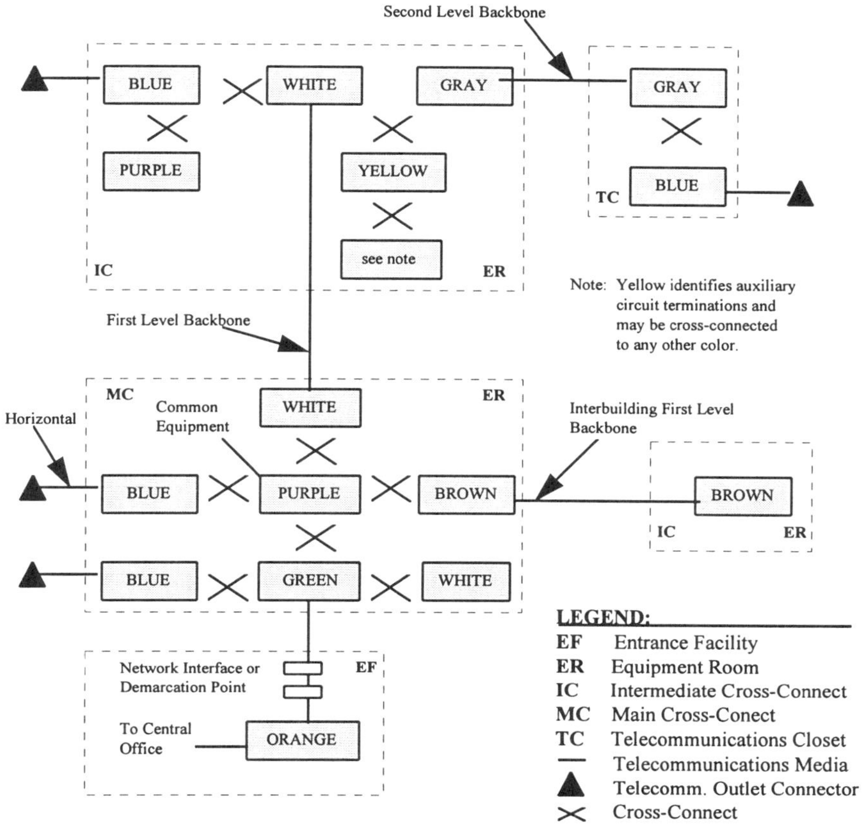

Figure 4.2: Example of administration-point label color coding.

Main Cross-Connect, the color white can be used to designate second-level backbone terminations. An example of color-coded labeling is given in Figure 4.2.

It is recommended that, for cables of different performance classes, you indicate the difference in performance through an "enhanced" color code or some other suitable marking. For example, high-performance station wiring terminations may be labeled with blue and white stripes rather than just blue, or else be marked with the performance category of the cable, such as Category 5, Cat 5, or C5.

That finishes the physical requirements of the TIA/EIA 606 Standard. It continues with four Annexes, like appendices, that are for information only. The first describes a paper-based administration scheme built according to the standard. In this scheme the space, pathway, and cable information is captured for the most part on diagrams and drawings, with conduit capacity, circuit information, and cable pair usage documented in tabular reports. A naming scheme is provided, although it is based on a rather simple example. It is emphasized that the naming or encoding scheme be fully documented and understood by all parties involved.

The second Annex addresses bridged taps, which are multiple appearances of the same cable pairs at several locations. Bridged taps are not allowed according to the TIA/EIA 568A Standard, so little emphasis is placed on their use. The 606 Standard recommends modifying the backbone cable record and the termination position record, but then simply gives the format of a report showing the termination locations of each pair count in the bridged cable and refers the reader to the backbone cable pair usage report for additional information. The example appears intentionally incomplete, and the administrator is instructed to develop their own method for bridged-tap recordkeeping.

The third Annex is a table of drawing symbols for the different types of infrastructure components. These include cable, termination devices, grounding/bonding components, pathways, spaces, splices, and miscellaneous components such as outlets, racks, and cabinets. A common understanding and usage of these symbols may help administrators use a vendor's design and engineering drawings for ongoing administration.

The fourth Annex provides a list of seventeen reference documents and the names of certain organizations which can be contacted for additional information. These include the corresponding Canadian standards CSA T527 through T530 and of course the TIA/EIA family of documents themselves.

For each element of administration, the 606 Standard specifies that work orders shall be used to document all changes. All records affected by the work order shall be updated, and the work order shall be kept on file. These changes could involve several components at once. Moving a patch cord, installing a conduit, and relocating an outlet box are given as examples of actions that must be documented with a work order. Further, the work order should list both the persons responsible for performing the work and for updating the records. There is no specific format given, but the required information for each infrastructure element is identified.

For changes affecting pathways, the pathway identifier, type, and associated space records shall be in the work order. Changes to spaces, wiring, terminations, and splices shall be documented in the work order by identifier and type. Work orders for all grounding/bonding changes shall include the bonding conductor and busbar identifiers and types. Additional information affecting the records may also be included.

The TIA/EIA 606 document is a very well-developed model for labeling and recordkeeping of the cable plant, and tries to close the loop by including requirements for using a work order and maintaining the records. It is a very detailed approach that would prove a challenge to implement with paper records for a large building or campus. Therefore, some type of computerized documentation and well-developed procedures are necessary. In a CMS, there are different table structures and relationships (the functionality that creates "linkages") that can adequately meet the 606 requirements. Throughout this book, we will be exploring the fundamentals with which you can evaluate the TIA/EIA 606 standard as well as more customized recordkeeping approaches suited to your specific needs and procedures.

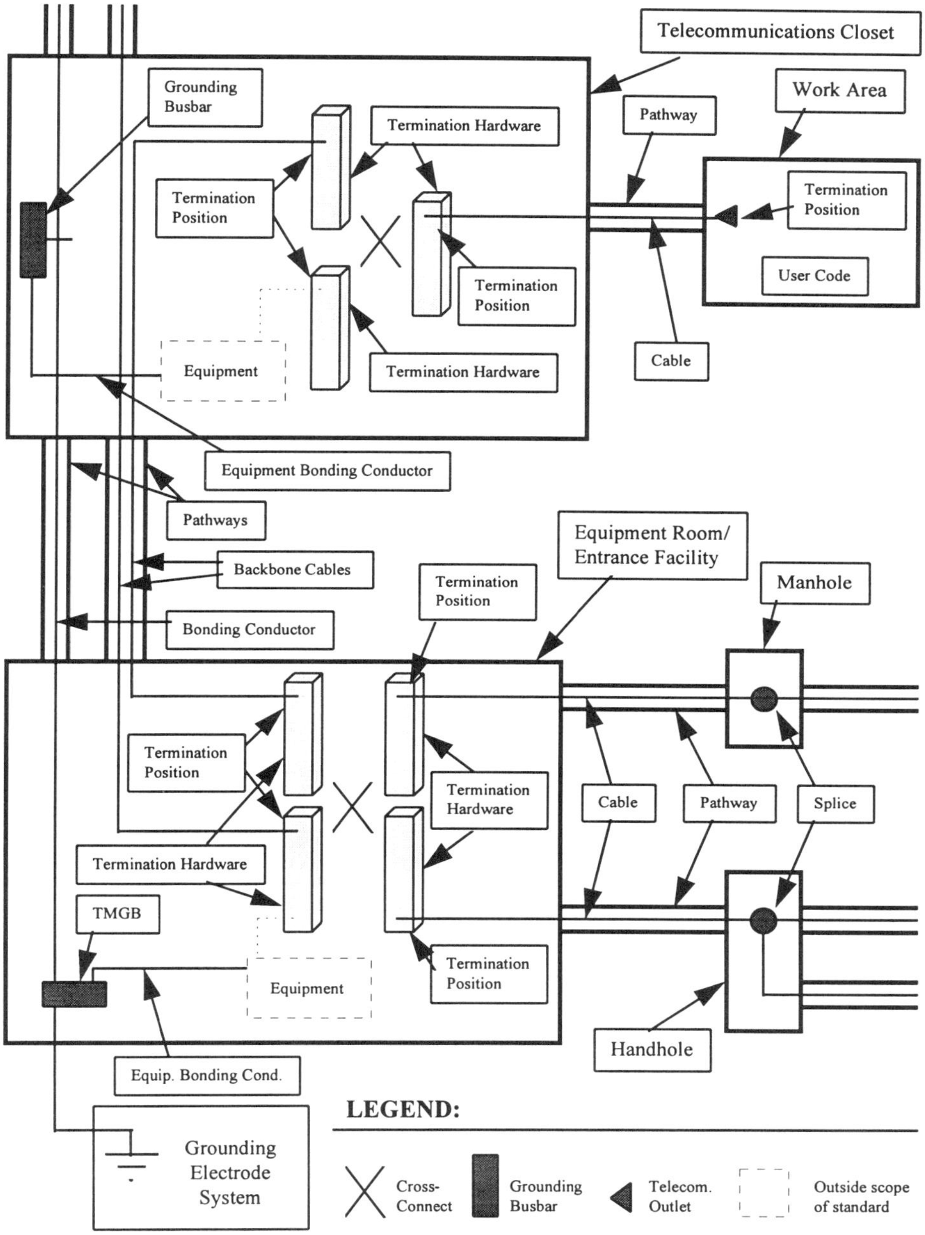

Figure 4.3: Telecommunications infrastructure items for TIA/EIA 606 administration-point labeling.

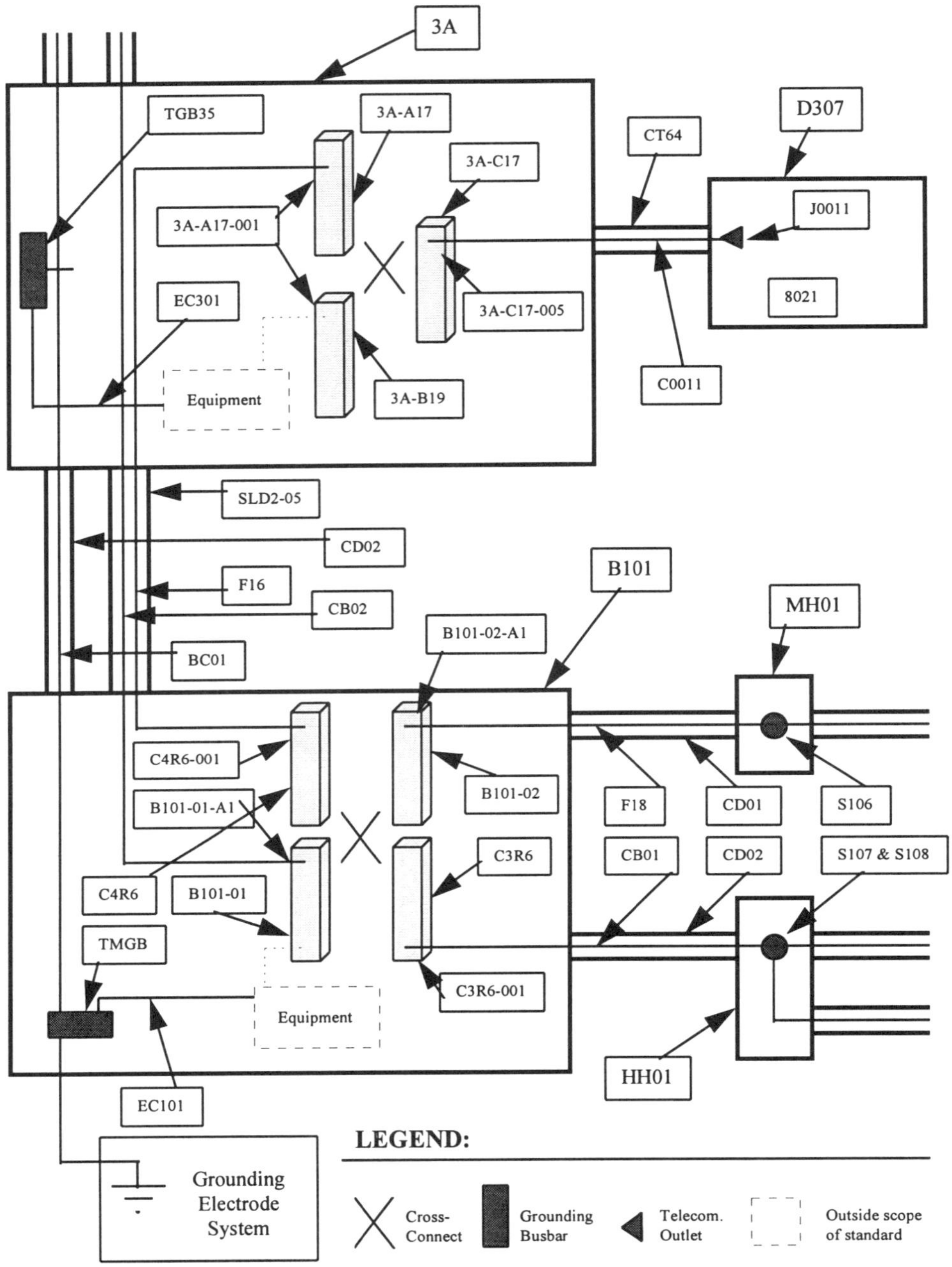

Figure 4.4: Telecommunications infrastructure items for TIA/EIA 606 administration-point labeling, with labels shown.

BENEFITS, VALUE, AND COSTS

5

So far, we have focused on the general concepts behind cable and connectivity management as well as the fundamentals of naming, labeling, and recordkeeping. We have emphasized that recordkeeping tools must be used and maintained in order to be of value, and have tried to show how records can help with specific tasks. But in general, what are the benefits of good connectivity management? Why would you want to do this in the first place? And how can you attempt to quantify the benefits, and the costs, of such a project? These are questions that management will want to have answered before you get too far along.

One of the most difficult aspects of cable and connectivity management is quantifying the real benefits that it can bring. Most of us would agree, in our hearts and our minds, that better management processes are always desirable, and that automated cable management would help us do our jobs better. But feeling good about it and getting it into the budget are two different things.

In many cases, cable management initiatives have been approved by a management team as part of an overall upgrade in management practice, without diving in to a rigorous justification. For example, a corporate quality-improvement program might mandate new processes for Customer Service or Help Desk functions. Measurable benefits occur as part of the overall strategy, but are not necessarily tracked back to a particular component like a CMS. Just as with the spread of personal computers, benefits are taken for granted and regarded as almost unmeasurable.

In other cases you may be required to develop a hard dollars-and-cents justification for expenditures, and may have to account for

- REDUCE LABOR COSTS
- IMPROVE LEVELS OF SERVICE
- FASTER RESPONSE TIME ON TROUBLE CALLS
- TRACKING OF TROUBLE HISTORY
- CAPTURE NEW CABLE PLANT RECORDS
- AUTOMATIC CIRCUIT ROUTING
- CABLE CAPACITY TRACKING
- PLANNING TOOL FOR LARGE MOVES
- IDENTIFY UNUSED CIRCUITS
- LESS TECHNICAL EXPERIENCE/TRAINING
- MANAGEMENT OF WORK ASSIGNMENTS
- DISASTER RECOVERY
- TRACKING OF CABLE PERFORMANCE CHARACTERISTICS

Table 5.1: Benefits of a good connectivity management plan.

improvements after implementation. At the very least, one needs to know the reasons and goals for any new management system. You will have to understand what can be quantified and what can only be judged subjectively, what you can know and what you must guess, and how you can make sound and reasonable assumptions. We must explore the benefits of a CMS process in a general way, but also must understand how specific tasks are affected in order to see how and why these benefits arise. After looking from both perspectives, we will offer some ways to quantify certain benefits and evaluate project costs. A list of the general benefits and improvements of connectivity management is given in Table 5.1.

Benefits

It helps to first consider the benefits of good cable management in a general sense. They may vary considerably from organization to organization. To begin evaluating benefits in a specific environment, start with some common sense items such as given in Table 5.2. First, what about the size of the organization? The budget and benefits of a fully-functional, commercial CMS exist when the user population gets between three hundred and five hundred people. Of course both the costs and benefits increase as the organization gets larger. Single sites under three hundred people may justify a CMS if people are spread out over a number of floors or buildings and are serviced by five or more wiring closets. Otherwise, good paper records or a spreadsheet and a labeling system are probably adequate.

The overall architecture and size of your facilities can have a great impact on CMS benefits. The more complex the cable plant, the more likely that significant time can be saved in the field using good, accurate records. But the same complexity will add to the cost of your CMS for inventory and ongoing upkeep. By complexity is meant the number of wiring closets, the number of backbone cables, the number of tie cables between closets, the amount of distributed control equipment, and the number of cable runs in an individual circuit. The nature and diversity of your communications systems works the same way - better records help manage multiple services.

The age, ownership, and life expectancy of your cable plant should also be considered. If your offices are in leased space, it would only pay to implement a CMS if you were going to be there for another three or four years at least. That's roughly a year to implement, and two or three years to get savings. If you own the facilities and expect to be there for a while, as with major corporate headquarters, universities, and hospitals, a system can definitely pay off. Older cable systems may be in need of replacement and difficult to inventory. If your cable plant has difficulty meeting current or proposed system requirements you'll have to replace it soon, but otherwise a full inventory and a good CMS can be less expensive than new cable. If you are thinking about or are in the midst of a new cable project, that is the best time to bring up a system and get your workflows in order. Over a full ten to fifteen year life cycle, a CMS will deliver its maximum benefits.

The rate of MACs within an organization can be a major incentive for a CMS. Some types of facilities and organizations naturally have a low rate of changes that might not warrant a CMS. The phones in a college dormitory don't usually move, nor do many phones in a hospital. (However, data services may be expanding rapidly.) In other environments, a voice or data connection may move, on average, more than once each year. In these places of high turnover, a CMS may be the only sane way to manage cable and connectivity, and each MAC benefits from recordkeeping and pre-assignment.

As was noted previously, a lot of cable management involves the management of people -

- SIZE OF FACILITIES
- COMPLEXITY OF CABLE PLANT
- AGE
- OWNERSHIP
- LIFE EXPECTANCY
- RATE OF MOVE AND CHANGE (MAC) ACTIVITY

Table 5.2: Determining factors for CMS benefits.

so we have to look at the actions and efficiency of people to understand many of these benefits. Looking back at Table 5.1, you see that reducing your labor costs for both Moves, Adds, and Changes (MACs) and for troubleshooting is at the top of the list. This assumes that time can be saved while executing these tasks. You know your facilities and routines better than anybody else, so your imagination and judgment must be applied when considering the time savings. Looking at the CMS benefits for specific tasks, we could draw an almost endless number of scenarios for MACs where good recordkeeping can play a role, and likely some where it doesn't. Instead, we offer some general tendencies for you to consider. Certain factors affect the number of tasks involved in a Move:

- How many people are involved?
- Does all equipment move with the individual?
- Is it within the same building, floor, or Telecommunications Closet?
- How many different types of services and connections are affected?

For moves of one or two people within the same Telecommunications Closet, it's easy enough to switch cable pairs or data ports without relying on records. For small moves between Telecommunications Closets, records will become more and more useful as the number of cross-connections for each circuit increases. Searching for the next cross-connection and toning or tracing of jumpers is avoided. Estimates might range from five minutes to over an hour saved on each move.

For "circular" moves, where two or more users rotate among offices, a similar logic applies. Only now, the technician must decide whether existing circuits can be left intact while switching only the user devices. A convenient record of all user services (circuits in the CMS) can indicate whether this is possible and desirable, eliminating substantial on-site inspection. Ideally, users can simply be told to pack up their devices and plug them in to a certain jack at their new location.

CMS records can also help determine quickly whether additional horizontal wiring is required at a particular location. This can avoid the need for on-site inspection and prevent unnecessary delays in ordering new wiring, where required.

For larger moves, CMS records allow for evaluation and planning without extensive on-site inspection. The technician or Move Coordinator can work on space planning with the relocating group, using the existing drawings or tables to determine who is affected, what circuits need to be moved, and what facilities are available to support them at the new location. The technician can even use the CMS to plan connections and assignments, going straight to each wiring closet with a schedule of new cross-connections to make at that location. This information would otherwise have to be developed manually and through appreciable on-site evaluation. Therefore, the possible savings during Major Moves can be appreciable, also increasing as the number of cross-connections and the number of different services increases. Provision can be made to allocate different cable resources (cable pairs and outlets) to groups with different network responsibilities during the move, making for a smoother cutover.

Adding or modifying backbone circuits can also be easier with cable records, but generally only when there is more than one run of backbone cable to consider. In other words, if there is only one cable between the two pieces of backbone equipment, then your choice of cables is self-evident. However, when there are two or more runs of backbone cable to consider, when there are different path options, or when there are cable length or performance considerations, then the CMS records will be most helpful. Selecting a cable route over a large campus or building could otherwise take from several minutes to several days, depending on complexity.

We can consider the benefit of cable records during troubleshooting in a similar fashion, where you again must consider the nature of your own facilities. For example, the technician responding to a trouble call from a user terminal must take several steps to correct the problem. If the technician is given an office number, and has no other record of cable or pair, the technician must first visit that office to determine the outlet number. The technician must then trace their way back along the circuit, visiting each wiring closet and equipment room to check the connections and circuit. After identifying the problem, they must return to the office to determine that the circuit is operable. There may be disruption for the user, such as when furniture must be moved to identify the jack number. In older facilities, the jack may have to be toned to find its location on the frame. At each cross-connect, the technician must use a toner or manually trace each jumper or patch cord. The search may lead all the way back to a Main Cross-Connect location.

On the other hand, if circuit information is available, the technician can either work outwards from the Main Cross-Connect location or go directly to the Telecommunications Closet. The initial office visit can be avoided, unless the user device itself is suspect. The cross-connection at each wiring closet is known, so no toning or tracing is required. Therefore, five to thirty minutes can be saved for such a trouble call. Obviously, the actual time saved will depend on the number of cross-connections in the circuit.

LAN circuits typically run directly from the office to the wiring closet, so the most beneficial information is knowing the jack numbers for each office and the user device attached to each. Little else about the wiring is immediately helpful, since visits would likely be made to both the office and the Telecommunications Closet anyway. In the wiring closet, it is helpful to know the outlet to port connection, so that this can be identified without tracing. Time savings for star-wired LAN user connections is only several minutes per trouble, unless there is some particular difficulty in identifying the jack number or port number of the circuit. However, daisy-chained LANs are different, and tracking a problem from a terminal far out on the chain may require tracing of the entire chain. Additional time would be saved if the path of the daisy-chain was known in advance.

It is with troubleshooting backbone circuits that circuit records could be most beneficial, for many of these circuits connect widely-separated equipment over several different backbone cables. Campuses and large office buildings are good examples. Identifying the complete circuit can involve going from wiring closet to wiring closet to wiring closet, identifying each leg one step at a time. Over a large facility, this can take several hours. But the real problem here is not just the time spent tracing the circuit, but the penalty for having the circuit down. Loss of a backbone circuit usually affects many users, sometimes in a critical fashion. For example, if a portion of the accounting department loses their connection to a host computer, some of those people are going to be sitting around doing very little until the circuit is restored. Incoming orders may be lost, or other business functions disrupted. This is a very difficult cost to quantify, but a very significant cost to avoid.

Along with faster troubleshooting comes the ability to track the trouble history of cable and network equipment. For example, if an underground cable has pairs going bad at a faster rate than other cables, this may mean some testing or inspection is in order to determine if the cable has been damaged or gotten wet. By tracking equipment history, the user can look for the failure patterns that emerge and use these as guides for troubleshooting similar problems later. Equipment with high trouble rates or low reliability can be avoided in the future.

When engineering new cable facilities, some type of survey is done to determine what is in place and what needs to be added. This survey may have to include tracing individual jumpers on a frame and determining which are active and which are dead. Otherwise you will not know your actual available capacity. Leaving dead jumpers in place is the norm rather than the exception. Many times, surveys must be repeated every year or every other year, incurring a substantial cost. With CMS records, you would expect much less time spent on cable surveys, since you always know capacity and availability of installed cable. This also includes a count of bad or defective cable pairs and equipment.

If plan drawings are available in the CMS, these can be used for engineering as well. Locating possible pathways for cable routing will save time on new installations. Knowing the availability or location of outlets in an office may avoid the expense of unnecessary horizontal wiring. Sometimes older, possibly abandoned cable such as RG-62 coax can be used for new services if its distribution is known.

Use of cable and connectivity records for new assignments might allow you to use MAC personnel with less technical experience or familiarity with your facilities. With proper documentation, often including CAD layouts, personnel can be given basic instruction regarding the connections they must make and the locations to make them. This could allow you to achieve reductions in hourly labor rates and make it easier to manage a third-party vendor working on site. Contractor personnel may change from time to time, or only be required on an as-needed basis, in which case they may not develop the familiarity with your facilities that you would expect of your own staff.

The process of inventory usually uncovers many voice and data circuits which are active but not being used and identifies cable with dead jumpers left in place. Freeing this cable and equipment up for needed services can save you the expense for additional capacity. Maintaining a CMS properly will also help keep this problem from recurring in the future. A typical survey would uncover five to ten percent of installed circuits are active but not being used. At a cost of $100 to $500 per port, this can add up to quite a bit of money. Other aspects of inventory management can save additional dollars by reducing spare inventory requirements, avoiding out-of-stock delays, and preventing theft.

Time management provides the ability to track work assignments and the time and materials spent on each service order or trouble ticket. This is not suggested as a yoke around the poor technician's neck, but as a source of management information that can improve your planning, staffing, budgeting, and overall efficiency. As we shall discuss, many times these costs are not currently tracked nor well understood. Supervisors can assign work and set priorities more effectively. Of course, this assumes your CMS includes the proper functionality to do so. Many times, a CMS can provide a billing platform for these services, allowing proper accounting and allocation of costs for various MAC activities. More and more, companies are not lumping these costs together as general overhead but are trying to allocate costs to individual business units so that the real operating costs for each business unit are better understood.

Connectivity management records can offer other "high-level" benefits. These might include improving Level of Service, or the turnaround time for MAC orders and problem resolution. A better Level of Service makes the organization look good, feel good, and receive the proper respect of upper management. A CMS also provides records for Disaster Recovery in case of a cable cut, water main break, natural disaster such as flood or fire, or other situation where restoration of service is required on a large scale. It is again difficult to estimate the value of records for this purpose since there is no value until the disaster strikes, at which time the records become almost invaluable. Most organizations consider this issue on the basis of the business cost of lost service, rather than the actual costs for restoration.

Another area where CMS records can provide benefit is in the tracking of the cable characteristics themselves, in terms of either fire safety and code requirements, electrical performance, or as-installed test data. Given the recent trend towards higher and higher bandwidth requirements, the identification of performance category or level can be important for planning and designing new circuits. Keeping a set of as-built test results can help with predicting circuit performance and diagnosing cable problems that arise. The costs for down time and restoration are both reduced.

Other less direct benefits can be added to this list, but they probably get even more subjective and difficult to analyze. And you won't necessarily get all of these benefits just because you buy a CMS - you must make each a clear objective of your project and build them into your management process every step of the way. Of course there will be offsetting expenses for maintaining the system.

Barriers to Analysis

Ideally, you could now take each benefit, analyze its impact on your organization, and estimate its dollar value. When you have quantified the benefits of each project goal, you could apply standard methods of financial analysis to evaluate the merits of the overall project. However, there are significant barriers to this type of analysis for connectivity management. These barriers are listed in Table 5.3.

Each CMS implementation is very specific and circumstantial. Like the connectivity management process itself, actual benefits are unique to each organization. Examples of specific factors include the annual rate of MACs, the complexity of the cable architecture, the types of terminal and control equipment you wish to track, whether you use in-house and/or outside staff for MAC work, and numerous others. The specific nature of a cable management project tells us that we should be wary of using generalizations as justification for a project. There have been sweeping claims that cable management will save you thirty, forty, fifty percent or more of your costs for Moves and Changes. While this may be a realistic projection in some organizations, it represents the best case more so than the norm. Few vendors (or users for that matter) have the studied case histories to back these results up, let alone represent them as the general case.

It is possible to use general formulas for estimating potential savings for certain benefits, provided your cable management plan is designed to deliver these benefits. For example, you can estimate savings in time and labor for MACs and troubleshooting. Basically what you need to know is the number of annual MAC or trouble calls, the proportion of these that would benefit from connectivity management records, and the average time spent on each. After that, you must approximate the time you can save for each relevant MAC or trouble, which allows for the individual circumstances in each organization. Otherwise, generalized numbers do not apply.

Another obstacle to your analysis lies in the fact that many potential benefits are simply unknowable ahead of time. For example, suppose you agree that your implementation of a CMS will uncover a certain amount of cable or equipment (capacity) that is really spare. You can avoid incremental costs for this new capacity, thereby generating a very real savings through your CMS project. But if you knew where this capacity was today, you wouldn't need a new project to find it! So the inherent benefit here is acquiring information that you don't have, and therefore the benefit is otherwise very difficult to estimate ahead of time.

A similar situation occurs with potential benefits related to your willingness to assume risk. Different people and different organizations manage risk in different ways. Consider the benefits of a CMS for disaster recovery. Some businesses would be greatly affected during a major outage such as a cable cut, and these businesses are more willing to spend money to reduce the chance of failure and fix the problem as quickly as possible. Banking and financial institutions, for example, typically have more extensive disaster recovery plans than most businesses. Each organization must therefore gauge the impact of a major outage and define the role of a CMS in troubleshooting and restoring service.

Even where you may have a reasonable formula for savings, you may ultimately be limited in your analysis by a lack of baseline information about service orders, troubles, and current costs. It's hard to see where you're going if you don't know where you are today. You need to have reasonably accurate information about your rate of MAC and trouble calls, and even breakdowns by type and location. If your staff cost for MACs and troubles gets lumped into an overhead cost with many other activities, then you really don't know what costs can be affected by connectivity management. For example, labor time for PBX software changes will probably not be affected by a CMS. If costs for both hardware and software changes are included in the cost for MACs, and if you apply some savings factor to the total, your savings estimate could be way off. The same goes for differentiating between routine MACs and trouble calls.

In all cases, you must take the time to analyze the information you do have. Perhaps some simple recordkeeping practices could be adopted to gather baseline information for a six to twelve month period. More formal time-motion studies or benchmarking can be performed if required. Ultimately, there are areas where you can make reasonable assumptions and realistic

- EACH IMPLEMENTATION IS UNIQUE
- INFORMATION IS UNAVAILABLE
- WILLINGNESS TO ASSUME RISK VARIES
- LACK OF BASELINE COSTS

Table 5.3: Barriers to estimating the value of CMS benefits.

$$T \times N \times P/100 \times C = \text{Annual Savings}$$

WHERE:

T = Time Saved per Field MAC, in Hours
N = Annual Number of MACs
P = Percent of MACs Requiring Field Actions
C = Average Technician Cost, Dollars per Hour

EXAMPLE:

T = 0.75 Hours (45 Minutes)
N = 2000 MACs per Year
P = 75% Involve Connections
C = $35.00 per Hour

0.75 Hrs/MAC x 2000 MACs/Year x 75/100 x $35/Hr = $39,375.00

Figure 5.1: Savings formula for Moves and Changes.

calculations that are specific to your circumstances, and there are things which will remain very subjective or unknown.

Estimating Actual Value

With that as background, you can now try to quantify some of the individual benefits of a CMS. To begin, we'll look at the MAC process itself. In perhaps an overly simple fashion, suppose you can estimate the average time saved during a routine Move or Change by having a complete and precise record of the current and proposed circuits. If you then know how many MACs to apply this to, and your average cost (labor plus overheads) per hour or day, you can multiply these factors together and get a hard savings number, as shown in Figure 5.1. But you will probably have to break it down a little further.

Estimating the amount of time saved per MAC depends on issues we have discussed previously. For small, simple facilities, this might only be a few minutes per MAC. For facilities with many wiring closets and several cables and connections per circuit, this could be up to two hours or more. It can vary between voice and data services. If the technician has to search for the right path, the right pairs, and the right port in an existing and/or new circuit, then substantial time can be saved with a CMS. Preferably, this circuit information would be in a service order.

Note that your savings only apply to MACs that involve a technician making or breaking connections. To estimate this number, it may require study or an audit of all service order data to separate out orders that were only software changes, simple changeouts of terminals or set types, directory changes, etc. You will probably find that fifty to seventy-five percent of all MACs require some actions in the wiring closet. For dedicated wiring systems, many connections are permanent and therefore even fewer MACs require changes. So if you multiply the total annual number of service orders by the percentage affecting field connections, you can estimate the number of MACs affected by a CMS.

$$\mathbf{T} \times \mathbf{N} \times \mathbf{P}/100 \times \mathbf{C} = \text{Annual Savings}$$

WHERE:

T = Time Saved per Trouble Call, in Hours
N = Annual Number of Trouble Calls
P = Percentage of Trouble Calls from Cables, Connections, and Ports
C = Average Technician Cost per Hour

EXAMPLE:

T = 1.5 Hours
N = 5000 Trouble Calls per Year
P = 30% Involve Cables, Connections, and Ports
C = $40.00 per Hour

1.5 Hrs x 5000 Calls/Year x 30/100 x $40.00/Hr = $90,000

Figure 5.2: Savings formula for trouble calls.

As an example, consider a smaller organization with 500 MACs per year, and that 85% of them involve connections (they don't tolerate many frivolous changes). Studies of their work habits suggest that one half-hour can be saved on each MAC affecting connections when complete circuit information is available beforehand (there are only seven wiring closets). The cost per hour for this work is a very reasonable $20.00 per hour. Multiplying these together gives an annual savings of $4,250.00.

In the example of Figure 5.1, there are 2000 MACs per year, of which 75% involve connections. A savings of forty-five minutes (three quarters of an hour) per MAC is estimated, and the labor cost is $35.00 per hour. Your calculated savings turns out to be $39,375.00 annually. Finally, consider a case of 20,000 MACs per year, of which 60% involve connections. A time savings of one hour is estimated for each MAC because there are many wiring closets and several connections to make for each circuit. Many of the changes are assisted by an outside vendor, so the average cost is $45.00 per hour. When these factors are multiplied together, the potential savings is over one-half million dollars per year. Remember, such savings are potential and not automatic. Also note that the typical rate of savings increases faster than the size of the organization, since the benefits of recordkeeping become more significant as the facilities get more complex and the rate of MACs increases.

You can perform a similar analysis when applying CMS records to the troubleshooting process, as shown in Figure 5.2. In this case, you must determine the amount of time saved on a typical trouble call and how many of these calls involve cables, ports, or connections. You can multiply these factors as shown and calculate an additional hard savings. Again, you may have to analyze your patterns of all trouble calls to estimate these factors. Estimates range higher and lower, but we have found that roughly fifty percent of problems can be traced back to a cable, connection, or port. It depends greatly on what you lump in as "trouble calls".

$$N \times P/100 \times V = \text{Savings}$$

WHERE:

N = Total Number of Active Circuits
P = Percentage Active, But Not Used
V = Annual Value Per Circuit/Port

EXAMPLE:

N = 5000 Active Circuits
P = 5% Active, Not Used
V = $375.00 per Circuit/Port

5000 Circuits x 5/100 x $375.00 = $93,750

Figure 5.3: Savings formula for unused active connections.

As an example, all trouble calls are logged through a central help desk of some sort. After weeding out problems with software and hardware, since a CMS would not routinely help you with these, you find that forty percent of your 1000 annual trouble calls involve connectivity. You can then perform a study that shows an average savings of one hour for each of these calls. Using an average cost of $35.00 per hour, calculations indicate a potential savings of $14,000 each year. This does not consider the costs of down time and lost productivity.

In a larger organization (the example in Figure 5.2), 30% of the 5000 annual trouble calls have been related to cable and connections. The complex and sophisticated network has been difficult to troubleshoot, so there is a savings of an hour-and-a-half for each connectivity problem. At an average cost of $40.00 per hour, the annual savings would be $90,000.00.

For both the MAC and troubleshooting estimates above, you can argue the assumptions and estimates, but the basic model is sound. The better you can make these estimates, the better your assessment will be. You might have to do them separately for voice and data services. Dig into your records, and do some experimentation yourself. Remember, there is just as much wrong with an overly conservative estimate as an overly optimistic one.

Another benefit that can be estimated with some confidence is the cost avoided when spare capacity is identified and managed. During a CMS implementation, substantial inventory is uncovered which is active, but no one is using. We have consistently found between five and ten percent of all facilities, voice and data, to be active but unused. At an average value of $100 to $500 per circuit, this represents a substantial cost to replace. As shown in Fig. 5.3, if you take even five percent of 5000 voice and data circuits, each worth $375 apiece (the price of a typical voice port), you have a total value of $93,750.00.

The other aspect of spare inventory involves the cable pairs themselves. To clean out the dead jumpers, perform a frame inventory, and verify circuit use may cost between $1000 and $2500 per thousand cable pairs at a single frame, and will probably have to be done over in a couple of years. Clearly, these capacity costs can be avoided with proper management through a CMS. As shown in Fig. 5.4, to figure a savings benefit here, estimate the percentage of your cable pairs requiring frame inventory each year (25-50%).

$$N \times P/100 \times E \times C = \text{Annual Savings}$$

WHERE:

N = Total Number of Cable Pairs
P = Percentage of Cable Pairs Requiring Inventory
E = Ends of the Cable (2)
C = Inventory Cost Per 1000 Cable Pairs

EXAMPLE:

N = 20,000 Cable Pairs
P = 25% Requires Inventory Each Year
E = 2 Ends of Each Cable
C = $1,500 Per 1,000 Cable Pairs Inventory Cost

$$20{,}000 \text{ Cable Pairs} \times 25/100 \times 2 \times \frac{\$1{,}500}{1{,}000} = \$15{,}000$$

Figure 5.4: Savings formula for frame inventory.

Determine how many pairs this represents, multiply by two, then multiply by a nominal $1500 per thousand pairs. You must multiply by two since each cable has two ends on two different frames, each of which must be checked.

If you can make each of the previous estimates, add them all up to approximate the hard benefits that a CMS can bring. Do your best to put a value on some of the more subjective benefits, and include them as well.

Costs of a CMS

On the cost side, a CMS implementation can be a challenging thing to analyze as well. It's one thing to consider the out-of-pocket costs, yet another to understand the internal costs. Meanwhile, the true cost of implementing a CMS usually depends more on the inventory and implementation requirements than on the cost of the software. It is very similar to the cable plant itself - over one-half of the cost of the cable plant is the installation labor. The inventory of equipment and connections can be time-consuming and requires proper planning and control to ensure the records going into the CMS database are accurate and complete. There are a number of factors which will affect the cost of your CMS implementation, and each of these is are covered in the following paragraphs.

There will likely be some hardware costs for installing a CMS unless a single-user system can handle your needs on an existing PC. The scope of your facilities and the amount of data to be recorded and maintained will determine the size and cost of the hardware required for your individual implementation. For a single-user system, the cost may be relatively small where only a PC needs to be purchased with plenty of disk space, memory and speed. For a large size facility with multiple users a separate file server may be required (with sufficient disk space), which can be added to the LAN to allow records to be accessed from multiple points. Other PCs or workstations may need to be added, as well as network cards, cables, modems, a printer, plotter, and/or tape backup hardware. A list of possible

SERVER	$5,000 - 15,000
WORKSTATION	$3,000 - 10,000
PERSONAL COMPUTER	$2,000 - 4,000
NETWORK CARDS (NIC)	$50 - 500
CABLES	$50 - 250
MODEM	$150 - 500
PRINTER	$500 - 1,500
PLOTTER	$1,000 - 5,000
TAPE BACKUP	$200 - 5,000

Table 5.4: Possible hardware costs for a CMS implementation.

hardware requirements is given in Table 5.4, indicating a typical cost range for each item. Note that hardware prices and features can change very rapidly in today's market. You may not have a clear picture of hardware costs until you actually select a CMS.

If you have determined you need a multi-user CMS database which will be installed on your LAN, you may need to purchase copies of the computer operating system (DOS or UNIX) and the Network Operating System to be installed on your server, unless the CMS package can be installed on an existing server already on the network. Depending on the CMS database you select, there may be additional software requirements such as the database software licensing (Oracle, Informix, SyBase, Ingres, etc.) and DOS or Windows software for user PCs. If you select a system with CAD capabilities, it may require purchase of a CAD software package in order to maintain, revise, or plot the drawings. If a separate CAD package is required, be sure to check the equipment specifications to insure it will operate adequately on the hardware you purchase.

Some CMS packages also warrant the purchase of SQL software or a third-party report generator for creating additional and custom reports. It is valuable to determine if an additional report writing package is needed, so that you are sure you can retrieve the records in a way that provides the desired benefits and results.

If remote access will be required, a communications software package (such as PCAnywhere or Remote2) may be necessary. This is helpful if access to the records may be required where there is no connection to the network, as well as providing a method for the CMS Vendor to provide remote support for your system.

Of course, there will be the cost for the CMS software itself. Make sure you know what is included in this price, and do not duplicate the purchase of items that may be included with the CMS. All of the above software items must then be considered and totaled for the project. A list of

OPERATING SYSTEMS	$100 - 1,000
DATABASE LICENSES (PER USER)	$100 - 500
CAD SOFTWARE	$2,000 - 4,000
CMS SOFTWARE	$3,000 - 75,000
COMMUNICATIONS SOFTWARE	$150 - 500
SQL/REPORT GENERATOR	$200 - 2,500

Table 5.5: Possible software costs for a CMS implementation.

possible software requirements is given in Table 5.5, indicating a typical range of cost for each item.

The concept of recordkeeping implies some type of inventory of cable and equipment. The inventory process can be complicated and extensive, depending on the nature of the facilities and the types of records you want to keep. The inventory could be a major undertaking including an inventory of each office, each cross-connect frame, and all backbone cable and equipment. It could be as simple as noting cable and pair connections in the local wiring closets.

It's easiest to capture information while a group is first moving into a facility, or by compiling "as built" information during cutover or installation. Older records may have to be audited and verified before then can be included in a new system. If some records already exist, it may save a percentage of the expense and time for inventory. You will need to determine how accurate and complete the records are, and whether they can be used for loading your CMS database. It is also important to make a note of the format in which the records exist. If they are in some type of electronic file, be sure to consider this when looking at different CMS products, and check with the vendors to see if they can handle file imports from existing flat files in order to utilize the existing records.

If an inventory is required to gather and record the information for your CMS database, you have two options on who will actually do the physical inventory and data entry: in-house or contractor staffing. There are several advantages for using your own staff for the CMS implementation. Using in-house personnel may avoid out-of-pocket costs, but remember the cost is still there nonetheless. Using in-house staff can be very valuable if the inventory personnel will be the same individuals who will be working with the CMS records ongoing on a daily basis. This

Inventory Activity	Estimated Time Required
Office Inventory: Information Outlet, User Name & Desktop Equipment	5 Minutes/Office
Voice Cross-Connects: Average of 2 Cross-Connect Points (Local Closet and Main Equipment Room)	10 Min/Extension
Data Connection: Station Wiring to Local Port Connection	2 Min/Port
Total Estimated Time Per Office:	17 Minutes

1000 Offices X 17 Minutes = $\frac{17{,}000 \text{ Min}}{60 \text{ Min/Hr}}$ = 285 Hours

Overhead at 25%: 285 Hours X 1.25 = 356 Hours (45 Days)

Inventory Cost: 356 Hours X $35/Hour = $12,460

Figure 5.5: Method for estimating inventory costs.

Data Entry Activity	Estimated Time Required
Total Includes: Backbone Cables & Control Equipment Voice Circuit Phone Extension, Information Outlet, 2 Separate Cross Connects Data Circuit Desktop Equipment, Information Outlet Port Connection User Information Name, Location, Department	12 Minutes/Office

Note: CAD Systems will require additional loading time.

$$1{,}000 \text{ Offices} \times 12 \text{ Minutes} = \frac{12{,}000 \text{ Minutes}}{60 \text{ Min/Hr}} = 200 \text{ Hrs.}$$

Overhead at 20% 200 Hours X 1.20 = 240 Hours

Data Entry Cost: 240 Hours X $15/Hour = $3,600

Figure 5.6: Method for estimating data entry costs.

provides an incentive to ensure the records are correct and complete. It also gets those individuals involved in the project at an early stage so they become familiar with naming and labeling schemes and the format of the records as they were recorded for the CMS.

Using clerical personnel for data entry who will also be involved with the ongoing CMS administration (Customer Service or Help Desk) is also a big advantage since they will become familiar with the nomenclature and set-up of the CMS database. They will gain a better understanding of how the records are created, making it easier for them to later update the database with changes resulting from trouble calls and routine MACs.

If using in-house personnel, keep in mind that these individuals may still have other work assignments, and special arrangements must be made to handle their normal workload. This may also mean that the inventory takes more time, since these individuals may not devote 100% of their time to the project. They may spend 50% of their time on the project, and the remaining on their current work assignment. Sometimes there are other demands on their time such as routine weekly meetings or special committees. If in-house personnel are paid on an overtime basis for after-hours work, the costs can quickly approach or exceed rates for contract project services. These individuals may not be experienced or trained in the most efficient methods of conducting an inventory, which may result in a higher rate of errors or incomplete information. It's important the data entry personnel be able to easily read the information which has been recorded so they can correctly enter the data into the CMS.

There are several reasons you may choose to use contractor personnel for handling your inventory requirements. The use of outside personnel allows for the inventory work to proceed at a faster rate, since they are not involved with daily activities and routine work assignments.

Project Management Activity	Estimated Time Required
Project Administration: Planning, Scheduling, Training, Inventory Form Preparation, Inventory Management, Auditing	2 Days for Every 5 Days of Inventory
Backbone Inventory:	1 Day for Every 5 Days of Inventory*

Total Estimated Time: 45 Days Inventory X $\frac{3 \text{ Days}}{5 \text{ Days}}$ = 27 Days

Project Management Cost: 27 Days X $500/Day = $13,500

* Note: A better estimate is obtained by using 2 days per wiring closet or equipment room.

Figure 5.7: Method for estimating project management costs.

Trained inventory personnel can accomplish the task in a shorter amount of time since they have experience with an inventory process which provides quick results with a minimum amount of errors. In fact, a shorter project cycle may result in lower costs overall. There is little disruption to the normal work schedule for moves, adds and changes, and inventory work can even be conducted after hours so there is little inconvenience to employees. In-house staff can concentrate on routine business, and begin using the new CMS records as soon as one area is complete and entered into the database. Any additional information which is required is identified at an early point, so the inventory staff can adjust the inventory procedures as required.

The last area which will impact the cost of the implementation is the scope of the project. This includes the number of buildings, floors, information outlets/jacks, circuits, cross-connects per circuit, equipment, and employee information to be recorded. For each item you want track, additional time is required to record the information and to enter it into the database. This is why it is important to determine the value of tracking each bit of information. If it does not provide value, it is not worth the cost to record, enter and maintain the information.

The inventory process can be broken down into three main component costs. These are for personnel conducting the inventory (costing between $15 and $45 per hour), personnel for data entry (costing between $12 and $25 per hour), and a project coordinator or administrator (costing between $450 and $600 per day) that can oversee the effort and perform the backbone inventory. The actual time required for each step may vary greatly depending on how much information is recorded on the desktop equipment (ID, Manufacturer, Model, serial number, etc.), the wiring structure, the number of outlets and connections per office, the number of cross-connects per circuit, and the condition of the cross-connect panels. Further, duration and costs will be affected by the experience and pace of the in-house inventory team. Things will go much faster as they get used to the routine. Some information about desktop devices can be obtained through the network via special software discussed in Chapter 16.

Hourly rates do not always indicate productivity, as a lower rate person will sometimes get more done than a higher-paid individual. On the other hand, low wage personnel with little experience in networking or telecommunications will likely have higher error rates and cause more rechecks. As with any effort, setting the pace to get it done right the first time is the most cost-effective approach. The overall pace of the inventory is highly dependent on the skills, experience, and motivation of the project coordinator.

Examples are given in Figures 5.5, 5.6, and 5.7 to show some sample costs associated with the implementation of a CMS database. They are based on an implementation covering 1,000 users or offices, and could represent out-of-pocket costs for outside services or the overhead cost for using in-house staff. These costs are also based on using trained inventory personnel who are devoting full time to the inventory process. The costs are broken down into inventory personnel, data entry and project management, with the assumption that the project administrator will manage the project and perform the bulk of the backbone inventory. The examples are also based on a typical office environment with a phone and PC in each office, with an average of one data connection per office.

The inventory and labeling of backbone cables and equipment usually falls to the project administrator, since at this point they are most familiar with the requirements for naming cables and devices which are being developed. A reasonable estimate is to allow up to two days per wiring closet or equipment room for this activity, and it must be done prior to a frame inventory. Over time, other personnel involved in the inventory may be familiar enough with the catalog to perform some of these tasks.

Keep in mind that the times and $/Hour costs are averages, and that the initial estimate of inventory time must include an additional 15 to 30% time for overhead such as lunches, meetings, training, rechecks, locked doors, confusion, etc. Efficiency will increase as the team gains experience. Actual time for inventory on a per-office or per-employee basis may range from 15 to 30 minutes, while the data entry time on the same basis may range from 10 to 20 minutes per office. Running cables and locating icons on a CAD drawing may take additional time.

The numbers used here are only examples, and your individual implementation will be different. The best way to determine the average time for your project is to inventory a sample of 15 offices, the backbone cable and equipment, and all the cross-connects for each circuit while making a note of the time required for each step. Then substitute the proper numbers in each calculation to get a better idea of what your implementation costs might be.

Once you have estimated costs for hardware, software, and project services you can put them in a summary similar to Figure 5.8 on the next page. Ongoing costs for CMS operation and maintenance will be covered in Chapter 7.

Summary Statement of CMS Implementation Project Costs
ABC Company

I.	Hardware	
	Database Server with Tape Backup	$ XX,XXX
	3 IBM Model PS2/80 Computers w/ DOS and Windows	X,XXX
	2 Ethernet Interface Cards	XXX
	Interconnecting Cables	XXX
	2 9600 Baud Internal Modems	XXX
	HP Laserjet 4 Printer	XXX
	Plotter	X,XXX
	Hardware Total:	$ XX,XXX
II.	Software	
	Server Operating System (Unix)	$ XXX
	Network Operating System and Modules	XXX
	Oracle 8-User License	X,XXX
	Oracle SQL Tools	X,XXX
	AutoCAD Rel. 13	X,XXX
	3 copies PC Anywhere	XXX
	CMS Software	XX,XXX
	Software Total:	$ XX,XXX
III.	Project Services	
	Inventory	$ XX,XXX
	Data Entry	XX,XXX
	Project Administration	XX,XXX
	Project Services Total:	$ XX,XXX
	PROJECT TOTAL:	$ XXX,XXX

Figure 5.8: Summary statement of project implementation costs.

CMS FUNCTIONALITY

6

Due to the extent and complexity of the records involved, the relational database is the tool of choice for building a Cable Management System (CMS). When you include the components of cable and equipment which make up your communications infrastructure, there is simply no better way to bring all this information together - both as lists of items and a representation of relationships between items that defines a "circuit".

Trying to describe, in a general way, how a CMS works is somewhat like describing all the ways to build a building. There are so many ways to approach the problem that you could spend forever on the subject and still not know all there is to know. But you could look at certain functions and features that are common to all buildings, for example, and make comparisons as to how a certain function or feature is designed. For example, most buildings must have a roof, walls, doors and windows, and some way to support their weight. Buildings usually have heat, light, electrical power, running water, and drainage plumbing. You could compare how buildings differ in their style of roof, or how they are strong enough to stand up, or how the ventilation system operates. But even after understanding how each requirement can be met, you can only imagine the different combinations that could actually be used to build a building.

Designing a building can be a very creative process, but it is ultimately limited by the constraints of money and space - there are only so many dollars to spend, and the building can only be so large. On the other hand, a building should be easy to use and effective in meeting its purpose or intent. Choosing the specific building methods means living within those constraints, even trading off among them. Once set, these features make certain things possible or not possible, convenient or inconvenient, for the inhabitants. Consider a simple choice: Should a particular building have windows that open, air conditioning, or both? If sealed windows and air conditioning are used, you will always be kept comfortable but will never feel a "fresh breeze" blowing through your office.

With CMS software, your first impression is often that they cover essentially the same set of records and therefore they must naturally be built in a nearly-identical or at least similar fashion. However, like the building previously mentioned, a CMS can achieve certain functions in a broad variety of ways. Unlike the building, people may reasonably disagree as to what those functions should be in the first place. And when choices are made and structure is set, there are certain things that will seem convenient, other things that will be inconvenient, and the remainder will probably go unnoticed. Much of your judgment comes from your personal preferences, your background, and the circumstances of your interaction with the system.

In this chapter, we will consider the common features and differences among the various approaches to CMS software. Not every possible approach, or even existing approach, will be considered. Our intent is not to judge, but to describe the mechanics and try to show the strengths and weaknesses of the different approaches. Only your specific needs and preferences can determine what is good, better, or best.

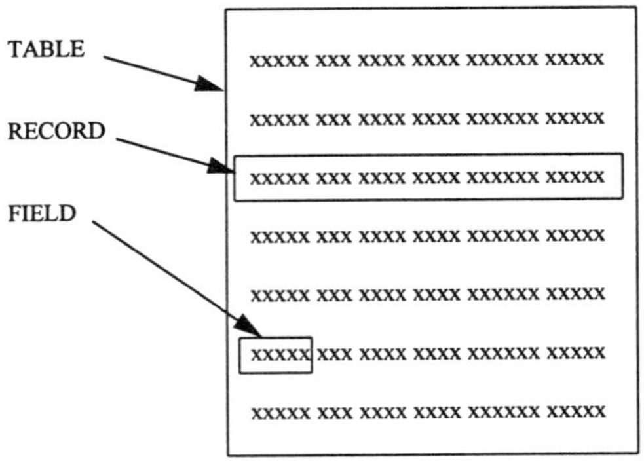

Figure 6.1: Representation of a database table, showing a record and a field.

The Relational Database

Our discussion will start with the operation of the relational database, since the database functions are the basic building blocks of the software developer for structuring, ordering, sorting, and relating information. The discussion is general and probably oversimplified, and a specific proprietary database may use different terminology. The basic building blocks of the database are shown in Figures 6.1 and 6.2.

The most fundamental unit of the database is the *field*, which holds a single string of characters and represents one piece of information such as your first name or a numerical value. Each field has a name, a specific format, and other characteristics assigned by the software code. The field formats could be all alphabetic, alphanumeric, numeric, integer, and even sophisticated formats such as date or telephone number. One or more fields are defined and included in a *record*, which represents a collection of information usually about or concerning one principal, important thing. A record is a sequence of related fields, such as your first, middle, and last names, which can be treated as a single unit.

One or more records, each using the same field sequence and formats, can be listed in a *table*. A record can now be thought of as a row of the table, while fields make up the columns of the table. Each record receives a unique identity, usually called an *index* or index number, by which it is known within the table. The table, as a file structure, stores its records by index number but can display them on a screen or a report in different order. This process is called *sorting*, and involves putting the records in increasing or decreasing order based on the value of a particular field or fields, much like a spreadsheet would do. Fields which can be used for sorting are often called key fields. Sometimes all of the fields in a record are key fields, sometimes not.

Imagine the list of names, addresses, and telephone numbers of your ten best friends. Each name and address on the list can be considered a record, using an identical set of fields to contain similar information. First name, last name, street address, city, state, zip code, and telephone number are each fields in the address record, with the entire list representing a table. If you number the addresses 1 to 10 in the order you wrote them down, then each record also has an index. Now imagine that you can sort the list (put the records in order) by last name, city, state, or zip code, in which case these four items would be key fields. Putting the list in a certain order might help you find a particular address, or give an indication of how many friends live in a particular state.

In a database the original table usually stays in index order, Record 1 through whatever. The table in sort order is placed in a separate table or file, where it can be displayed on the screen or printed to a report. If you sort the original table, you can *reindex* the table in sort order. If you reindex a table in the sort order that you use the most, it will take the least amount of effort, on average, to use the table. Sorting your list of friends by last name would probably serve you best. You might then rewrite your list, again numbered 1 to 10, in alphabetical order by last name and use this as the primary copy. You have now reindexed your list.

There are two ways commonly used for a database to display a table. The first method is as you would normally think of a list, with the records displayed down the screen in some order, and with the fields for each record displayed across the screen in a row. With this format it is easy to scroll up and down the list, or highlight a particular record on the list, but it becomes awkward if the fields of each record pass off the right of the screen. In that case, you can never see a whole record at once, even if you can scroll right and left. This is also a more limited and difficult format to edit or change directly.

The other way to display a table is one record at a time on a special screen, where the table can still scroll forward or back. The special screen can contain a much larger record and have the record's fields labeled with their names. Each field can have special software code that looks at the string you are entering and determines whether it is valid before it is accepted. A simple example is enforcing the basic field formats. The software can also insist that certain fields be filled in before the record is saved, making them *required fields*. Other code may validate the entry against a list of acceptable entries, the process of validation, or it could display another table with entries for you to choose from, often referred to as a *look-up table*. Information that is commonly used in many fields or in many tables, such as building names, floor numbers, and department codes is often put in support tables for those fields for the convenience and control of validation and look-up.

Overall, the database can be viewed as a group of tables containing lists of records, with each record made up of the character strings in a defined set of fields. The architecture of each table and its records we will generally refer to as the *dataset structure* of the database and CMS.

If your only concern is lists of things or items, you could stop there and have a fairly simple structure to work with. But the problem of connectivity recordkeeping, and the beauty of the database, is relating things on different tables together in new and different ways. The concept of a circuit as a string of elements requires the database to relate or associate items together in a specific sequence. The database uses a combination of software code and other tables listing these relationships to accomplish this job.

The CMS software developer has a good deal of freedom in structuring tables, defining possible relationships as new tables, and writing the additional code to tie these together in a meaningful way. How they are presented on the screen and how additions and modifications are handled in each table must be addressed. As was stated earlier, there are trade-offs made which make certain operations easier and certain operations harder in any given system. Ultimately, the structure and operation of a CMS reflects the operations and limitations of a database more so than it necessarily follows the real world. There is no conceivable database that would be perfect in all situations.

The process of selecting or retrieving specific records from a table is called a *query*. The process of querying a certain table in a CMS means going to the screen which displays a single record for that table and entering strings in one or

TABLE: AN INDEXED LIST OF RECORDS WITH IDENTICAL FIELD STRUCTURE
(LIKE A LIST OF NAMES AND ADDRESSES, OR A PHONE BOOK)

→ RECORD (ROW): A STRING OF ONE OR MORE FIELDS (LIKE A COMPLETE ADDRESS)

→ FIELD (COLUMN): A STRING OF ONE OR MORE CHARACTERS OR A VALUE (LIKE A LAST NAME)

Figure 6.2: Descriptions of the basic building blocks of a database.

more fields. This sets the search *criteria*. The search then asks the system to display records with fields matching those entries. The fields have to match exactly with respect to characters, spaces, upper or lower case, punctuation, etc. If more than one field criterion is used, the selected records must match all of them. To select records with matches to any field (as opposed to all fields) or to select records with matches over a range of values usually requires special software called SQL (Structured Query Language) to create custom queries and reports. More information about working with SQL and report generators is given in Chapter 32.

Most CMS databases allow the use of characters, called *wildcards*, to be used in the criteria strings during a query. This is similar to the asterisk (*) used in DOS file names. The wildcard can mean any character or character string within a given criterion field. For example, use of the wildcard character alone in a field would select all records, since the wildcard can be anything. Entering "A*B" would match any record where the same field string started with "A" and ended with "B". The use of wildcards gives great flexibility to the queries available within a CMS. As before, multiple field entries with wildcards must match the entries in all fields of the selected records.

Some databases also allow the use of operators such as >, <, =, ><, etc. to broaden the search criteria. For example, if you enter ">50" in the cable length field, the query will produce all cable records for cables with a length greater than 50 feet.

While each record would always be unique by index number, most tables require that the string in at least one field be unique - that is, different from all other entries in the table for that field. This is sometimes referred to as the primary key field, and is usually the name of whatever is described by the fields in the record. If the record described the features of a telephone set, for example, then the primary key field would likely be the phone number or extension, the name we would naturally give. By making the primary key unique, the entire record becomes unique and duplicate records are avoided. This eliminates a number of potential problems and allows the CMS functionality to operate on the records without

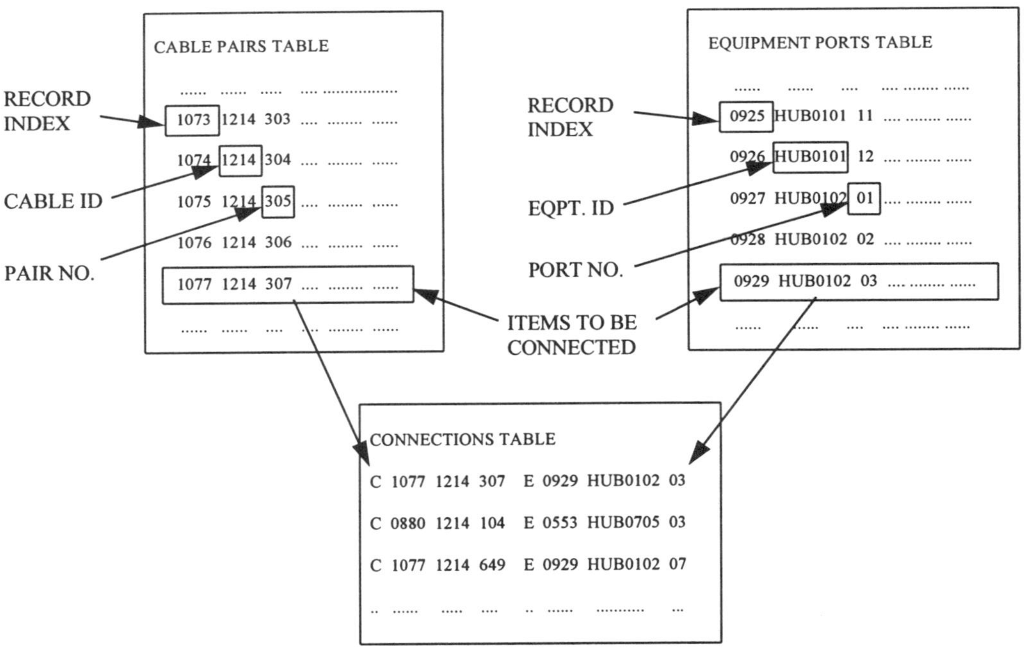

Figure 6.3: Example of connection records tracked in a separate table.

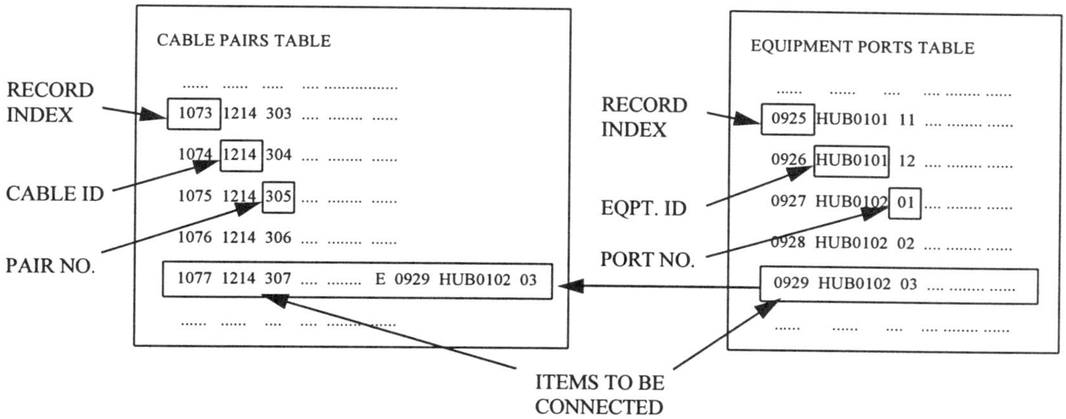

Figure 6.4: Example of connection records tracked in an item table.

getting confused, especially when relating different records together. Primary key fields are usually required fields as well.

Each table in the CMS usually has a screen display (or several) showing all the fields in a single record. This screen is used for entering or modifying data in a single record, and is usually accessed by a menu selection. However, the task of making or breaking connections, the relationships created between different records, is usually on a separate screen that only shows some of the information about each record being related. For example, connecting a cable pair to a port might indicate the cable ID and pair number rather than the entire record of cable pair characteristics.

This should be kept in mind when developing cable and device configurations, since you would wish to see all relevant information displayed on the screen when making the connection. In other words, you don't need to see the cable's installation date but you may want to see its type or class. Information that is significant for making connections should only be embedded in naming conventions, when required, for fields displayed on the connection screen. Information embedded in other names will not be available when making connections, and therefore has no value for this purpose.

The actual functionality and table structures used for connection relationships varies greatly from CMS to CMS. The nature of connection records will be very dependent on the fundamental item table structures in the database. How the CMS relates cable pairs and ports will depend on how cable pairs and ports are defined in the first place. Usually the connection relationship is tracked in another table, where the field ID and record index of each related record are stored as shown in Figure 6.3. There may be a single table for listing all connection relationships, or there could be several for listing different types of connection relationships. For example, cable-to-cable connections and cable-to-equipment connections may be in different tables. Otherwise, connections may be tracked in the original item table, such as when the table of cable pairs itself tracks the record index of a connected port as shown in Figure 6.4. This structure can be simpler but more limiting, since it usually places more restrictions on the types and quantity of connection relationships allowed.

The table structure for tracking connection relationships has two significant effects on operation. The first is the flow and number of data entry steps. Where connection relationships of any type are stored in the same table and are accessed on the same screen, making a set of circuit connections flows quickly. If different circuit connections are done in different tables and on different screens, the user has to move from screen to screen in proper sequence to create the circuit. On the other hand, trying to make all connections fit a single table or screen may be limiting. Reporting on connections would be similarly affected, since either one or several tables must be accessed for connection reports.

The second major effect of connection records structure is in the development of automatic circuit creation routines. These routines are very structured themselves, and operate in conjunction with the connection tables. Proper operation may require that cables and equipment be defined in specific ways, and that initial connections such as cable-to-termination position be properly set up. This has a strong impact on how the catalog is developed, and must be considered during the cataloging process.

Differences in table structures will have other minor effects on operation, such as how validations work, how menus and commands are presented, how the circuit trace logic functions, etc. We do not suggest that any one approach is good or bad, but rather that it should be understood in some detail in order to find the best match for your requirements.

Computer-Aided Drafting (CAD)

The addition of CAD functionality to a CMS database offers certain advantages, but also creates some special needs as discussed in Chapters 17 and 31. By use of the terms "CAD" and "graphics", we mean those systems that use some type of architectural rendering of a floor plan, even if it is done in a very simple fashion. An example is given in Figure 6.5. Some CMS software offers a representation of sequential circuit components in the form of a schematic, more like a block diagram as shown in Figure 6.6. The creation of schematic diagrams is sometimes referred to as graphics or a graphic representation. This type of illustration we will refer to as "schematic", and is not the subject of this chapter. The use of schematic diagrams is useful, but in a different fashion than actual plan drawings. There are even CMS products which offer both CAD (graphic) and schematic diagrams.

Each of the CMS products with CAD/Graphics capabilities have somewhat different procedures and methods for interfacing with a database, but there are basically two different types. First, there are systems that can be considered fundamentally CAD-based with some database features added. These products start with fully functional CAD software, making the CAD

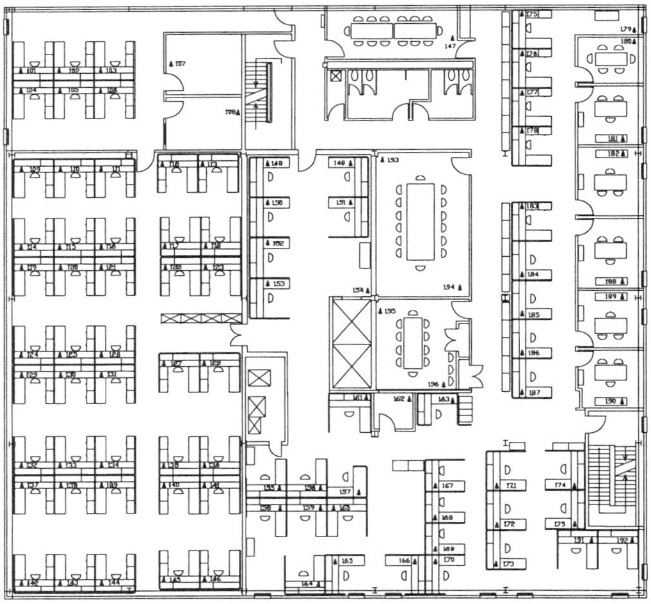

Figure 6.5: CAD architectural plan drawing.

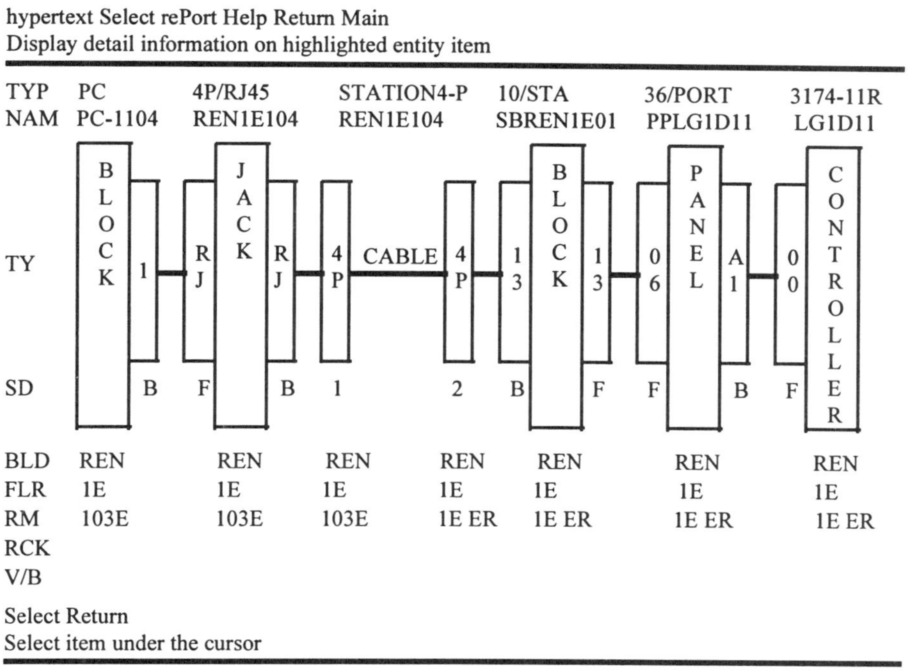

Figure 6.6: Schematic representation of circuit (*Courtesy of CCAM from CDS, Inc.*).

"native" to the CMS. They include all the required tools for drawing modification (move, add, erase, edit), printing/plotting, and import/export capabilities.

This first class of CMS usually has a database associated with them for tracking additional information about each of the elements/icons placed on the drawing, and they can usually keep tabs on spare/available cable capacity and ports on controlling equipment. In some systems, database tables are created which can track and report on cable connections and device connections as well, but the sophistication varies. The location of a database element is usually tied to a grid location on a drawing, and therefore *plan drawings are required* in order to utilize the database. CAD-based systems usually require more training since users must operate in a CAD environment and be able to navigate around the drawings to perform queries and routine tasks.

The second type of system is fundamentally a CMS database with CAD features added or somehow linked to the database. Usually with a more complex and robust set of tables and associations, they tend to do a better job of tracking communications devices and relational information such as cable connections and circuit information. Tracking of cable utilization and reporting of spare capacity are often standard features. The use of floor plan drawings is optional, therefore *CAD drawings are not required* in order to use the database.

These systems usually require less formal training, since CAD capabilities are usually limited to Zoom and Pan features, with limited printing/plotting abilities. Database elements often have to be located onto the drawing through an additional data entry step. These systems normally can add the CAD module later, so the CAD portion does not have to be implemented at the same time as the database portion.

CAD systems can also be used to create drawings of actual equipment racks and patch panels as shown in Figure 6.7. This type of drawing can take significant time to prepare and maintain, but may have already been prepared during the design phase of new facilities. There are CMS systems which have actually tied database information in to this type of drawing, such as clicking on a port location to retrieve

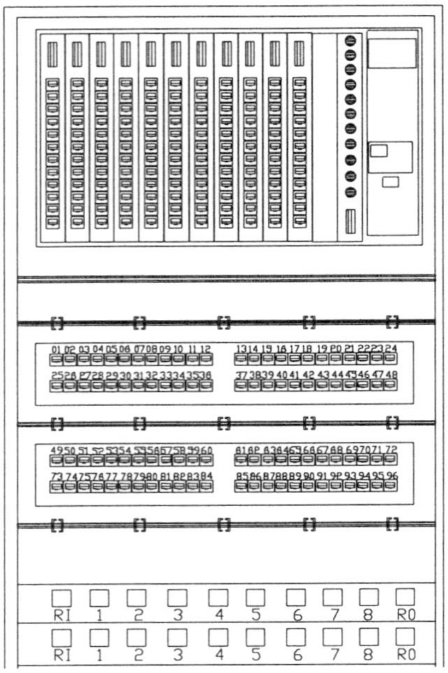

Figure 6.7: CAD drawing of an equipment rack showing patch panels and network hubs.

information about the port, or actually making patch cord connections between two patch panel positions. This type of functionality is not usually found in the major systems since it requires that all connections be managed in this fashion, which may not be appropriate for the entire organization.

The initial preparation of drawings requires that the proper drawing scale be used and that information be segregated into different layers of the drawing. Each drawing may have to have an origin, or spatial reference point, assigned so that the CAD system knows the spatial relationship of one drawing to another. How to layer the drawing is up to you, but it is most often done so that common elements of one system (such as power distribution or plumbing) are on separate layers. For additional information on layering, you may wish to refer to the *CAD Layer Guidelines* available from the International Facility Management Association (IFMA), Research Publications Department. The *CAD Layering Guidelines* recognize that there is no universal method for structuring and naming layers, but that a hierarchical approach is typically most useful. The highest level breakdown of drawing layers is shown in Table 6.1.

Common CMS Functions

Despite major differences in structure, almost every CMS database attempts to provide certain functionality for the user. Of course, there have to be tables with a record for each cable.

ID	GROUP
A	ARCHITECTURAL AND INTERIOR
S	STRUCTURAL
M	MECHANICAL
P	PLUMBING
F	FIRE PROTECTION
E	ELECTRICAL (INCL. COMMUNICATIONS)
C	CIVIL AND SITE WORK
L	LANDSCAPE ARCH.

Table 6.1: Major groups for drawing layering.

• SUPPORT TABLES
• CABLE AND EQUIPMENT TYPES AND CONFIGURATIONS (CATALOG)
• CABLE AND EQUIPMENT ITEMS
• CONNECTIONS AND CIRCUITS
• SERVICE ORDER
• TROUBLE TICKET OR WORK ORDER
• REPORTS

Table 6.2: List of common CMS functionality.

There will be some way to break these cables down on a pair or strand basis. There will be tables for listing termination devices, control devices, and desktop equipment. There will likely be some tables for listing users (a directory) or Circuit IDs. And there will be some functionality to link, associate, or relate items from these different tables into a circuit description. Some of the differences in table structures and circuit relationships are covered in the next section.

In terms of basic tables, each CMS includes supporting tables, the tables of cable and equipment types, the tables of cable and equipment information, and possibly tables for tracking connections. Additional tables may exist for functions specific to each database. The table or data set structure is unique to each database, as discussed in the next section.

Most CMS software has some higher level functions to operate on these tables, more in accordance with the way you do your job. Most CMS software comes with a Service Order function, covered in more detail in Chapter 9. A key distinction between systems is whether the connection information can be manipulated directly through the Service Order, using a pending or reserved status, and whether the connection changes are done through the database or through the CAD interface. If the Service Order does not provide this functionality, the circuit changes must be made manually outside of the Service Order. The Service Order also tracks some level of administrative information, such as due date, technician assigned, and work charges.

In a similar fashion, a Trouble Ticket function is usually included, either as a separate module or combined with a service order as a more general Work Order. The Trouble Ticket also may or may not allow direct display and manipulation of circuits. It usually tracks administrative information such as problem type, resolution, technician assigned, work charges, etc.

Each CMS comes with a built-in or "canned" set of reports which provide a screen display, electronic file, or paper document listing selected records from a table. Additional "ad hoc" reports can be obtained through a report generator, sometimes included with the CMS, or through the SQL (Structured Query Language) tools available for the specific database. Reporting is covered in more depth in Chapter 18. A summary list of the common CMS functions is given in Table 6.2.

Specific CMS Structures

A significant difference among CMS software, of course, is the type of database being used. Traditionally, a CMS was married to one database, such as Oracle, Informix, SyBase, etc. For some systems, this remains true today. Other systems are now offering portability among databases through the use of an Application Program Interface (API) based on the Open Database Connectivity (ODBC) standard. If the CMS supports ODBC, then it can utilize any database with an ODBC interface. Such a system will be easier to integrate and support using the organization's existing, guidelined database.

As mentioned, the biggest thing that gives each CMS a unique character is the table structure describing the basic cable and equipment components. This table structure dictates how you will track information about each element, and almost more importantly, how each of these elements can be connected or related to other components. The table structure captures the component's *configuration*, whereby you define and name the connection points on that type of cable or device. A simple example would be defining each cable by pair versus defining the same cable by conductor - while a cable conductor could be connected to a pin on a jack, a cable pair would be connected to a pair of pins. They represent two different configurations. A CMS table structure could allow conductors only, pairs only, or allow either one or both in the same table.

The process of creating the configurations of cable and equipment, called cataloging, is covered in Chapters 21 and 28. At first this can be approached with no constraints to develop the desired structure, but ultimately the catalog must be made to conform with the table structure of the chosen CMS. One of the challenges of cataloging is to develop a consistency among connected items. If you define cables as conductors, then you should define termination positions as pins. If you define cable as pairs, then you should define termination positions accordingly. If you define a device to have a set of ports (connection points), the port might connect to one, two, three, or four cable pairs, and the database must allow for this.

Defining cable and device connection points usually must follow some conventions on directionality, as discussed in Chapter 3. Cables must have a Start and an End, or an End 1 and an End 2. Each end must have a configuration, and the cable may have different configurations on each end. Device connections may have to be specified as to whether they are on Side 1 or Side 2, Front or Back. Directionality keeps connections more organized and allows the circuit trace and automatic circuit creation to work properly.

If one type of port connects to one type of cable, while another type of port connects to a different type of cable, how would this be indicated? Does it even need to be indicated? With the features available in a CMS for pre-assignment and Service Orders, the answer is yes - compatibility and consistency of connection must be evident. The identifier for each port should give an indication of the type of cable or connector to which it can be attached. In other words, the records should clearly indicate that the connection points for the cable and the device are compatible, appropriate and allowed. Connecting the two particular items is therefore legitimate.

Most CMS software will allow cables to be defined as either pairs, conductors, or strands. In a few cases, copper cables are defined by pair while a separate table is used for optical fiber, so it can be defined on a strand basis. Some software even has a separate table for horizontal copper cables, where these cables can be defined as any combination of one, two, three, or four-pair units. For example, the TIA/EIA 606 Standard allows a 4-pair horizontal cable to be defined as four individual pairs or a single four-pair unit. However, if it is described as a four-pair unit, and if it is carrying more than one service, you can not distinguish which service is on which pair. On the other hand, if the cable is described as four individual pairs, and all pairs are used for one service (whether active or not), then there are four connections required to indicate this one circuit.

The table structure and configuring of devices can naturally be more complicated than configuring of cables. Defining connection points on a device presents two unique problems. First, the configuration of the device may change over time, depending upon the cards or modules which are added or removed. This suggests that the device configuration must account for this population of cards currently in place, and the total number of cards which can be added. Sometimes, there are restrictions on how the cards can be configured in the device. Imagine the cards in an expandable network hub or PC, for examples.

Capturing this information is possible and desirable, especially for interfacing with network management systems, although it makes things more complicated. Some software is structured to allow device connections to be defined at both the card and port level, while other systems would require the cards themselves to be devices related back to some base device. Working with this structure is part of the cataloging process, and consistency between device and card connection points must be maintained.

The second problem with devices is that connection points often have a physical description as well as a logical or functional description. Each description is useful under different circumstances. For example, the port

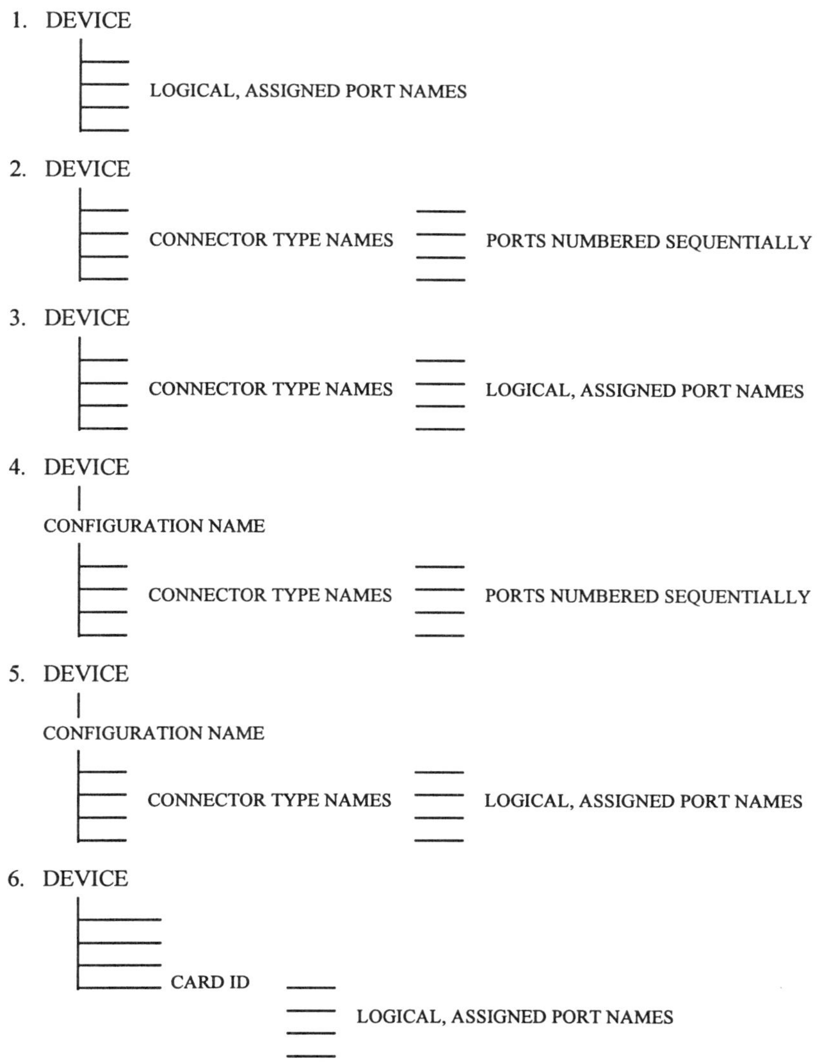

Figure 6.8: Sample approaches to device configurations and connection-point naming.

(connection point) on an Ethernet Hub card may be an RJ45 connector and might also be known by a Port ID, whether this be a network address or simply Port Number 4. The physical connector type is useful to assure compatibility with the assigned wiring connection, but the Port ID is useful when identifying the connection on the hub or diagnosing a network problem.

The table structures in different CMS packages reflect a variety of approaches to this concern, with several different structures shown in Figure 6.8. In some systems, the connection points on a device are simply treated as a Port ID. While it is not identified as to type of connector, it can be connected to cable pairs, conductors, or strands from the cables table. It is possible, in some cases, to include the connector type or functional name in the Port ID. For certain communications systems, such as SNA networks, the logical port name might be preferred and the connector reference cannot be included. In other cases, the CMS may assign a number to the port with no modifications allowed. Sometimes each numbered port has its own record where the physical description can be included.

Another approach to device connections is to treat them in the opposite fashion - all connections are described as physical connectors numbered one through whatever. While this can produce great accuracy in the physical description, it also eliminates the possibility of using any logical descriptions besides. The correspondence between physical and network information may be lost. There are cases where something more than a connector type and number is useful. The IBM 8228 has 10 data connectors, but they are Ring In, Ring Out, and eight user ports. Clearly, it is helpful to differentiate between these connections.

A useful alternative developed by some CMS systems is to structure the device configuration tables so that each connection point is defined as a connector type to which a unique name can be assigned. If the name format is flexible enough, actual logical or network names or Port IDs can be used. This is the best of both worlds, provided that the right characteristic of the connection point (physical and/or logical) is presented on the screen or report at the right time. This type of configuration is the most complex to create and maintain, and takes a little extra time to set-up and test in the database. In the long run, the flexibility and usefulness of this CMS records structure will outweigh the extra time spent to understand and develop an effective catalog.

A final distinction applies to this latter case, where connector type and ID are given. In certain systems, the initial definition of the device type will request that you define the connector types and names directly. The entire list of connection points must be entered for each device. This takes some time, but gives some flexibility when modifying the configuration later. As a convenience, other systems allow you to actually define a configuration of connector types and names, and then assign that configuration name to the device when the device is defined. This means the same configuration name can be assigned repeatedly to similar equipment types, making the creation of devices simpler and faster. Use of named configurations can add another level of complexity to the database and sometimes get a little awkward to modify later, but they represent a very powerful and flexible approach.

The configuration structure of cables and devices is part of the overall method for handling connections. The device configuration determines what types of things can be connected to what other types of things, and more importantly for the database, what types of records can be related to other types of records. This point and the points from the earlier section about how connection records are tracked will dictate what a circuit can look like, how the circuit trace logic will work, and what the circuit display and report will look like. The steps involved in creating or disconnecting a specific circuit will therefore vary from system to system, usually trading off between a complex, highly flexible approach versus a simpler, less flexible approach. The more detailed and structured the configuration is, the more complex and flexible it will be.

Other considerations are the ability to make multiple connections to a single connection point, and the logic that would apply when tracing through points with multiple connections. In some CMS software, connection points can have only one connection. For example, several cable pairs cannot be connected to a single port. Other steps are taken to handle this case, such as modifying the cable configuration to create a single unit from several pairs, or by requiring termination of the cable pairs on an intermediate device (such as a wiring block or patch panel) which would then allow one-to-one patching between port and termination. Tracing through such a connection is straightforward in either direction.

Other CMS software allows for multiple connections at a connection point, which gives the most flexibility in configuring cables and equipment and for representing complex circuits. However, tracing through the circuit may require choosing a particular connection to trace through, and this can sometimes produce strange results. Proper cataloging is required to present the right information for selecting the appropriate connections, wherever they may branch, and thereby tracing along the proper path.

Finally, some CMS software will require that a particular component be included as part of a circuit. This might be a User ID, a Circuit ID, an outlet, a piece of desktop equipment, or a network port. Each approach has advantages and disadvantages. If your selected CMS has a requirement like this, you must be prepared to treat each and every type of circuit on this basis.

CMS ADMINISTRATION

7

Using CMS software for cable and connectivity management requires you to properly support and administer the CMS system. Administration can cover both the routine use and maintenance of the records themselves as well as the ongoing support of the software application. We will discuss approaches to the overall use and maintenance of the records, with the goal of documenting this process in a formal way. Additional aspects of daily administration are frequently influenced by other management systems as discussed in Chapter 8, and the Service Order and Trouble Ticket functionality in the CMS which will be covered in Chapters 9 and 10 respectively. Finally, the maintenance of the CMS application and supporting software will be covered here.

Putting the CMS to Use

Putting a CMS to good use also requires that you consider how to use and maintain the system during the course of everyday activities. Many people in different groups may use and affect the information in a CMS for different purposes. The more types of information that you track in the CMS, the greater the number of people involved. Therefore it is essential that you consider the steps required to perform different jobs, and adapt these workflows to use and maintain the CMS effectively

Your organization's adaptation to a CMS cannot be taken lightly for several reasons. First, the nomenclature (naming of things) in a CMS reflects and defines your environment. All participants must understand and describe the environment in the same terms, and these terms must be global - that is, they must be consistent across all different types of cabling and networks. A LAN technician who refers to a "LAN Cable" and a voice technician who refers to "D-Inside Wire" may be talking about the exact same thing. The CMS will likely issue instructions in one fashion or the other, but not both.

Secondly, a CMS is only as accurate as its maintenance allows. Compliance with reporting and updating requirements is mandatory, not optional, for anybody affecting this information. This presents a management challenge to monitor the flow of information and activities, and to mandate the required compliance at the highest levels of the organization. CMS maintenance must be tied in to job performance for the system to be used effectively.

Finally, the benefits and burdens of a CMS are not always evenly distributed. Having complete and accurate connectivity information can improve Help Desk response significantly, reducing the time for diagnosis, user counseling, and hands-on troubleshooting. The burden of supporting these records falls on the technical and administrative staffs. While technicians can also benefit from these records for Moves, Adds, and Changes, many see this recordkeeping as a chore that only adds to their workload. They may balk at using the records in the first place, feeling that they lose some control over their own activities.

Just as wiring exists solely for the benefit of end users, so recordkeeping exists to benefit the entire support organization rather than any particular individual. Management must be committed to explaining this principle and holding people accountable. A formal approach to planning, documenting, and managing workflows is the necessary ingredient.

Exactly how CMS records are used will depend on the structure of the organization as well as how these records are part of the bigger telemanagement and network management processes. When you set out to design a CMS process, you must identify: 1) who will use particular information, 2) for what purposes they will use it, and 3) who will maintain each piece. Nowadays, it is possible to buy integrated software which can handle cable records, connectivity, user directory, and asset management. Service order, trouble ticket, and billing functions can be considered. Interfaces to network management systems are also available.

The development of workflows must therefore account for all management functions and all recordkeeping systems supported, provide for the flow of information among these tasks at the proper time, and define roles and responsibilities for individuals in different groups. For example, a simple change to a voice circuit may require you to use and change PBX records, connectivity records, asset records, and directory information as well as cutting a service order document.

There is great variety in the organizational structures and methods used to handle these tasks today. Groups function with different degrees of integration and centralization. It would therefore be impossible to cover all cases. We will consider three general scenarios that can be applied to communications management to illustrate how a CMS might be used and maintained. The first case is where every individual assumes direct responsibility to use and update the records. The second case is based on one or more key individuals in each group maintaining the database. The third scenario is based on a truly centralized administration, where all work requests are funneled through a central group and all work is done via Service Order or Trouble Ticket.

In the first situation, every individual that acts on or affects the communications transport infrastructure (cable and equipment) must use and update the CMS records according to the actions they take. If they move a PC or install a new hub, they themselves must change the database. These updates must be done regularly, preferably at the end of the day when the actions were taken. In this case, there really is no formal workflow to document, since it simply becomes a matter of individual habits.

Such an approach is self-managed, and requires a high degree of discipline from every individual. It is difficult to provide oversight in this situation, since nobody but that individual knows what was done. If an individual neglects to update the database properly, the information will deteriorate and ultimately affect other persons working with the system. This approach is best suited to a small group of two or three persons who are each willing to take responsibility for the database.

Beyond two or three persons, this method usually begins to break down. Unfortunately, our human nature begins to create politics, conflicts, and self-interest that get in the way of each individual assuming responsibility. If there is no oversight or accountability, there may be little incentive for individuals in large groups to maintain the discipline required. Frankly, even managers can become parochial and fail to sustain the level of commitment and cooperation necessary for successful recordkeeping.

The second method of administration relies on a number of key persons to be responsible for database usage and upkeep. These individuals are usually considered Lead Technicians or Group Leaders that are technically involved but also oversee the work of others. They may each represent one of the groups involved in the CMS, with responsibility for a particular type of network service, a particular site, building, or physical domain, or a particular part of the cable plant. Each Lead Technician receives work requests directly, manages the connectivity within their domain, uses the database records to assign tasks within the group, and assumes responsibility for making database updates within that domain as shown in Figure 7.1.

An example might be a facility using network administrators, a Lead Voice Technician, and a Facilities Coordinator to manage the database records. The network administrators, each with responsibility for a different network (or areas of a common network), receive all Data MAC requests and problems. They may deal with each request directly or assign personnel to the task. Updates to network management, asset

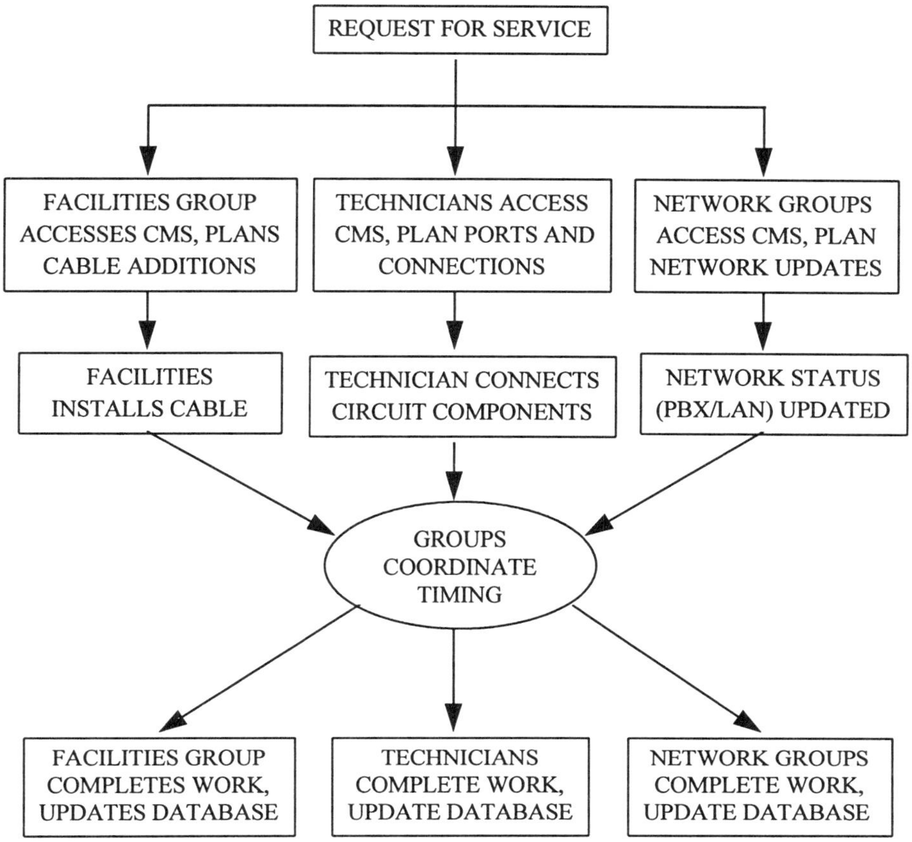

Figure 7.1: Flow chart of administrative procedures based on key individuals in different groups having a direct interface with the CMS.

management, and CMS systems are coordinated or handled by the Group Leader. Similarly, changes affecting voice circuits go through the Lead Voice Technician, who coordinates activities and updates PBX and CMS software directly. Changes involving new cables or pathways, architectural modifications, or other space planning functions are handled by the Facilities Coordinator, who is responsible for maintaining CAD drawings and related database updates.

This approach places the responsibility for CMS recordkeeping with several designated individuals, each in a good position to know what is going on and to make sure the records are updated. The key issue is coordination between groups, since once a work request is received they operate fairly independently and maintain an independent interface to the CMS. When and how they choose to use CMS records is up to each group, provided that each is maintaining an agreed-upon base of information in a timely fashion.

Preparing a workflow diagram for this circumstance is generally a case of showing how a particular type of work request is distributed to and within each group. Beyond this, there is little exchange of paper documentation between groups. For example, a typical Move involves relocating both voice and data connections concurrently, but they can often be performed independently. Internal procedures can also be documented for each group, but there is great flexibility here as long as each Group Leader stays abreast of the changes being made and makes the database

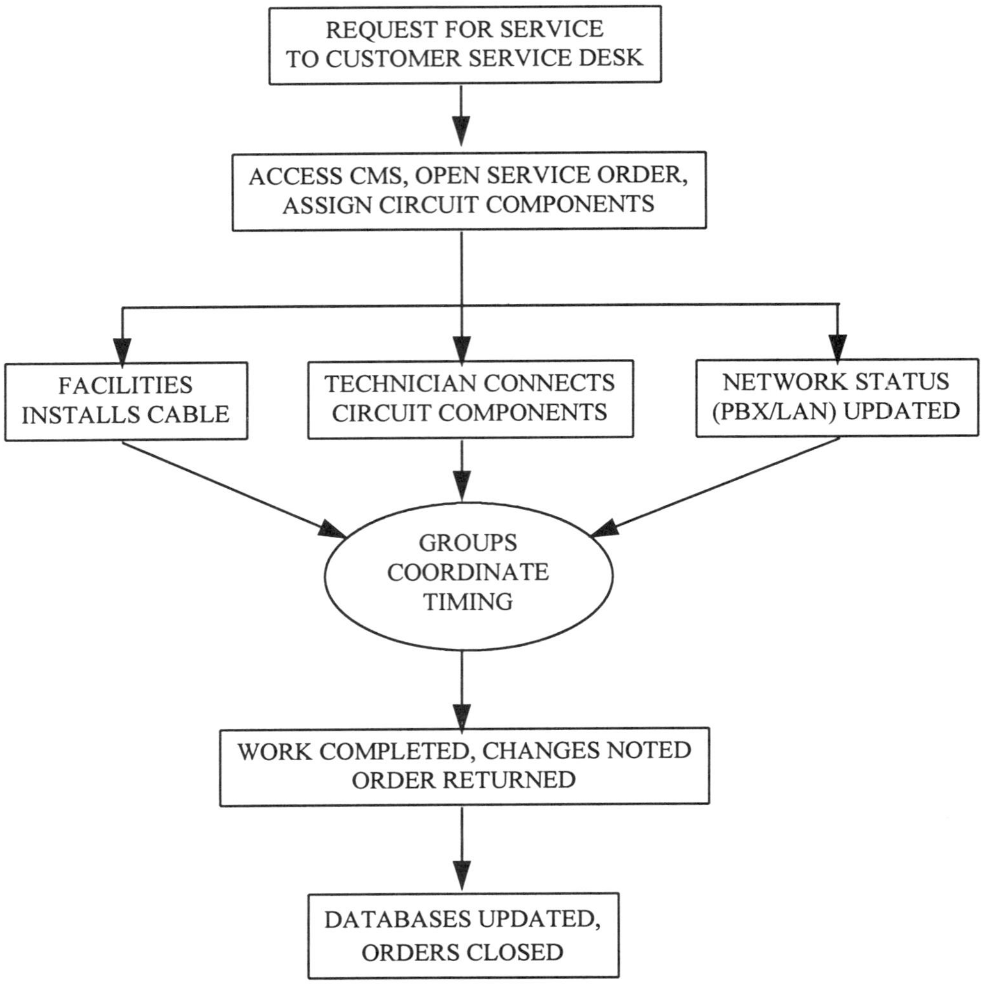

Figure 7.2: Flow chart of administrative procedures based on centralized CMS access through a Service Order.

updates immediately upon completion of a task. Certain work requests may require coordination between groups as to task timing as well as the sequence of CMS updating. If a building is renovated, the CAD drawings must be updated first, then new cable and network equipment defined, then new connections entered.

The third process for using a CMS is based on centralized administration through Customer Service and/or Help Desk functions. In this case, all MAC requests come through a single group which prepares a Service Order and distributes this order to the appropriate persons for completion. All Trouble Calls come through a Help Desk which creates a Trouble Ticket and distributes this document to the appropriate persons for problem resolution. All interfacing with the CMS is done by the Customer Service and Help Desk groups as part of their administrative operations, as shown in Figure 7.2. The Service Order and Trouble Ticket are paper or electronic documents provided to the technician. Cable pair and port assignments may be made ahead of time and made part of the document, thus instructing the technicians to perform very specific connections. Any changes to the assigned connections must be reported by the technician in proper format when the Service Order is returned.

If no pre-assignment is made, or if the tasks are more complex, the technician must report the actions that were taken, again in the proper format. Upon return of the order, it can be closed out in the CMS. Any pre-assigned connections will be automatically updated in the database.

With this approach, the glue that holds the whole process together is the Service Order. The Service Order must capture all activities, and all activities must be managed through the order. If actions are taken outside of this process, they will result in incorrect database information. It is imperative that the Service Order, or Trouble Ticket, not be closed out until the database is updated.

The Service Order can be issued to all appropriate groups at the same time, allowing each group to evaluate the required tasks and coordinate their actions. Cable, equipment, and connections can be installed in that order, followed by relocation of desktop equipment and software changes to the network or PBX. It is more difficult for things to slip through the cracks this way. If new equipment needs to be ordered, or if major reconfiguration of network devices is required, there will be plenty of lead time to put the pieces in place so that connections can be made. The user(s) can be kept abreast of progress, and notified of successful completion. This is particularly important when resolving user problems.

This process also provides for accountability. Key milestones can usually be indicated on the order, such as for the installation of a new network hub. Individual technicians or groups must perform these key tasks by the time indicated, or else risk jeopardizing the entire schedule. Actions can be taken to expedite the key steps rather than allow them to slip because no one was aware of a critical date.

As mentioned, the details of these workflows will be different for each organization, depending on the types of networks involved and the structure of the current organization. A separate Facilities or Building Services group may or may not keep their own records, but if they are installing new cable, then they must provide information to help maintain the CMS. Sometimes the management of desktop assets (PCs, terminals, modems, etc.) is handled by the user organization, in which case they need to be part of the workflows or else recordkeeping will be circumvented. In other cases, all or part of the management process is done by third parties, whose agreement and understanding is necessary to bring them "into the loop". An outsourcing provider must also be required to report connectivity changes accurately and completely.

You can see that the typical CMS does not exist in a vacuum, but must be integrated with other recordkeeping and other processes in order to be viable. A formal process of evaluating current workflows and developing new ones is critical to the successful implementation of cable management. Each organization may therefore be required to adapt its structure or methods in order to support, and be supported by, a fully functional cable management system. The opportunity exists for management to hold individuals responsible for their work and raise the level of services while bringing down overall cost.

Maintaining the CMS Applications

Like any piece of significant enterprise software, the CMS and its supporting applications must be properly maintained to operate smoothly and effectively. It is beneficial to give one individual or group the responsibility for this maintenance so that certain tasks are performed routinely. This individual can become familiar with all aspects of CMS operation and serve as the principal trainer, troubleshooter, and administrator for the system. If the CMS is managed through Customer Service or Help Desk groups, it is possible that an individual from one of these groups can be given the role of system administrator.

The tasks shown in Table 7.1 are broken down into two main areas. The first is system administration, where daily user access and routine operation of the system must be sustained. These tasks include managing proper operation on the network, monitoring files and directories, setting up new users, managing user authorization levels and passwords, helping individual users with questions and problems, and organizing formal training when required. The administrator must be an expert on the database catalog, able to address questions on new and existing cable and equipment configurations.

SYSTEM ADMINISTRATION TASKS
• MONITOR NETWORK COMMUNICATIONS LINKS
• MAINTAIN FILES AND DIRECTORIES
• SET UP NEW USERS
• MANAGE AUTHORIZATION LEVELS AND PASSWORDS
• PROVIDE USER ASSISTANCE
• PREPARE TRAINING
• NEW CATALOG DEVELOPMENT
MAINTENANCE TASKS
• DATABASE BACKUP AND RESTORATION
• ARCHIVING SERVICE ORDER AND TROUBLE TICKETS
• FILE INDEXING AND TABLE MAINTENANCE
• INSTALL AND TEST NEW RELEASES
• INSTALL AND TEST NEW SUPPORT SOFTWARE
• CUSTOM REPORTING

Table 7.1: System administration and maintenance tasks for ongoing CMS support.

The second area of dedicated support is for ongoing maintenance tasks. These will include routine database backups and restorations, reindexing database files, installing and debugging new CMS software releases, installing and debugging new versions of operating systems, the database application, or a CAD application, and perhaps doing other database table maintenance or creating custom reports through the Structured Query Language (SQL) tools available. Archiving of old Service Orders and Trouble Tickets should also be performed routinely.

While this may seem to be a long list, the actual time to administer the CMS itself is fairly low - perhaps one day a week, if that. During periods of training or high levels of MAC activity, this time will increase. If new areas are being brought in to the CMS, the administrator may play a key role. Testing new releases or version upgrades to the CMS, the database software, or the operating system will also create temporary demands on the administrator's time.

The position of system administrator represents an ongoing cost to the organization. Other recurring costs for the CMS will include maintenance fees, usually 15 to 20 percent of the initial system price. A Maintenance Agreement often includes free upgrades to new versions, otherwise the upgrades represent a possible ongoing cost. Upgrades to the database software and operating systems may also be required, since many software vendors will only offer support on the latest two releases. Hardware upgrades for future additional users should be considered.

Major building renovations may require significant CAD drawing updates, or even completely new drawings. This may represent ongoing out-of-pocket costs, or could be handled in-house. Of course, bringing new areas into the CMS means additional inventory work as well.

INTERACTIONS WITH OTHER MANAGEMENT SYSTEMS

8

The overall process of communications management involves a number of related but distinct functions, including that of connectivity management. During the course of everyday activities, several different functions or systems can be affected by a single change such as moving a telephone. If other software-based management systems are in place, then the relationship of connectivity management to these other functions must be considered.

For any given system, there is a choice - whether to combine its functions within a CMS, build software interfaces or provide file transfers between systems, or manage their interface on a manual basis. We can look at a number of these systems, describe their general purpose and functions, and consider how a CMS might interact with each. Specifically, this chapter will consider how a CMS might be "bridged" or combined with various other management software. These systems include software for administering a PBX (Private Branch Exchange), for tracking and billing of phone usage, for billing of other services through the organization's general ledger, for managing the health and performance of the communications network, and for space planning and management of the physical facility. Asset Management, the tracking of "things" for purposes of accounting, depreciation, maintenance, and trouble history, will be discussed in Chapter 11.

PBX Administration

A traditional requirement for managing telecommunications has to do with operating a PBX. One or more PBXs can be the heart of an organization's voice network, and as a general rule they have been available much longer than modern Local Area Networks. Therefore, administration of the PBX is an established organizational need that has helped to shape the management of the cable plant as well.

The PBX is a computerized switching system designed to provide dialtone, connect different phone lines to one another, and provide access to outside trunks through an interface with the public network. The PBX has a central processing unit (CPU) and peripheral devices, an operating system, and software code that runs the

- PBX ADMINISTRATION
- CALL ACCOUNTING/DIRECTORY
- BILLING (GENERAL LEDGER)
- NETWORK MANAGEMENT
- FACILITY MANAGEMENT

Table 8.1: Examples of other management systems or interfaces.

moment to moment operation of the switch.

The typical PBX has a population of connection points for circuits on line cards and trunk cards, which operate in either an analog or digital format. Each port on a line card (or its equivalent) needs some software definition or configuration to operate properly. For example, each port may have to be configured with a set type, user name (for display purposes), extension number, button configuration, and features configuration, as well as storing the call forwarding, call pickup, and hunt groups for each line. The exact nature and format of this software is usually proprietary and varies from manufacturer to manufacturer.

The PBX software is fundamental - the switch cannot operate without it. As such, the CMS software cannot replace it, although it may duplicate information or have an active interface such that the port/line information can be accessed through the CMS. However, from the standpoint of connectivity, the important information is the relationship between port number, extension, and cable pairs that define the individual circuit. This may even include sequential backbone cable pairs and the office outlet number.

Some PBX vendors have introduced PBX software which includes cable and connection tracking features, so that the PBX software itself acts like a CMS. Most of these systems use a table structure and connection functionality that is oriented towards voice wiring, and they are not necessarily adequate for capturing data circuits and equipment on an enterprise basis. Information on pathways, termination hardware, and controlling equipment may be limited or non-existent. Similarly, the PBX software may have a work order function to initiate updates, but it is probably not robust enough to use as an all-purpose Service Order document. If you are managing a voice wiring system alone, such an approach may do just fine.

As a more general case, the CMS will need to operate side-by-side with the PBX management software. This raises several issues. First, it must be determined where certain information will reside, what information must be duplicated, and what key information can be used to manually relate records (information) between the two systems. For example, both systems will likely require the extension number and the port number, and both of these fields can be used to cross-reference information between systems if they are in an identical format. The CMS needs to know the cable pairs and outlet for a particular line, and you must decide whether to duplicate this information in the PBX software. Conversely, the PBX software may track set types and features, and you must decide how much of this information to duplicate in the CMS. It is recommended that you minimize duplication since it represents extra work, leads to favoring one system over the other, and can become a source of conflicting information between systems.

If you follow this approach, your second concern must be the sequence in which information is used and updated between the two systems. The PBX software must be used for port assignments, even if they are also tracked in the CMS. Therefore, a CMS Service Order cannot pre-assign PBX ports but can only have the port connections updated when the order is closed. The same might be true for a new extension number. If connections to PBX ports are changed for reasons of maintenance or balancing, the CMS must be manually updated.

Call Accounting/Directory

Call Accounting is the process of tracking telephone usage and billing the charges for each line back to a user, client, or cost center. These charges may be internally generated, from a PBX for example, or be received from a local access provider or long distance carrier in some file format. This file or tape is often called the Station Message Detail Record (SMDR). In addition, monthly charges for individual phone sets, particular line features, or additional hardware may be included in an inventory or asset management module. Often, as part of the call accounting or PBX management functions, an Employee Directory is maintained for use by company operators and for publishing an internal directory of telephone extensions.

Call accounting may be one part of a larger system designed to support voice services commonly referred to as telemanagement software. How different functions break down on a module-by-module basis may vary from system to system. The call accounting software usually requires that each telephone extension be defined in a table, along with a user name and cost center. Additional features of the line may be defined for billing purposes in cost allocation and/or inventory modules, and many times the telemanagement system uses a real-time or batch transfer interface to the PBX to obtain this information. Many call accounting systems allow for definition of trunk circuits or special circuits between sites, as they also may have charges associated with them.

Since the call accounting module includes each extension or line circuit, and have provision for other circuits, it was a natural outgrowth of telemanagement systems to include some information about the circuit connections themselves. Many of these packages can track backbone cable pairs and outlets for each extension, similar to the PBX software. But again, with a few recent exceptions, the telemanagement systems tend to look at the world in terms of voice circuits and architecture, and do not always handle cable and circuits on a general basis. For example, if a circuit is not associated with an extension, then it must be associated with a Circuit ID (how a circuit from the phone company would be identified), with no provision for defining the equipment at the end of the circuit. Pathway and termination hardware information may also be limited. There are systems which have now broken with tradition and have merged the call accounting functionality with a robust capability for defining cables, equipment, and connections.

If call accounting in a telemanagement system and a CMS must operate in parallel, it must be determined what information will reside in which system, and what information must be duplicated between systems. If the call accounting system is tracking phone set types and features through an inventory or asset management module, and handles all monthly charges for each set through a cost allocation module, then the CMS should track a minimum of information about each phone set. The CMS need only track each set by extension number and whether it is analog or digital. Tracking additional detail about set types, such as model number, features, or additional hardware may have merit for technical reasons, but will quickly duplicate information in the telemanagement system or PBX software. Distinctions between set types should only be made where there are differences in connections, such as a plug, adapter, power source, number of cable pairs required, etc. On the other hand, there is now little reason to track cable pairs or outlet ID in the telemanagement software.

The extension number field in each system should correspond for a particular telephone. A phone known as 2323 in the call accounting system should be identified the same way in the CMS. The specific set 2323 can therefore be found in either database. Modems and data modules should be so identified in the CMS, since the call accounting system may not have this information. Special circuits identified in the call accounting system by a Circuit ID may be identified as such in some CMS software, thereby tracking the cable pairs and connections of a special circuit. In other CMS systems, the Circuit ID can be used to name a fictitious device or piece of equipment so that the correspondence is maintained between systems.

A call accounting system and a CMS can operate independently, yet both may depend on the PBX software for an extension number assignment. Cable pair and outlet assignments can be made from the CMS, while set type and features are entered into the cost allocation module if required. Both systems should be updated in a timely manner so that correspondence of phone sets between systems is maintained.

The Directory function, while frequently derived from call accounting or PBX information, can also be a separate system designed to provide easy look-ups of extension numbers by user name and department. It may include home phone, emergency numbers, and a backup phone number for each person. This system may be also be linked to personnel files maintained by a Human Resources group. Many CMS packages have fairly well-developed Directory functions that can be used for this purpose, with the benefit that Directory information is automatically updated during the course of routine CMS administration.

A separate Directory, being dependent on extension assignments from the PBX software, is usually updated manually by paper document or by electronic file transfer from the PBX. Its relationship to a CMS is similar to the call accounting system, in that they are not interdependent. The CMS should not duplicate information from the Directory other than user name and phone extension. Both should be updated with changes in a timely fashion. For batch transfers to the Directory, you will need to determine whether the PBX, the call accounting system, or the CMS will be the most complete and timely source of information.

Billing (General Ledger)

Assuming that call accounting and cost allocation is handled in a separate system, the CMS may still require an interface with the general ledger for chargeback of Service Order or Trouble Ticket costs, monthly equipment charges, or equipment depreciation. This type of interfacing is usually handled through a batch file transfer rather than on a real-time basis. Some CMS software already comes with the ability to generate a file or report on current charges that may be imported to the ledger. Otherwise, this may be custom functionality added onto the CMS.

The job of the CMS, as the origin of the batch file, is to gather and format the proper data. The format of the file can vary, but must be readable by both the CMS and general ledger software. Charges are usually posted monthly, whereby the CMS queries records in the Service Order or Trouble Ticket tables to find any charges entered since the last report. This can be done strictly based on date of entry, or can use additional fields or flags in the record to indicate new charges versus those previously posted.

Different CMS packages offer different levels of charge detail. This could be as simple as a single charge entry, separate charges for labor and equipment, or a detailed list of individual labor and/or equipment charges. Your need for this detail will depend on your current billing practices. Using the CMS for monthly chargeback can be accomplished by using an existing or modified field on the equipment or device screen to enter the monthly charge, and then generating a monthly report for all device records indicating a charge. Certain CMS software may already come with this field, and allow the monthly charge to be set in the catalog as part of the definition of this type of device. When a device of this type is created, the monthly charge is automatically entered into the proper field.

Monthly chargeback is a simple case of asset management, and is not uncommon in a CMS. Additional detail that covers depreciation, maintenance agreement costs, or other recurring charges is more sophisticated. These charges can also be provided to the general ledger via batch file. More detailed asset management will be covered in Chapter 11.

The key fields that are common to the CMS and the general ledger system are the equipment identifier, and the cost center or cost code of the responsible department. A field for cost center is usually provided for each device in the CMS, or can be easily adapted by the Vendor.

Network Management

The idea of network management can represent a number of different methods for tracking the configuration, operation, and performance of network devices. The earliest software systems for network management involved monitoring a specific set of similar devices and capturing error messages or alarms from these devices. The management system was specific to the hardware and the communications protocol, and many times had to be manually configured as to the logical or physical arrangement of these devices. These systems have been referred to as "element" managers, since they only monitor one "element" of the total network.

Each type of communications protocol has traditionally used a different format or language for communicating alarms and problems over the network. A packet switch, an IBM SNA network, modems on dedicated leased lines, and a particular Local Area Network (LAN) may all use different signaling schemes to indicate an alarm, and may even differ in what the alarms are or mean. Since modern networks are comprised of a number of elements, there needed to be a number of different element managers to gather the required information. A problem arose whenever different elements would signal alarm conditions that might be related to the same point of failure, yet there was no way to correlate these alarms and use them for quick diagnostics.

The need for integrated network management has led to two useful developments. The first is an attempt at standardization of alarm conditions and signaling through a common language such as SNMP (Simple Network

Management Protocol) or CMIP (Common Management Information Protocol). These developments are still underway, although significant progress has been made.

The second development has been the creation of higher level management systems, often referred to as enterprise management systems or as a "manager of managers." HP Openview, Cabletron SPECTRUM, IBM Netview/6000, and SunNet Manager are examples of such systems. These systems can track, theoretically, any network component using a common management protocol, and pass alarms and messages from this component to the network administrator. The component is usually identified by a network address, whose specific format may vary from network to network.

Network management systems track logical information about the network - what is connected to what. Many systems are even smart enough to "look and see what's out there" and create their own database of equipment and interconnections. However, these systems cannot automatically know where a particular device is located or how it is physically connected. Such information can be retrieved manually from a CMS, or else interfaces can be built to do this automatically. Several newer CMS packages have done just that, and can operate with a network management system to provide the physical information necessary for locating troublesome equipment and connections.

To do this manually requires a correspondence of names between the CMS and the network management system. Since the network manager knows everything by network address, somehow the CMS must use the network address to name the corresponding equipment or port connection. Doing this at the port or card level is preferred, since this allows the hub or chassis to be physically named for asset tracking.

When the CMS and network management system are linked together, the network manager can identify the name or network address of equipment that is broadcasting an alarm or is not responsive. The network manager can then call up the CMS to show the location of the faulty equipment, usually displayed on a CAD drawing. Additional information about the device, such as outlet number, user name, and office can be displayed. Physical connections can be obtained if they have been defined in the CMS. This information, in textual and drawing format, can be provided to a technician to assist with finding and fixing the problem device. In addition, an alarm status detected by the network management system can be passed to the CMS to generate a Trouble Ticket and track the problem status.

Linking to a network management system has been done with CAD-based CMS software. These systems enjoy the same strengths and weaknesses as CAD systems in general. They work best with changes that involve moving an icon, such as the desktop device in the office. They are particularly suited to LAN environments with a dedicated port connection to each outlet.

We expect to see continued development of interfaces between CMS and network management systems. In this usage, the CMS can be populated with a minimum of cable and connection information, yet still serve its purpose for network management. Over time, additional physical and connection information can be added to the CMS provided that information can be developed and maintained properly.

Facility Management

Just as with connectivity management, building managers are frequently looking for automated tools to document and track information about the physical facility. These tools are extensively CAD-based, since a plan drawing best conveys the core information about building layout, space occupancy, building utilities, etc. It is likely that the interests of connectivity and facility management will continue to converge over time.

The facility manager is responsible for planning and administering the physical environment. While there are numerous tasks that must be done, we can discuss three basic areas of facility management that relate directly to managing connectivity and communications. The first is space planning, whereby the need for building space is forecast and space is allocated and inventoried. In other words, the space planner wants to know how much space is required to support organizational needs, how much is available at any given location, and how much is occupied by a particular group or department.

FUNCTION	CAFM	CMS
ALLOCATION	SPACE	OUTLETS
INVENTORY	FURNITURE FIXTURES	DESKTOP EQUIPMENT AND CONNECTIONS
MAINTENANCE	PLUMBING HEATING/COOLING ELECTRICAL SECURITY	BACKBONE CABLES PATHWAYS NETWORK EQUIPMENT GROUNDING SYSTEMS

Table 8.2: Correspondence of recordkeeping functions between Computer-Aided Facility Management (CAFM) and Cable Management System (CMS).

Frequently, charges or costs associated with leasing and maintaining space are allocated back to the individual department. For these functions, plan drawings can easily indicate space occupancy and provide a total square footage by group. CAFM (Computer-Aided Facility Management) tools are designed to track this space and report this information automatically. Therefore both CAFM and CMS tools need to show the user assigned to each space, with the user related to square footage in the first case and to communications outlets in the second case.

Both CAFM and CMS functions must go a step farther and track interior planning and installation. The facility manager desires to know the furniture and movable partitions in each space (and assigned to each user), while connectivity management seeks to know the communications equipment and connections in each space. Both groups manage an inventory of their respective equipment which can be managed in a database. This inventory must be tracked with regard to location, cost, maintenance, trouble history, etc. Finally, both facility and connectivity functions are involved with the movement and relocation of this equipment as a result of the routine Moves and Changes that occur in each organization.

Finally, both groups are concerned with routine maintenance and operations of the building infrastructure. For the facility manager, this includes building functions such as plumbing, heating, air conditioning, painting, carpeting, security, etc.. For the communications manager, this may include maintaining backbone cable, pathways, outlets, grounding systems, and distributed network equipment. The CAFM and CMS tools track the physical arrangement of these utilities on CAD drawings, allowing for improved maintenance capability and fewer conflicts or interferences between building functions.

If either the facility or communications groups chooses to develop CAD-base drawings for allocation, inventory, and maintenance functions, as described in Table 8.2, it makes sense for the other group to participate as well. A number of software vendors provide tools that combine the management functions of each group in a meaningful way. Otherwise, facilities groups may have a key role to play in maintaining drawings for a CMS. The CMS should be able to import drawings and drawing updates in a format compatible with existing CAD systems. Updates to architectural information (walls, floors, partitions, etc.) are fundamental and should be done prior to updates on cable runs and connections. Both groups should coordinate their efforts during major moves and relocation's, and share information as required to expedite changes.

For additional information on the functions of automated facility management, we recommend you obtain *Computer-Aided Facility Management* by Eric Teicholz (McGraw-Hill, 1992).

THE SERVICE ORDER

9

As has been previously suggested, utilizing a Service Order process can be the glue that helps keep track of, and control over, all the individual activities at work as networks are changed and reconfigured. At first, it might seem beneficial to forego a rigorous service order process, since staff can appear to respond more quickly and spontaneously to needs that arise. Over time, however, lack of structure in managing these changes can lead to breakdowns, errors, and oversights which could end up costing more time and money in the long run. Further, keeping connectivity records requires discipline on the part of every individual to update and maintain the information in a complete and timely way - and this is much easier to accomplish when there is some pattern and consistency to how requests are handled, what actions must be taken, and what records are affected by the changes.

Different approaches to CMS administration are discussed in Chapter 7. When a Service Order is used, it must capture all changes initiated as a result of that order before the order can be closed. If the information is not available, the order cannot be closed, and remains an open item in the system, as a reminder to someone that updated information is needed. If all actions are tracked through an order in this fashion, then all required updates will be flagged, and there is less chance that something will fall through the cracks. Perhaps more importantly, requiring a Service Order can help assure that information is made available and utilized before any actions are taken, which speeds up the change process, and provides the maximum benefit from the connectivity records.

Service Order Functionality

The typical Service Order in a CMS must accomplish several things, and may require several screens of information to do so. First, it must serve as a general record of the initial request and the actions taken. The order should track to whom it is assigned, when it needs to be completed, and other administrative information we will mention in a moment. Second, it may be required to generate a charge to some department or cost center for the service provided. This must be done through a billing interface with the general ledger, described in the previous chapter. Third, it should directly manage the changes and additions to circuit and connection information. Pre-assignment of circuit connections will speed up the technician's job and help to keep the records up-to-date. The Service Order should indicate a Pending or Reserved status on all affected components and connections until the order is closed. Finally, the Service Order document must convey the proper information to the technician, so the instructions are clear and unambiguous as to what changes are required.

The Service Order serves as an administrative record of requests that are received and actions taken. A sample screen showing potential information tracked in the Service Order is given in Figure 9.1 If these fields are used consistently, then it is possible to query, sort, and retrieve orders according to these fields. One might determine all the orders due on a given day, or assigned to a certain technician, or received from a particular department. Overdue orders are easily identified. If orders are tracked by type, patterns will emerge over time that can be useful for long-range planning and management of the Customer Service function. The Service Order can be used to avoid problems and make your organization look good to your internal clients.

Note that a charge or charges for completing the order can be tracked against a charge code or cost center. Even in organizations that simply treat charges as corporate overhead, or allocate total MIS or IT costs on an annual basis, this informal accounting can be helpful for specific business units or groups to see the costs that each has incurred. Otherwise, where direct chargeback is not used, it is easy for departments to take MACs for granted and then complain about the overheads they are burdened with.

The Service Order in different CMS software handles connectivity information to different degrees. There are systems with no interface between Service Order and connections, and all changes must be tracked and updated manually. In this case, extensive textual description should be included to indicate the nature of the change and any components reserved for this order. Some systems will have limited functions for changing connections, but not for adding or deleting new cables or equipment. Other systems will allow for complete access to inventory functions, adding and deleting equipment as required and making or breaking connections between cable and devices. Including this in the Service Order usually makes for the most effective use of connectivity information, and provides the greatest assurance that the database will remain accurate.

An example of a service order request might be where a user calls the Customer Service Center and requests to have their phone and computer moved from their current office on the first floor to an office on the third floor. The Customer Service Representative would open a new Service Order and record the user's name, telephone number, current room number, department, cost code and the room number of the new office, as well as the date when the service needs to be completed. While information on the user's phone number and room number should be available from the CMS records, it is beneficial to request the information from the user so the information can be verified to make sure the CMS records are accurate. The Customer Service Representative would then complete the remaining fields on the Service Order screen, such as type of service order (move), assigned technician(s),

SERVICE ORDER #: 940312 DATE ENTERED: 03/06/94 DATE COMPLETED: ________

STATUS: O PRIORITY: 3 REQUESTOR NAME: SANDY THOMPSON

PHONE: 111-1321 DEPARTMENT: ENG EXPENSE CODE: 163-071

APPROVED BY: W. HALEY S.O. TYPE: MOVE DUE DATE: 03/15/94

DESCRIPTION OF WORK: MOVE EXTENSION 1321 AND PC1000 FROM OFFICE 105E IN BUILDING REN TO OFFICE 316E IN REN. USER NEEDS SAME TYPE OF DATA SERVICES AS SHE HAS IN HER CURRENT OFFICE

ASSIGNED TECHNICIAN(S): SWEENEY ________ ________

INCLUDE CIRCUIT INFORMATION WITH PRINTED SERVICE ORDER: YES (YES / NO)

CIRCUIT INFORMATION: EQUIPMENT / CIRCUIT NUMBER: PC1000

TYPE:	NAME:	CONNECTOR:	BLDG:	FLR:	ROOM:	STATUS:
DEVICE	PC1000	2	REN	1E	105E	IN USE
OUTLET	REN1E101V	P3	REN	1E	105E	
CABLE	REN1E101V	P3	REN	1E	105E	
CABLE	RC-REN1E	27	REN	1E	1E ER	
PANEL	SNAPP01	14	REN	B	B22E	CIRCUIT INFORMATION:
CTRLR	LG1D11	TA-08	REN	B	B22E	PAGE 1 OF 3

Figure 9.1: Sample screen showing administrative information in a Service Order.

priority, etc.

Depending on the CMS product, the Customer Service Representative would then retrieve the current circuit information for the existing location, and check the availability of outlets or horizontal cabling in the new office. The Customer Service Representative would then mark the existing circuits for the phone and computer to be disconnected and assign the new connections for the new office. This information would then be issued to the appropriate technician(s) with all the current and new circuit cross-connect information.

Once the technician received the Service Order, the circuit cross-connect changes would be made, and the completed service order would be signed, dated and returned to the Customer Service Center with any additional changes noted. The Customer Service Representative would then update the CMS records with any changes, and close out the order with the completed date and any appropriate charges for the move. The records in the database would then reflect the new circuit connections, and the outlets and cables in the old office will now be vacant and available for new service.

There are situations that may be awkward to handle within the Service Order, and this may depend on the specific system. One case is the circular move, where two users are switching offices. If the outlet and/or user devices are allowed only one connection, then all connections may have to be broken before any devices are moved and reconnected. In other words, you cannot disconnect one device and connect it at the new outlet while the old device is still there - the old device must be disconnected first so the outlet is free to use. This may simply take one or two additional steps, but must be considered before making any changes through the Service Order. The same logic may apply when several users are switching offices or moving to a location that is already occupied.

The second situation that may require special treatment is a large Move, usually ten or more people at a time. In this case, it is better to print reports and track changes on paper than worry about defining every move in the Service Order. It may flow more quickly to disconnect all devices at once and then reconnect them all at once in the database, as long as someone takes responsibility for updates and they are completed before the Service Order is closed. An intermediate step of making the new connections on a pending or reserved basis can frequently be done manually. This avoids having to modify the Service Order if changes are made during the Move, as is frequently the case.

Finally, the printed Service Order document must convey the proper information to the technician, especially if connections or equipment is pre-assigned. This means indicating the specific user device, outlet, cable pairs, and port to be used for each service. Proper configuration (cataloging) of cable and equipment will help make sure this is done, and the Service Order document should be reviewed before the final catalog is completed. The technicians must be trained to read and understand these instructions, and clearly know their responsibility for updating the circuit information if it is changed in the field. In this manner, the actual connection information is what is used to update the database.

Additional information can be provided with the Service Order document, specifically a CAD floor plan and/or equipment rack elevation. These drawings can assist the technician with finding locations and equipment quickly. Even an E-size drawing (3' x 4') printed on 8 1/2" x 11" paper can sometimes be useful. Depending on the CMS, drawings could be generated directly through the Service Order or as a manual step after the order is opened.

Using a Service Order

It should be determined early in the project whether the CMS database is to utilize a Service Order/Work Order system. This will impact the workflows and procedures, as well as how the database and records are configured. If there is a current service order tracking system, determine if it can be replaced by the CMS database, or if it will work in conjunction with CMS records. The following questions will help to direct the discussions towards the best method for your organization.

- What is the volume of requests per day?
- How are requests currently tracked?
- How does the technician get the request?

- o How are current requests for changes received? (paper form, call to Customer Service, Electronic Mail, Etc.)
- o What circuit information, if any, is currently provided to the technician?
- o Are completion dates or status tracked?
- o How will the technician obtain current Circuit Information and available information for the new circuit?
- o How will the new/revised circuit information get updated in the CMS?
- o Should the technicians have access to the database or should all information be provided with the service request?
- o Are the technicians comfortable with using computers every day? Will they?
- o Do technicians have time to update records? Will they? Should they?
- o What is the cost of a technician vs. a clerical database administrator?
- o If a CMS Service Order is not used, how will records be kept up-to-date?

The above list of questions should help to determine if a service order system is needed to track user requests and provide technicians with a document which includes the circuit information.

Note that the use of a Service Order assumes there is a clerical Customer Service group available to take incoming requests and manage the Service Order. They can be given the proper training to do this on their own, or in conjunction with a Lead Technician or some other form of technical support. Persons at the Help Desk and in Customer Service are often more comfortable working with a computer every day, and have demonstrated the ability to master a CMS with proper training and guidance.

A Service Order can be particularly valuable when using third-party or outside contractors for MACs or troubleshooting. Documenting the work request and transmitting specific instructions will help assure quicker and more reliable service, and help you keep your records straight when others are working in the wiring closets. If the Service Order provides connection information, and you wish to pre-assign circuit components, it will be necessary to provide some fundamental training for the contractor in reading the Service Order and reporting back any changes.

Interfacing with an Existing Service Order System

If an existing service order system is currently used and must remain in place, it will be necessary to create a link between the current service order system and the CMS database. There are a few options on how this might be approached, depending on the type of service order system and the platform it utilizes.

One method is to have the existing service order electronically linked to the CMS database and retrieve the circuit information based on a Circuit ID. This could be the telephone extension number or a network port. This information could then become a part of the service request which is issued to the technician.

Another option is to provide the service order from the current system to the technician, and the technician would then be responsible for obtaining the circuit information from the CMS database. This approach might be best where a technician receives all of their requests for the day at one time, where they could look up all the required circuit information before leaving to perform the connectivity changes. With this method, it would likely be the responsibility of the technician to update the CMS database after making the circuit changes.

Another approach might be to have the Administrator who works with the current service order system access the CMS database and retrieve the circuit information. This could then be attached to the original service request before being given to the technician. With this approach, the technician could then report back changes in the circuit to the Administrator who would be responsible for updating the CMS records manually.

Any combination of approaches might be considered. This will greatly depend on the type of service order system currently being utilized, how the requests are distributed to the technicians, and also what information is recorded and provided to the technician. The most important point is to determine how the information will be provided to the technician and how the changes will be updated in the CMS once the request has been completed.

TROUBLE TICKETS AND PROBLEM MANAGEMENT

10

Like a good Service Order system, the Trouble Ticket can be used to track and control the individual actions taken to resolve problems encountered on voice and data networks. The challenge of any problem management process, however, is that problems arise in many forms and in many areas. Inevitably, people in several different groups have to respond to a variety of problems which might range from incompetent users to software bugs, to hardware crashes, to loose cable connections, to defective ports on a controlling device. Managing a centralized Help Desk can be a difficult chore, especially when information for problem diagnosis is in short supply or is contained in multiple databases in several different software systems. Frequently, different groups utilize their own trouble ticket systems, each oriented to their particular system or area of responsibility.

Using the Trouble Ticket or Trouble Report functionality in a CMS cannot be expected to meet every conceivable need. However, connectivity information is often very useful in troubleshooting, especially the locations of equipment and individual connection points, and can often help avoid long periods of downtime. To be able to access circuit information through the use of CMS records can be very beneficial. Therefore the CMS deserves serious consideration as an enterprise Help Desk tool. Interfaces are being built between CMS software and network management systems so that physical network information can be linked with information in the network management system to assist in locating problem equipment and troubleshooting network alarms.

Trouble Ticket Functionality

There are various approaches to structuring a Trouble Ticket or Trouble Report in a CMS. Like the Service Order, the ticket must serve as a general record of the problem and the actions taken. The ticket should track to whom it is assigned, what priority it assumes versus other current problems, and other administrative information. Second, it may be required to generate a charge to some department or cost center for the service provided. This must be done through a billing interface with the general ledger, as described in Chapter 8. Third, it should provide a textual or schematic description of the affected circuit to assist with troubleshooting and provide a document which can be marked with changes for updating the database. Finally, it must convey the proper information to the technician on the paper document so the problem and affected components are clearly indicated.

The Trouble Ticket serves as an administrative record of the problem reported and the actions taken. A sample screen showing potential information tracked in the Trouble Ticket is shown in Figure 10.1. If these fields are used consistently, then it is possible to query, sort, and retrieve tickets according to these fields. One might determine all the problems reported in a particular building, or assigned to a particular technician, or related to a certain network service. Outstanding tickets are easily identified. If tickets are tracked by problem type or problem resolution, patterns may emerge over time that can be useful in diagnosing future problems.

Some CMS software can track a trouble history of particular components. By compiling this information, it is possible to determine whether a particular type of equipment or device is prone to problems, allowing you to replace these devices or avoid them in the future. Other problem management systems, often working in conjunction with a network management system, allow correlation of problem resolution with problem type and store these correlations as a set of guidelines or recommendations which can be applied to future problems of a similar nature.

In some cases, a charge or charges for resolving the problem can be tracked against a charge code or cost center. Even in organizations that simply treat charges as corporate overhead, or allocate total MIS or IT costs on an annual basis, this informal accounting can be helpful for specific business units or groups to see the costs that each has incurred.

The Trouble Ticket in different CMS software handles connectivity information to different degrees. There are systems with no interface between Trouble Ticket and connections, and all changes must be tracked and updated manually. In this case, extensive textual description should be included to indicate the nature of any changes made in resolving the problem. Some systems will have limited functions for changing connections, but not for adding or deleting new cables or equipment. Other systems will allow complete access to inventory functions, adding and deleting equipment as required and making or breaking connections between cable and devices. Including this in the Trouble Ticket usually makes for the most effective use of connectivity information, and provides the greatest assurance that the database will remain accurate.

As a general rule, full circuit functionality is less common in the Trouble Ticket than the Service Order. Due to the unpredictable and varying nature of problems, connection or component changes cannot be made ahead of time. Manual database updates, prior to closing the ticket, are usually required.

An example of a trouble call might be where a user calls the Help Desk and reports a problem with their connection to the LAN. The Help Desk Administrator would open a Trouble Ticket and record the user's name, telephone number, room number, and a description of the problem. While information on the user's phone number and room number should be available

TROUBLE TICKET #: 94-132 DATE OPENED: 02/19/94 DATE COMPLETED: _____

STATUS: O PRIORITY: 1 REPORTED BY: MARILYN MCCARTHY

PHONE: 111-1123 DEPARTMENT: ENG TROUBLE TYPE: NDT

PROBLEM EQUIPMENT: 111-1123 PREVIOUS TROUBLE TICKET(S): TT93-476

DESCRIPTION OF PROBLEM: TELEPHONE GOES DEAD FROM TIME TO TIME WHERE USER DOES NOT HAVE ANY DIAL TONE. SHE HAS HAD A SIMILAR PROBLEM LAST YEAR.

ASSIGNED TECHNICIAN(S): HUTCHINS

CIRCUIT INFORMATION: EQUIPMENT / CIRCUIT NUMBER: 111-1123

TYPE:	NAME:	CONNECTOR:	BLDG:	FLR:	ROOM:	STATUS:
PHONE	111-1123	2	REN	1E	105E	IN USE
OUTLET	REN1E101V	P1	REN	1E	105E	
CABLE	REN1E101V	P1	REN	1E	105E	
CABLE	RC-REN1E	132	REN	1E	1E ER	
CABLE	1214-REN	432	REN	B	B22E	
CTRLR	PBX-A2	A20311	ADM	B	B01	

Figure 10.1: Sample screen showing administrative information in a Trouble Ticket.

from the CMS records, it is beneficial to verify the information. The Help Desk Administrator would then complete the remaining fields on the Trouble Ticket screen, such as the problem type, the trouble device, the assigned technician, etc. The CMS system would automatically enter the date and time the trouble call was received.

Depending on the CMS product, the Help Desk Administrator would then retrieve the circuit information for the problem device and any previous trouble history, and then dispatch the Trouble Ticket with the circuit cross-connection information to the appropriate technician. The ticket can be printed locally, printed at the technician's printer, transmitted electronically via E-Mail or workflow software such as Lotus Notes, or transmitted via paging system. Some CMS software tracks technician qualifications, and automatically assigns the problem to the proper person. A schedule of each technician's time can be maintained.

Once the technician received the Trouble Ticket, each cross-connect point of the circuit and the port connection would be checked. If there were no problems with the circuit, the user device would then be checked to identify the source of the problem. Once the trouble source has been identified and corrected, the technician would note any actions taken to resolve the problem, signed off on the trouble ticket, note the completed date and time, including the amount of time spent on the problem, and return the ticket to the Help Desk Administrator. The Administrator would update the CMS records with any changes and close out the ticket with the completed date and time, and indicate the problem resolution code. The records in the database would then reflect any changes to circuit connections and would indicate the nature of the problem and its resolution for future reference.

Finally, the printed Trouble Ticket document must convey the proper information to the technician, especially if a circuit description is provided. Proper configuration (cataloging) of cable and equipment will help make sure this is done, and the Trouble Ticket document should be reviewed before the final catalog is completed. The technicians must be trained to read and understand these instructions, and clearly know their responsibility for updating the circuit information if it is changed in the field. In this manner, the actual connection information is what is used to update the database.

Additional information can be provided with the Trouble Ticket document, specifically a CAD floor plan and/or equipment rack elevation. These drawings can assist the technician with finding locations and equipment quickly. Even an E-size drawing (3' x 4') printed on 8 1/2" x 11" paper can sometimes be useful. Depending on the CMS, drawings could be generated directly through the Trouble Ticket or as a manual step after the ticket is opened.

Using a Trouble Ticket

It should be determined early in the project whether the desired CMS should have a Trouble Ticket system. This will impact the workflows and procedures, as well as how the database and records are configured. If there is a current trouble ticket system, determine if it can be replaced by the CMS database, or if it will work in conjunction with CMS records. The following questions will help to direct the discussions towards the best method for your organization.

- What is the volume of trouble calls per day/week?
- How are troubles reported? (call Help Desk, Electronic Mail, etc.)
- How are trouble calls tracked? Are they tracked?
- How does the technician get trouble ticket information?
- Are trouble tickets tracked with closing or completion information?
- How will the technician obtain current Circuit Information from the CMS?
- How will any wiring or connection changes be updated in the CMS?
- Should the technicians have access to the CMS database?
- Are the technicians using computers every day? Will they?
- Do technicians have time to update records? Will they? Should they?
- What is the cost of a technician vs. a clerical database administrator?
- If a Trouble Ticket module is not used, how will records be kept up-to-date when

changes are made as a result of a trouble call?

The above list of questions should help to determine if a trouble ticket module is needed to track trouble calls, and provide a means for capturing changes resulting from trouble calls. The process for handling trouble calls will become important in establishing new workflows and procedures for working with the CMS database records, to insure maximum benefit is received from the records, and that they are maintained with the highest degree of accuracy.

A Trouble Ticket can be particularly valuable when using third-party or outside contractors for troubleshooting. Documenting the work request and transmitting specific instructions will help assure quicker and more reliable service, accurate billing, and help you keep your records straight when others are working in the wiring closets. It may be necessary to provide some fundamental training for the contractor in reading the Trouble Ticket and reporting back any changes.

Interfacing with Existing Problem Management Systems

If an existing trouble tracking or reporting system is currently in use, determine if the existing system must remain in place, or if it can be replaced by the new CMS record keeping system. As with the Service Order process, there are different options for handling the links between the CMS and an existing problem management system.

One method is to have the existing trouble tracking system electronically linked to the CMS database to retrieve information on the problem device or circuit. This circuit and device information can be included with the Trouble Ticket when provided to the technician. This would require that there be a specific circuit name or device name indicated in the trouble ticket which would correspond to the information required from the CMS.

Another option is to provide the Help Desk or user of the current trouble ticket system with access to the CMS, where the user could query the CMS database for the appropriate circuit information and attach it to the trouble ticket before issuing it to the technician. This approach would require the technician to indicate any changes on the trouble ticket and returning it to the Help Desk so the CMS database could be updated.

An alternate approach would be for the technician who receives the trouble ticket to query the CMS database for the required circuit information prior to resolving the problem. This would require the technician to be near a terminal or PC where he could retrieve the information when needed, and would also require that the technician then update the CMS records with the changes which were made during the resolution of the problem.

Any combination of these approaches might be considered. The method that will work best for your organization will be dependent on the type of trouble ticket system currently being used, how the tickets are delivered to the technician, and whether the technicians would be responsible for updating the CMS records based on changes resulting from the resolution of the problem. It is best to consider each approach, and walk through each process to determine which option provides the highest probability of success

ASSET MANAGEMENT

11

Asset management, in the broadest sense, is the tracking of information about things. Knowing what you have, where it is, and what its value is can be a great benefit in many areas. Tracking of cable is a form of asset management, for example. Just about any thing of value owned by the organization can be managed in this fashion.

More narrowly, asset management can be regarded as the tracking of items for purposes of accounting. This could be for billing, for inventory control, for depreciation, or simply knowing book value. Many times, asset management is required to comply with standard accounting practices as well as state and federal tax laws. It can apply to a broad variety of items - computers, telephones, desks, and chairs, for example. Therefore the interest in asset management can come from a number of groups, including Telecommunications, Information Systems, and Facilities or Building Services. Since a CMS is usually required to have a list of equipment and devices, it makes sense to consider this list as the basis for tracking the asset information about each item.

General Considerations

As mentioned in Chapter 3, one important aspect of asset management is that the "things" being tracked must have fixed, physical names as opposed to logical names. For example, tracking a PC for asset purposes requires that the PC be given a permanent ID rather than use the network address of an embedded network interface card. Tracking a phone set requires a fixed ID rather than just an extension number. If the equipment is replaced, the new equipment might take on the old logical name (network address or extension) and there would no longer be a record of the original equipment. If equipment must be tracked as spare, or under repair, it needs to have a permanent ID. Reconciling equipment against a purchase order or a maintenance agreement also requires a permanent ID.

In cases where these devices require both physical and logical names, the physical (permanent) ID must be used as the primary identifier. The logical name can usually be included in a field of the device record. Often there is a separate component that can use the logical name. In the case of expandable network hubs, the hub (chassis) itself requires a physical ID, while each card or port connection can be assigned a logical name to retain the cross-reference to a network management system.

Sometimes the item serial number can be used as the equipment ID. However, serial number formats vary from manufacturer to manufacturer, even for similar devices, and are often placed in an obscure spot. Sometimes the serial number is 15 or more characters and may begin with a string of zeroes. In order to keep equipment IDs in the same format and of a fixed length, it is useful to assign the equipment ID and track the device serial number in a field of the device record. Of course, items with no serial number must have an ID assigned. If IDs are assigned, they should each be noted on the device in some permanent and readily visible fashion. Bar-coding should be considered.

The MIS, Information Technologies (IT), or Computing Department usually tracks computer room devices and major network components. These devices include CPUs, Front-End Processors, disk storage devices (DASD), high-speed printers, card readers, etc. These devices are usually expensive and are covered by maintenance agreements in case of failure. Many times these are capitalized items that must be depreciated rather than expensed. Therefore, there must be a certain annual charge for depreciation over the specified lifetime of the equipment. This lifetime is set for different types of equipment by the Internal Revenue Service (IRS). Depreciation may be handled automatically in the accounting system, or could be derived annually from an asset tracking system.

Much of the growth in networking and personal computers has occurred at the departmental or group level. Purchases and expense accounting have also been handled at this level, and are not necessarily tracked over the entire organization. More recently, businesses appear to be establishing more central control over network, PC, and software purchases, since MIS or IT has been given the problems of maintenance and compatibility among these various devices.

With this trend, more groups are trying to implement asset control and tracking of PCs and network equipment. Some network management systems can track this information, but not always on an enterprise basis. The network management software may be limited to certain networks or devices. If so, the CMS can be a good home for this information since it can cover almost all types of network systems and devices.

Major telecommunications equipment has traditionally included the PBX and its supporting hardware, which have not typically relied on asset management systems for tracking. It was unusual to have specialized lists of analog phone sets, whose quantity could be roughly inferred from the number of active ports on the PBX or incoming phone lines. With the introduction of high-priced digital phone sets and auxiliary devices, asset tracking becomes more important. It is also helpful to have an internal list of any sets which are leased, under warranty, or under maintenance agreement. Frequently the PBX software can include serial numbers and set types. It is not necessary to duplicate this inventory in the CMS, thereby allowing sets to be easily tracked in the CMS by extension number only.

Other telecommunications equipment for T1 multiplexing and other high-speed circuits should naturally be tracked in the CMS for purposes of showing connectivity. If not resident elsewhere, its asset information could be included in the CMS as well.

Facility Management groups have a similar opportunity to track various inventory components, including furniture, utility and maintenance equipment, specialized building fixtures, etc. Space itself is a desired component of a facility inventory, readily tracked on plan drawings. Just as with communications items, there is a question as to what level of detail is appropriate and beneficial. Tracking every table, chair, and desk may not be practical or of much value, considering that these items are not very expensive to begin with.

The facility inventory items, particularly furniture, tend to be more portable than communications hardware and more easily relocated from office to office. There is also a greater spare or uninstalled inventory of furniture to consider. (Very few organizations keep a roomful of spare PCs under lock and key.) So the management of facility inventory is somewhat different than for communications, requiring a greater level of inventory control. Use of bar-coding can be helpful in this regard, since a person with a portable bar-code scanner can inventory a location very quickly.

Asset Information

Asset information about any particular thing begins with a unique name or identifier for that thing. Additional information to consider is shown in Table 11.1. In the case of communications hardware, many or all of these devices will be tracked anyway for the purpose of showing connections, and asset information can simply be included as part of the original device record. Many CMS packages include user-defined fields which can be modified for this purpose. If asset information is tracked in a separate system, the names or identifiers should be the same in each system so that information can be cross-referenced from one system to another. It is possible to build batch file transfers between systems to move information back and forth as appropriate.

EQUIPMENT NAME OR ID	NETWORK ADDRESS
TYPE OR MODEL	LOCATION
SERIAL NUMBER	USER ID
STATUS	DEPARTMENT
MANUFACTURER	COST CODE
MANUFACTURER'S PART NUMBER	
PURCHASE ORDER NO.	MAINTENANCE CONTRACT NO.
VENDOR	SERVICE PROVIDER
PURCHASE DATE	CONTACT NAME
PURCHASE PRICE	CONTACT PHONE
MONTHLY CHARGE	START DATE
COMMENTS	END DATE
WARRANTY PERIOD	COST
	MONTHLY CHARGE
	COMMENTS

Table 11.1: List of potential fields in a device record for tracking asset information.

Additional information should include the equipment type or model, and the actual serial number. You may consider tracking the manufacturer and the manufacturer's part number as well. It is useful to indicate the device status, such as In-Use, Spare, or Reserved. For communications components, the presence of an active connection to something else may provide sufficient indication of status. It also may be necessary to indicate the current network address of active devices. Many CMS databases will automatically fill in the location and a User ID (from validation tables) when the device is installed and connected. Certain network-related information, such as network address, serial number, BIOS version, PC type, etc. can be obtained via special network software that polls and investigates a logged-on PC.

If financial or maintenance information is desired, then you will need to track the original purchase order (PO) number, the vendor, the purchase date, and the purchase price. Some management systems, including certain CMS products, can actually generate a Purchase Order, or at least a Bill of Materials. Maintenance information should include the agreement number, the service provider, the cost for the total maintenance agreement, and the starting and ending dates of the agreement. A contact name and telephone number for the service provider may be helpful. A comments or notes field is useful to indicate what is covered, what is not covered, and any other special conditions or charges that apply. The table should be able to be queried to list all items associated with a particular purchase order or maintenance agreement.

Monthly maintenance and depreciation charges can be provided to the general ledger by way of a batch file. A department or cost code must be indicated on each item record, consistent with the codes used in the general ledger. This type of accounting can be handled in a variety of ways, but there are two we will consider here just to illustrate the process. The easiest way, of course, is to simply indicate the monthly charge for depreciation (or cost) and/or the monthly charge for maintenance in individual fields. Some CMS software will allow a cost to be specified for the particular type of equipment, and this cost appears as the default cost (modifiable) on each item screen. The batch routine can then identify these monthly charges and create a file which can later be uploaded to the general ledger. The problem is, there is no immediate indication of how much these charges have accumulated over time, especially if the charges are meant to be of limited duration, say two years. It is possible to have additional fields indicate the accrued charges, and have these fields increased monthly by the proper amount.

Another approach, which is a more complicated but more accurate method, is to indicate the total amount to be charged, the number of monthly installments, the starting date, and the total accrued charges. The monthly charge is then calculated to be the total charge divided by the total number of months. Each month, a charge is calculated and posted, and the total accrued charges are increased. When the number of monthly installments has been met, no further charges are posted.

For purposes of annual depreciation for tax accounting, most accounting software has provision for tracking items and accruing depreciation against expenses. In an asset management system, you can use an approach similar to the previous, using annual installments for allocating the original cost rather than monthly installments. Be sure to check with your accounting personnel to determine the proper method.

As you can imagine, this type of billing routine is fairly sophisticated and is more typical of what is available in asset management software. It would not be a typical approach in a CMS unless it was developed through software customization.

The purpose for attempting to consolidate asset management into the CMS is to avoid redundant recordkeeping, since device information will always be required in the CMS in some format. If a separate asset management system is desired or in place, both the CMS and the asset records will need to be updated in a timely fashion. If common equipment IDs are used, a batch program can be written to automatically compare the two equipment lists and report any discrepancies.

PART 2: PLANNING FOR AND SELECTING A CMS

THE PROJECT TEAM

12

A Project Team should be created when you begin to define the needs for your Cable Management System. The strength of the Project Team will have a definite impact on the success of implementing a recordkeeping system. A team provides the opportunity to generate a more complete list of ideas and concerns which must be addressed when changing current procedures for documenting connections, as well as the activities of those individuals making the changes.

No matter how much a single individual may know about what's needed, an enterprise CMS will impact other individuals and groups. It's important to include the comments and input from these groups early in the project, in order to gain the maximum level of support and commitment. This approach also assures that you have considered all the possible requirements for your CMS recordkeeping, and identifies any records which are currently being maintained. In a smaller organization or at the site level, this ultimately may be only two or three individuals.

The Team Approach

First, make a list of the organizations and individuals who have access to the wiring closets, computer rooms, LAN rooms, and main distribution rooms. These are key individuals who can make changes that affect your database, so it's critical to get their input and commitment to the project.

Make a second list of the organizations and individuals who are involved in employee relocations, major moves, or building renovations. These are also key individuals who know when changes are being made which will affect the CMS database records. These organizations are usually the first people to know when an employee (therefore, a phone) will be moving or that new services will be required. Their knowledge and experience will not only be an asset to your team, but will help to provide an additional source of information for keeping your CMS records up to date. Their involvement in routine activities will help to define the current procedures and workflow processes which are in place, and where modifications to procedures will need to be made in order to support the CMS database records. They will also be valuable in identifying areas where information or changes could be performed that might not go through the normal channels.

Add any other organizations and individuals to the list who could have an interest or involvement in your project. These individuals could be part of the Help Desk, Facilities or Building Maintenance, Employee Directory,

- IDENTIFY TEAM MEMBERS
- INFORM SUPERVISION
- DETERMINE RESPONSIBILITIES
- ESTABLISH MILESTONES
- RECOGNIZE ACCOMPLISHMENTS

Table 12.1: Elements of the team approach.

Human Resources, etc. This will help to ensure you develop a good base of information to define your CMS requirements, identify what record keeping tools are used today, and what is needed for the future.

Contact the Managers of each of these organizations and inform them about your CMS project and how it will impact their organization. Ask for their commitment and involvement by assigning one member of their organization to participate in your CMS Project Team. This can be done individually with each manager, or in a group. It will be the responsibility of team members to represent the collective ideas and needs of their respective groups and to keep their own organizations informed of the CMS Project progress and initiatives.

When preparing presentations to be given to managers or key individuals in each group, include the specific interest areas and benefits which apply to their organization. Target the key issues and concerns which the CMS project will address. For example, when addressing Help Desk personnel, indicate the additional information available regarding the caller's equipment and connections which can make the Help Desk more knowledgable and provide a higher level of service.

Roles and Responsibilities

Once you have identified the members of the project team, it is important to provide them with an understanding of their roles and responsibilities as a member of the team. They need to be an active member, where their contributions are not only valuable but are necessary to build a good foundation for the CMS and the processes for maintaining these records. Their responsibilities should at least include those provided below in Table 12.1. Depending on the resources available, individual team members may have responsibilities and tasks assigned in addition to those provided below.

Each team member must represent the views and ideas of their organization or group, and report CMS project goals and activities back to their respective groups. This includes keeping their supervisor or manager informed of the CMS project progress and initiatives in order to maintain their continued support and commitment. Each member must attend all requested meetings or send a designated representative, and complete all assigned tasks pertaining to the CMS project which are agreed upon, pending approval from their respective supervisor/manager.

The Project Leader has primary responsibility for the project. This individual needs to be committed to the concept of cable and connectivity recordkeeping and should be very familiar with either the wiring and communications infrastructure or the move, add and change process. They need to be highly motivated, capable of making decisions, and demonstrate good leadership skills as well as good people skills. This individual needs to be capable of presenting ideas and needs to management, and compiling the necessary documentation to convey the project goals, planned activities, and resource requirements. They will be the key individual interfacing with management personnel in the various organizations impacted by the CMS project, and it is their responsibility to ensure that

- REPRESENT THEIR ORGANIZATION'S NEEDS AND IDEAS
- KEEP THEIR ORGANIZATION & MANAGEMENT INFORMED
- ATTEND PROJECT TEAM MEETINGS
- COMPLETE ASSIGNED TASKS
- BE A CONTRIBUTING TEAM MEMBER

Table 12.2: CMS team member responsibilities.

- PRIMARY PROJECT RESPONSIBILITY
- DECISION MAKING
- DEMONSTRATING LEADERSHIP
- PROJECT DOCUMENTATION
- MANAGEMENT PRESENTATIONS & COMMUNICATION
- SECURING MANAGEMENT SUPPORT & COMMITMENT
- PLANNING & GOAL SETTING
- CONFLICT RESOLUTION
- DEVELOPING PROJECT COST AND SAVINGS ESTIMATES
- MAINTAINING TEAM ENTHUSIASM & COMMITMENT

Table 12.3: CMS team Project Leader responsibilities.

the necessary management commitment and support is clearly demonstrated.

In addition to the primary responsibilities of driving the project forward, the Project Leader or their designate should schedule all meetings, notify team members of meeting dates and locations, and set goals and objectives for each project and team meeting. They must address concerns raised by team members and resolve conflicts which arise during the course of the project. And of course, the Project Leader is the principal spokesperson for obtaining management commitment and resources. The support and commitment demonstrated by management is the key to a successful project.

The planning and scheduling of project activities should be coordinated through the Project Leader. Documentation of the project justification and estimates of project costs will also likely be their responsibility.

As you begin to discuss the requirements for the CMS, keep your options open and at least give the ideas presented by team members a chance to be discussed. Make a list of all the ideas which are presented. This will become your initial "Wish List", which will later be prioritized as to which items will be critical requirements for the CMS. After going through all the steps outlined in this manual, it will become apparent which items will be a low priority on your list of requirements. It's important to make all team members feel their input and opinions are important. This helps to keep the necessary enthusiasm, ongoing support, and commitment required for a successful project.

Milestones

Every project should establish milestones, with anticipated completion dates. This helps to keep the project on schedule, and provides a means for measuring success. Make the milestones significant but keep them within reach, so that they are achievable, demonstrate success and renew the team commitment to the project and to other team members. Be sure to identify milestones which are reached early in the project, so the team knows they are making progress and can feel good about themselves and the project.

Make a special effort to celebrate the successes and milestones as they are reached. Rewards and celebrations do not have to be expensive, but something that represents the "Pat on the Back" will go a long way to keeping team enthusiasm and commitment. A pizza lunch meeting, doughnuts for morning meetings, even little things like bringing candy, popcorn or snacks to add some energy to afternoon meetings can help to keep team members interested and enthusiastic.

Document and communicate the progress and success of the project. Include status updates in weekly or monthly reports or newsletters that might be issued by the various sections or departments. Distribute monthly reports to the managers or supervisors of the organizations who have representatives on your team. This helps to gain exposure and support for the project.

Some organizations have given the project a special name or logo to help promote visibility and support for the project. This could be a set of letters, abbreviations or even something simple like Project CMS. Project awareness can help create commitment from team members. You can even make it a contest for team members to come up with a project name, with a small prize or award for the winner.

Developing a good project team and maintaining a high level of enthusiasm and commitment to the project are keys to a successful CMS implementation. The project interest demonstrated by team members will spill over into the rest of the organization and help to promote the project goals and objectives. This will be a real asset as the project unfolds and more and more individuals become involved in the daily use of the records and information contained in the CMS database.

DEFINING PROJECT SCOPE

13

A CMS implementation project begins by determining the scope of the project in several ways. The project scope includes the domain of facilities (which buildings, floors, etc.), the domain of cable plant (which copper backbone cables, horizontal cables, fiber optic cables, pathways, etc.), and the domain of active devices (telephones, PCs, network hubs, etc.) which will be included in your recordkeeping. Knowing rates and patterns for Moves and Changes (MACs) can also be considered part of the scope of information affecting the project. After collecting some basic information about these domains overall, you will have to narrow each of them down to exactly which facilities, cables, devices, etc. you are going to manage. Together, these items will ultimately determine what types of items, records, and connections will be maintained in the CMS. Further, your chosen domains will also guide you in planning the administrative processes, or workflows, which will use and maintain the information in the proper fashion.

By considering the project scope in this fashion, you will have good information for estimating the size of the project, determining the types of information you want to track, developing the nomenclature for this information, and understanding who needs to be involved to make it all work. This is where a broad-based project team becomes valuable, since different individuals may know more information about specific areas. We will use our example of the ABC Company facilities, the Landisville Research Park described in Chapter 2, to illustrate the process.

- SITES
- BUILDINGS AND/OR FLOORS
- # OFFICES PER BUILDING OR FLOOR
- # OUTLETS PER OFFICE, BY BUILDING OR FLOOR
- # PAIRS PER OUTLET, BY BUILDING OR FLOOR
- # CIRCUITS PER OFFICE, BY BUILDING OR FLOOR

Table 13.1: Initial data for CMS planning.

Evaluating the Physical Plant

The first area which will impact the scope of the project is the size of the facilities, such as the number of buildings and floors involved, the basic nature of the cable plants themselves, and the quantities of cable pairs and outlets to be tracked. This general information, summarized in Table 13.1, can be compiled into several lists on paper or computer spreadsheet. The lists are broken down so you can follow in a clear, step-by-step fashion. After reading this section, you may choose to structure your information differently or dispense with some of the intermediate steps, provided that you have acquired the appropriate information.

To begin, make a list of the sites which will be included in the CMS project. You should include any standard abbreviation or designation currently in use for each site, as well as any

FOR: ABC COMPANY

SITE	ABBR	COMMENTS
HILLSIDE PLANT	HMF	MANUFACTURING
LANDISVILLE RESEARCH PARK	LRP	CAMPUS
DOWNTOWN OFFICES	ABC	HIGHRISE

Table 13.2: List of potential sites for CMS planning.

descriptive comments or notes that are appropriate. Sometimes it is also helpful to prioritize the list, which will be useful when putting together the implementation schedule. This information will also help determine if all the records can be maintained in a single database at a central location, or if separate files or databases may be required for different sites. An example of this list is given in Table 13.2.

The next step is to break things down to the building and floor level, since the cabling may vary from building to building and even from floor to floor. First, make a table listing the individual buildings at each site, one list for each site, including information on the number of floors in each building and any designations or abbreviations currently in use for each building. As with the site designations, you may want to consider these abbreviations for use in your recordkeeping system. An example for the ABC Company's facilities at Landisville Research Park is shown in Table 13.3. Remember that a table must be created for each site.

To continue getting a sense of quantity and size, you need to consider the typical average number of offices per floor and the average number of jacks or outlets per office. To have an accurate description from floor to floor, it is recommended that you make a new table of buildings with an entry for each floor. Use any current abbreviations or designations for each floor. Indicate the estimated number of offices or work locations on each floor along with any room numbering schemes in use. Leave some room on the right for additional columns of information, since you will need to expand this table with the required planning data. An example is given in Table 13.4. Again, remember that a table must be created for each site with all sites being considered for planning purposes.

In some organizations, the number of outlets or jacks per office may vary from building to building and site to site, depending on the when the current cabling was installed. Determine the approximate number of outlets per office for each building/floor listed, and add this information to the chart for each of the different buildings (Table 13.4). This should cover all types of outlets, such as modular, coax, fiber, etc. If the type of modular outlet is different from location to location, it may also be useful to note the type of outlet (i.e. simplex, duplex, or quad) and the number of pairs per outlet, if appropriate. For example, a single 4-pair station wire may be split to two different

SITE: LANDISVILLE RESEARCH PARK (LRP)

BUILDING	ABBR	FLOORS	COMMENTS
ADMINISTRATION	ADM	B, 1, 2	PBX MDF
FACILITIES OPERATIONS	FAC	1, 2	MAINTENANCE
RESEARCH/ENGINEERING	REN	B, 1, 2, 3	DATA MDF

Table 13.3: List of potential buildings for CMS planning.

SITE: LANDISVILLE RESEARCH PARK (LRP)

BLDG	FLR	ROOM NUMBERING	OFFICES PER FLR	OUTLETS PER OFFICE	TYPE/PRS	TOTAL OUTLETS
ADM	B	B01-B99	20	2	RJ45/4	40
	1	101-199	50	4	RJ45/4	200
	2	201-299	50	4	RJ45/4	200
FAC	1	G101-G140	40	4	RJ11/2, BNC	160
	2	NONE	20	3	RJ11/2, BNC	60
REN	BE,BW	B01-B99E/W	50	2	RJ45/4	100
	1E,1W	101-199E/W	100	2	RJ45/4	200
	2E,2W	201-299E/W	100	3	RJ45/4, DC	300
	3E,3W	301-399E/W	100	2	RJ45/4	200
		TOTALS	530			1460

Table 13.4: List of buildings, floors, and outlets for CMS planning.

jacks, 2 pair and 2 pair or 3 pair and 1 pair. Other station wiring may be only 2-pair or 3-pair to each jack. Your interest here is in the outlet connection itself, pairs per outlet, since additional information about the horizontal wiring will be compiled on another table.

The method for describing outlets or jacks will vary from organization to organization. Some organizations refer to each RJ11 or RJ45 as an individual outlet, while others may consider an outlet to consist of multiple RJ11 or RJ45 connections. The options for defining outlets and jacks are covered in more detail in Chapter 21.

Outlet information will be valuable when you look at naming and labeling procedures for outlets. For each building or floor, multiply the number of offices times the number of outlets per office to determine the total number of outlets. These can be added up for all buildings/floors at each site to get the total number of outlets to be

SITE: LANDISVILLE RESEARCH PARK (LRP)

BLDG	FLR	OFFICES PER FLR	OUTLETS PER OFF	TOTAL OUTLETS	VOICE CIRCUITS	DATA CIRCUITS	OTHER
ADM	B	20	2	40	1	1	
	1	50	4	200	1.5	2	
	2	50	4	200	1.5	1	
FAC	1	40	4	160	1	1	
	2	20	3	60	1	.5	
REN	BE,BW	50	2	100	1	1	
	1E,1W	100	2	200	1	1	
	2E,2W	100	3	300	1.5	1	
	3E,3W	100	2	200	1.5	1	
TOTAL		530		1460			

Table 13.5: Primary Planning Table.

tracked in the database.

It is often useful to determine the number of employees that will occupy those buildings and floors which will be included in the CMS. This information may be available through your personnel section, or by looking at the employee directory. Make a note of the number of employees at each site on the bottom of the Primary Planning Table for each site. It should roughly agree with the number of offices or work areas identified on that same chart. This number will be helpful in determining the size of the database and the potential number of circuits involved in routine Moves, Adds and Changes.

As a final step, it is useful to determine the average number of circuits per office - in other words, how many circuits and connections are actually active? All office outlets may not be in service, while other outlets may actually be carrying two or more services. These can be estimated by building and floor and added to Table 13.4 to create the Primary Planning Table shown as Table 13.5. Active circuits per office could range from 1 to 4, depending on the environment, and can be broken down into Voice, Data (LAN or Host connections) and Other. For example, some users may have a second voice line for a modem or answering machine, and may have one or more data connections such as Mainframe or LAN.

This information will help to identify the volume of records which will be stored in the CMS database and will help to indicate the extent of physical inventory required, particularly if you are interested in tracking control or desktop equipment. It is sometimes possible to get the total number of voice lines from the employee directory, or from the main voice technician. The total number of data circuits may be more difficult to determine since there may be several different types of data networks, with local controllers or hubs distributed among the various wiring closets.

Evaluating the Cable Plant

Having developed an overview of the nature and extent of the physical facilities, you

SITE: LANDISVILLE RESEARCH PARK (LRP)

CABLE	TYPE	COUNT	END 1 SPACE	END 1 TERM'N	END 2 SPACE	END 2 TERM'N
CAMPUS						
1214	UTP	900	ADM B01	CB01-18	REN B22E	CB01-18
1214	UTP	300	ADM B01	CB19-24	FAC ER	CB01-06
TRUNK		300	MTU CO		ADM B100	FB01-06
FIBER	62.5 MM	36	REN B22E	FC-RENB	ADM B100	FC-ADMB
FIBER	62.5 MM	12	REN B22E	FC-RENB	FAC ER	FC-FAC
ADM						
RISER 1	UTP	200	ADM B01	RB01-04	ADM 104	RB01-04
RISER 2	UTP	200	ADM B01	RB05-08	ADM 242	RB01-04
FIBER	62.5 MM	12	ADM B01	FC-ADMB	ADM 104	FC-ADM1
FIBER	62.5 MM	12	ADM B01	FC-ADMB	ADM 242	FC-ADM2
FAC						
NONE						

Table 13.6: Identified backbone cables for CMS planning (continued next page).

CABLE	TYPE	COUNT	END 1 SPACE	END 1 TERM'N	END 2 SPACE	END 2 TERM'N
REN						
1E	UTP	200	REN B22E	RB01-04	1E ER	RB01-04
1W	UTP	200	REN B22E	RB05-08	1W ER	RB01-04
2E	UTP	200	REN B22E	RB09-12	2E ER	RB01-04
2W	UTP	200	REN B22E	RB13-16	2W ER	RB01-04
3E	UTP	200	REN B22E	RB17-20	3E ER	RB01-04
3W	UTP	200	REN B22E	RB21-24	3W ER	RB01-04
FIBER	62.5 MM	12	REN B22E	FC-RENB	1E ER	FC-REN1E
FIBER	62.5 MM	12	REN B22E	FC-RENB	1W ER	FC-REN1W
FIBER	62.5 MM	12	REN B22E	FC-RENB	2E ER	FC-REN2E
FIBER	62.5 MM	12	REN B22E	FC-RENB	2W ER	FC-REN2W
FIBER	62.5 MM	12	REN B22E	FC-RENB	3E ER	FC-REN3E
FIBER	62.5 MM	12	REN B22E	FC-RENB	3W ER	FC-REN3W

Table 13.6, continued: Identified backbone cables for CMS planning.

should create a similar, reasonably complete description of the backbone and horizontal cables that are on premises. For each building you have identified on your Primary Planning Tables, make a table listing by name all of the backbone cables running within that building. Make an additional table listing all of the interbuilding (campus) backbone cables and feeder cables as well. For each cable, indicate the type of cable, the pair or strand count, and the termination location (space) and termination method at each end. Examples of these lists for Landisville Research Park are given in Table 13.6. If you desire, make a list of the backbone cable pathways at the same time, indicating which cables and pair counts are in which pathways as shown in Table 13.7. You can add more detail about the pathways later.

If you don't know all of the cables, then a quick survey is in order. A good method for identifying all your cables is to start at the source point, such as the telco demarc point, or at the main equipment room where the voice switch or main data center is located. Working from this central point out to individual wiring closets in the local office areas will help to make sure you have identified all possible cables, and quickly identifies where cable names are duplicated and may require modified naming or labeling schemes. Don't spend forever trying to track down the last missing items, but take good notes and sketches wherever unidentified cables appear. Note the cable type, pair count, jacket color, manufacturer's labeling, or any other characteristic markings to compare with other "unknown" cable appearances. It is helpful to tag or label the unidentified cable with a number and track it accordingly. Chances are, you will be able to identify these components by a process of elimination. Otherwise, such items should be identified during the full inventory.

You can now total up the number of backbone cables, cable pairs, and fiber strands that will become records in the CMS. This provides information on quantities of things, and helps to identify what information is missing and must be obtained through inventory. Note that you should have an indication of every administration space (wiring closet, equipment room, etc.) as well as the types of termination hardware to track. You may wish to make a separate list of administration spaces at this point.

The same process should be repeated for the horizontal cabling for each building and floor listed on your Primary Planning Table (13.5). While your primary table could be expanded to include this information, it is probably easier to start a new table like that shown in Table 13.8. For each building and floor indicate every type of horizontal cable in use, such as 4-Pair UTP, IBM Type 1, Category 3, RG-58, etc. Also make sure to identify the different wiring configurations, such as 568A, 568B, USOC, etc. and to verify that the wiring and configurations are consistent with the number of pairs per outlet on Table 13.4. This information will be important when defining the structure for your horizontal cable in the CMS database as well as how connections are made at each end. It does not have to represent your ultimate description of this wiring, since we will address nomenclature in more detail in Chapter 21. Most automated CMS products have some flexibility in how you define your station cables, which will impact the simplicity or complexity of making connections and tracing circuits during daily use and maintenance of the CMS records.

For each set of cables, indicate the naming/numbering convention used (if any) for labeling these cables and their terminations. It's a good idea to note the room numbering scheme on each floor as well. This helps identify duplicate or missing numbering schemes to be addressed in the final cataloging and labeling process.

As a final consideration of project scope, it is useful to determine the level of activity involved in trouble calls and Moves, Adds and Changes, since this will affect the time required to maintain and support the database ongoing. If there is a current Service Order or Work Order system, this will provide an estimate of the number of MACs processed over a given period of time. If requests are not currently tracked, determine the estimated number based on information from technicians who handle the changes.

If there is a current system used by the Help Desk or Customer Service Center, this should

SITE: LANDISVILLE RESEARCH PARK (LRP)

PATHWAY NAME	CABLE NAME	PAIR COUNTS
CWADM-1	1214	1-900
CWADM-2	1214	901-1200
CWADM-3A	FIBER/ADM	1-36
CWADM-3B	NONE	
CWADM-3C	NONE	
CWADM-4	TRUNK	1-300
CWREN-1	1214	1-900
CWREN-2	NONE	
CWREN-3A	FIBER/ADM	1-36
CWREN-3B	FIBER/FAC	1-12
CWREN-3C	NONE	
CWREN-4	NONE	
CWFAC-1	1214	901-1200
CWFAC-2	NONE	
CWFAC-3A	FIBER/FAC	1-12
CWFAC-3B	NONE	
CWFAC-3C	NONE	
CWFAC-4	NONE	
CWLRP-1	1214	1-900
CWLRP-2	1214	901-1200
CWLRP-3A	FIBER/ADM	1-36
CWLRP-3B	NONE	
CWLRP-3C	NONE	
CWLRP-4	NONE	
CWTEL	TRUNK	1-300

Table 13.7: Identified pathways for CMS planning.

SITE: LANDISVILLE RESEARCH PARK (LRP)

BLD.	FLR	TYPE	CONFIG	NAMING	ROOM NUMBERS	# PER OFFICE	TOTAL
ADM	B	4-PR/CA3	568A	101A,B	B01-B99	2	40
	1	4-PR/CA3	568A	101A,B	101-199	2	100
	1	4-PR/CA4	568A	101C,D	101-199	2	100
	2	4-PR/CA3	568A	101A,B	201-299	2	100
	2	4-PR/CA4	568A	101C,D	201-299	2	100
FAC	1	2-PR/CA3	USOC	101A,B,C	G101-G140	3	120
	1	COAX	RG-62	101D	G101-G140	1	40
	2	2-PR/CA3	USOC	101,A,B	NONE	2	40
	2	COAX	RG-62	101D	NONE	1	20
REN	BE/BW	4-PR/CA3	568A	101V-150V	B01-B99E/W	1	50
	BE/BW	4-PR/CA4	568A	101D-150D	B01-B99E/W	1	50
	1E/1W	4-PR/CA3	568A	101V-150V	101-199E/W	1	100
	1E/1W	4-PR/CA4	568A	101D-150D	101-199E/W	1	100
	2E/2W	4-PR/CA3	568A	101V-150V	201-299E/W	1	100
	2E/2W	4-PR/CA4	568A	101D-150D	201-299E/W	1	100
	2E/2W	TYPE-1	IBM-DC	1-50	201-299E/W	1	100
	3E/3W	4-PR/CA3	568A	101V-150V	301-399E/W	1	100
	3E/3W	4-PR/CA4	568A	101D-150D	301-399E/W	1	100

Table 13.8: Identified horizontal wiring for CMS planning.

provide an estimate of the number of trouble tickets which are handled on a daily or weekly basis. If no current method of tracking trouble calls exists, determine the estimated number based on information from the technicians who handle the calls involving problems.

At this point, you should have a reasonable estimate of the physical scope of the project, information which will be used for both project planning and system design. Information about control and desktop equipment will be discussed in the next chapter. Particular sites, buildings, floors, or cables which will not be part of the project should be noted on the appropriate table. You will continue your planning in the next chapter by looking at specific types of cable and equipment that you might track and the specific information required for each item.

The Project Scope Document

Once you have collected the appropriate information for each site and discussed the project scope, it needs to be put into a brief summary which describes the overall scope of facilities and activities that will become a part of your CMS record keeping. It is not critical that all of the information mentioned on the pervious pages be available immediately, but each additional bit of information you develop now will make project planning and execution easier. Remember, decisions on what information to track and how to manage it in the CMS database will continue through all project phases, even during final implementation. It is unlikely that you will anticipate everything in advance, and you should not feel limited to the items originally discussed and included at the beginning of the project.

The information in a project scope document will be important when developing estimates on timing and costs for gathering and inputting the data into the CMS database. The Project Scope could be included in the Request For Proposal to assist the Vendors in understanding the background and overall goals for your project. A sample Project Scope is provided in Figure 13.1 on the following page.

Connectivity Management Implementation Project Scope:

The Cable and Connectivity Management System (CMS) for ABC Company will be implemented for all ABC Company sites in the State of Maine. The sites include both industrial and commercial facilities which support approximately 2,000 employees. It is estimated that there are approximately 2,500 voice lines (including service from two PBX's as well as leased lines from the local phone company), and an estimated 2,500 data connections involving both an IBM Token Ring LAN and 3270 Mainframe connections. There are also an estimated 100 special type circuits utilized for alarm and security services.

All copper and fiber outside plant cable and information pertaining to utilization of these cables shall be tracked. All inside plant cable, including copper and fiber distribution systems within buildings shall be tracked. The scope of the project consists of 3 individual sites (manufacturing facilities, headquarters building, and research park) located in the State of Maine. There are approximately 20 buildings, served by an estimated 36,000 copper cable pairs associated with the outside plant and local inside cabling. Fiber cabling consists of 36-strand fiber between buildings, and 12-strand fiber cabling distributed within the buildings to each individual wiring closet. There are an estimated 30 wiring closets, and approximately 2,500 workstation drops/jacks consisting of 2-4 outlets each, served by these wiring closets.

Specific equipment to be tracked in the Cable Management System (CMS) shall include, at a minimum, phones, fax machines, answering machines, printers, 3170 controllers, modems, terminal servers, Token Ring hubs, bridges and routers.

The Information Technology group currently handles approximately 175 service requests and an average of 125 trouble calls per month in supporting the equipment and connections for all the Maine locations.

Figure 13.1: Sample Project Scope document.

DEVELOPING A RECORDS STRUCTURE

14

Having captured a general description of the facilities and components, there are a number of planning issues which must be considered when developing the ultimate "structure" for this information and choosing a CMS database to contain it. These issues include determining what information to track and how to organize this information into tables and relationships, which will be covered in this chapter. How to define the configuration of connection points for each component is addressed in Chapter 21. Preparing an RFP and comparing this structure against what is available in a particular CMS package is covered in Chapters 22 and 23. Finally, you will need to adapt this structure to fit within the selected CMS, as discussed in Chapters 21 and 28. The following discussion is written for the most general case, implementation of an electronic CMS, but applies to paper recordkeeping as well.

The first need is to determine exactly what information you wish to track about the individual components. This can be considered as the "scope of information" in the CMS database, and must consider the logical relationships between things as well as their characteristics and attributes. We will look first at cable itself and then the information necessary to describe circuits. By looking at your desired circuit descriptions, you can get a sense of how you will characterize and track active components such as controllers, hubs, and terminals.

The second consideration is how to form this information into a specific structure or format, based on how certain information is related and ultimately used. As discussed in Chapters 3 and 6, any records structure (paper or database) is good for some things and not as good for other things. It should fit your overall requirements and be easy to set-up and maintain.

Your proposed records structure then becomes the model or template against which you can evaluate different CMS software. This is your ultimate goal. As a planning issue, understanding records structure will help you see how each CMS is different, where each shines and where each falters. Further, what you get will be the CMS, not your proposed records structure. Not only will you have to select the "best" system, you will have to address any final adaptation of your plans that the "best" CMS might require. For all of these reasons, a good understanding of records structure is essential. You may wish to review Chapter 3 on Basic Recordkeeping and Chapter 6 on CMS Functionality before proceeding.

Creating this records structure involves identifying the specific tables required, the information or fields that will be part of each table record, and the relationships and desired functionality between items (fields) in these records. For example, the table of cables includes individual cable records, which then include certain fields for specific information such as cable type and end locations. A table of cable pairs includes a record of each cable pair and its type, and each record must be able to be related or associated to a device or termination record in a different table. The TIA/EIA 606 Standard is a good example of a specific table and record structure, but gives only the most basic requirements for defining relationships between records. In addition to tables and records, you should identify any master lists that are helpful,

S T	DATA ELEMENT	BACKBONE UTP	HORIZONTAL UTP	FIBER	COAX	OTHER
C	NAME/ID	1214	REN1E101V	F/ADM	FAC101D	
C	TYPE	1-900PR	4PR/CA3	F1-36	RG-62	
C	FROM BLDG	ADM	REN	ADM	FAC	
C	FROM FLR	B	1E	B	1	
C	FROM ROOM	B01	105E	B01	G102	
V	FROM TERM.	CB01-CB18	101V	FC-ADMB	NONE	
C	TO BLDG	REN	REN	REN	FAC	
C	TO FLR	B	1E	B	1	
C	TO ROOM	B22E	1E ER	B22E	FAC ER	
V	TO TERM.	CB01-CB18	SB01	FC-RENB	BNC	
O	OUTLET	N/A	101V	-----	NONE	
O	CONFIG.	N/A	568A	N/A	BNC	
V	SIZE	900 PAIR	4 PAIR	36 STR	N/A	
O	MANUF.	GEN-CAB	BELDEN	BELDEN	BELDEN	
O	PART NO.	GF-5900	1229A	T255336	E7103	
O	INSTALL DATE	2/5/87	4/16/87	2/15/93	5/12/87	
V	LENGTH	300 LF	150 LF	300 LF	150 LF	
C	STATUS	USED	USED	USED	SPARE	
O	PATHWAY	CWADM-1	CWREN1E-1	CWREN-3A	-----	
O	FIRE RATING	-----	-----	-----	-----	
O	LAST TESTED	2/5/87	4/16/87		5/12/87	
O	OTHER	-----	-----	-----	-----	

Table 14.1: Proposed list of data elements that describe cables.

such as Buildings, Floors, Item Types, Department Codes, etc.

Identifying Required Information

We suggest this process begin with two lists, one for cables and one for circuit information. From these two lists, you can get a good idea of what the overall records structure should be. To begin with cables, make a new table listing each of the major types of cables you identified in Tables 13.6 and 13.8 of the previous chapter, and then list the information which needs to be tracked about each type. One approach, shown in Table 14.1, is to list across the top of the page the various types of cable on site, such as horizontal UTP, backbone UTP (Riser), fiber optic cables, coaxial cables, and any other special cables such as EIA-232C interface cables. Down the left side of the table, list all the information and attributes that you want to track about each type of cable. You may refer to Chapter 4 for the specific requirements of the TIA/EIA 606 Standard if that method of recordkeeping is desired.

In each column (for each cable type), enter an example of the information for each field or element on the list. There may be fields required for one cable type and not others. Using the examples, you can determine the minimum field length for each element. Your list may be longer or shorter than the sample in Table 14.1. Start by listing every item suggested by team members. After the list and the examples are completed, discuss the importance of each item and rank them in order of importance, or add a code designation to the left of the field name. A simple set of criteria should be applied:

- o Do you have control over the information (the item and its connections)?

- o How much effort is required to identify and record the information?
- o What steps are necessary to maintain the information? Will it be hard to maintain?
- o How (during what tasks) will the information be used? What is the value?
- o Does the use and value of the information justify the effort?

Obviously, items and information over which you have little or no control should not be required, but they can be optional. The effort to gather and maintain the information must be weighed against the benefit provided. If a particular item of information offers little benefit and is difficult to acquire or maintain, then it should not be required. However, this must be considered from all points of view represented by the team, and a consensus reached. Once you have evaluated each of the items and information on the list, you can rank them in order of priority. As an alternative, you can simply identify the status of which elements are Critical (C), Valuable (V) and Optional (O) along the left side of the field, as shown in Table 14.1. This list now provides the essential recordkeeping elements for the cables and terminations in your facilities. We will return to it in a moment.

So far, your focus has been on the cable plant and cable records. However, it is likely that you will want to track some information about control equipment, and perhaps even expand your inventory to desktop equipment. Even more significantly, you will likely need to track something about these devices such as the physical port ID in order to have a reasonable representation of a circuit. As you did for cables, make a chart listing each of the different types of circuits/services which will be tracked in the database, such as Voice, Token Ring, Ethernet, etc. An example is given in Table 14.2. Down the left column, list the proposed data elements required to represent the different circuit types.

In each column, list an example for each field of information to be tracked for each type of circuit/service. For example, a Voice circuit may require telephone number or extension (111-2222 or 707-111-2222), the equipment class (Phone), the equipment model (D-7407), etc. There may be fields which are required for one system but not another, so you may want to make a note that these fields should be optional. This will provide

S T	CIRCUIT ELEMENT	VOICE	TOKEN RING	ETHERNET	SNA/3270
C	CIRCUIT NAME	111-1222	TH-REN1E01	TS1234	LG1D11
C	EQUIP. TYPE	PHONE	HUB	DECSERVER	CONTROLLER
V	MODEL	D-7407	SYN-3030	DEC-500	3174-11R
O	SERIAL NO.	00031426	723514	113245	263214
C	NET. ADDRESS	B20103	01102123	000021022321	TG1D1100
V	NO. OF PORTS	1	36	64	32
V	CARD TYPE	N/A	3502/3514	CXY08	TMA
O	USER	ADAMS, S.C.	N/A	N/A	N/A
O	DEPARTMENT	ENG	ITG	SLS	ENG
C	BUILDING	REN	ADM	ADM	REN
C	FLOOR	2E	1	2	3W
C	ROOM	214E	104	242	3W ER
C	STATUS	USED	USED	USED	USED
O	COMMENTS		SEGMENT 1		UPGRADE

Table 14.2: Proposed list of data elements that describe circuits.

you with an idea of what fields you will need in your database, and what the minimum field lengths need to be for this information.

Your list may be longer or different than the example given. As noted before, list every item suggested by team members. After the list is complete, discuss the importance of each item and rank them in order of importance, using the same criteria given for the cable information. Once you have evaluated each item, put the elements (fields) in priority order or add an indicator next to each field name (as shown to the left of the field names in the sample chart), indicating the priority such as Critical (C), Valuable (V), and Optional (O). Between the two basic lists of cable elements and circuit elements, you have defined your basic recordkeeping requirements. If you wish, you can stop right there and use these simple tables to consider various CMS products. For a more rigorous evaluation, and a better understanding of CMS software, you should now organize this information into a more specific format.

Considering the Records Structure

Once you have prioritized each list of information, you will need to go down each list to see whether a single table will be sufficient to describe all cables, and whether all circuit information can use the same basic format. Starting with cables in Table 14.1, if there are significant differences in information between different cable types, you might want to specify that they be tracked in different tables. If so, make a separate list for each proposed type. A specific record for horizontal cable, for example, may contain information about the outlet number, outlet type, and wiring configuration at the outlet. Otherwise, all cable records must have a field to indicate an outlet, whether one exists or not. The horizontal cable can usually be described either as a cable or as an outlet, with similar results.

At the time of CMS evaluation, be sure you understand the functionality that applies to each type of cable record, and that you are not somehow limited by this approach. In your example, if certain devices in the CMS database can only be associated to a horizontal cable record, you would want to know this in advance to determine if it will work properly. Examples of separate backbone and horizontal cable records are shown in Table 14.3 and 14.4.

If you are tracking information for usage,

DATA ELEMENT	SAMPLE RECORD
CABLE NAME	ADM01101C
CABLE TYPE	4PR/CA4
CABLE CONFIG.	PR1-4
OUTLET NO.	101C
OUTLET TYPE	RJ45
OUTLET CONFIG.	568A
FROM BUILDING	ADM
FROM FLOOR	01
FROM ROOM	113
TO BUILDING	ADM
TO FLOOR	01
TO ROOM	104
INSTALLED DATE	5/14/89
TEST RESULT	2.5 DB
STATUS	

Table 14.3: Example of a separate horizontal wiring record.

DATA ELEMENT	SAMPLE RECORD
CABLE NAME	RC-REN1E
CABLE TYPE	R-200PR
PAIR COUNT	1-200
LENGTH	60 LF
SHEATH	S1
TERM. HDWE.	66 BLOCKS
FROM BUILDING	REN
FROM FLOOR	B
FROM ROOM	B22E
TO BUILDING	REN
TO FLOOR	1E
TO ROOM	1E ER
INSTALLED DATE	7/13/89
PATHWAYS	CWREN-R1

Table 14.4: Example of a separate backbone cable record.

STATUS	DATA ELEMENT	SAMPLE RECORD
C	CABLEWAY NAME	CWREN-1
C	CABLEWAY TYPE	UNDERGROUND
C	SIZE	4"
O	MATERIAL	PVC
C	CURRENT FILL	75%
O	PERCENT LOADING	N/A
V	CABLES CONTAINED	1214
O	CABLE USE	VOICE & SPECIAL CIRCUITS
O	COMMENTS	BETWEEN MH-1 & REN-B22E

Table 14.5: Proposed list of data elements that describe pathways.

availability, or connections, then each cable must be broken down into individual pair or strand records, as discussed in Chapter 3. User device, control device port, or connected cable pair can be either indicated on or associated to this record, as discussed in Chapter 6. If you desire complete circuit information, then it is likely that user devices must be defined to include certain connectors or connection points. Similarly, control equipment must be defined so there is a record for each port or connection point. Circuit connections are now represented in a meaningful fashion, and usage and availability information can be maintained for each component. Determining this structure and suitable naming conventions are covered in detail in Chapter 21.

Next, you can review the cable record(s) to determine if additional elements listed in the record really need to be in their own separate table. For cables, these would likely be terminations and/or pathways. Simply listing the pathway for each cable, as a part of the cable record shown in Table 14.1, tracks the basic relationships between cables and pathways. You must then determine whether you need to know (and will use) anything additional about each pathway. If so, you can create the record of pathway information for the list of pathway types given in your Table 13.7. List potential data elements for each type of pathway and then determine which items, if any, are Critical (C), Valuable (V), or Optional (O) as shown in Table 14.5. Determine whether pathways must be associated to all different cable types, if there is more than one type of cable record desired.

Based on the specific use for this information, you may wish to consider whether it should be tracked on drawings, in the database, or both. You can track cable, pathway and outlet locations on CAD drawings alone, since they don't change as often as individual connections. This leaves the desktop devices, jacks/outlets, control equipment, terminations and connections to track in the database in a simpler format.

As discussed in Chapter 4, the TIA/EIA 606 Standard requires you to track termination hardware and termination position for each cable pair, fiber strand, or coaxial cable. This approach should be evaluated carefully since it adds complexity to the records structure and database. We recommend a full discussion of the routine benefits of tracking terminations in this manner, whereby these benefits can be clearly demonstrated. If there is minimal value identified, then it is suggested that you track cable on a pair or strand basis only, a method which is completely adequate for describing individual circuits. Be aware that some CMS software may require the inclusion of termination devices, such as blocks, patch panels and outlets in order for automatic pair assignment or circuit creation to work properly as well as for handling connections of cable pair to cable pair. In such cases, other benefits of the particular CMS may be sufficient to justify this approach.

If desired, information about termination hardware must be evaluated in the same way. You

STATUS	DATA ELEMENT	SAMPLE RECORD
C	HARDWARE NAME	RBRENB01
C	HARDWARE TYPE	66B
C	NO. OF CONNECTORS/POSITIONS	100
V	CONNECTOR POSITION NUMBERS	1-100
V	CONNECTOR/POSITION TYPE	PINS
C	MFGR. PART NUMBER	66M50
C	BUILDING	REN
C	FLOOR	BE
C	ROOM	B22E
O	COMMENTS	RISER CABLE TO REN-1E

Table 14.6: Proposed list of data elements that describe termination hardware and position.

can simply indicate a termination identifier as part of the cable record or as part of the individual cable pair record. If more information about the termination device is required, then make a list similar to Table 14.6 of potential data elements for termination hardware such as type, number of positions, type of positions, etc. and rank them as critical, valuable, or optional just as you did with cables, circuits, and pathways. As with the relationship of cable to pair, each termination position must become its own record, perhaps in its own table. It is unlikely that any additional information (fields) about each termination position will be required.

You may also want to consider whether information outlets or jacks require their own table record, or whether they can be tracked as a component of the horizontal cable. If the information to be tracked for outlets or jacks is similar to the elements listed for termination hardware, such as type, and number and type of positions, you may want to consider a tracking outlet records in the termination hardware table. If the outlet number, type, and configuration are tracked as fields in the horizontal cable record, a separate outlet table would not be required since this would be a duplication of information. If more information about the outlet is required, then make a list similar to Table 14.7 of potential data elements for outlets, then rank them as Critical (C), Valuable (V), or Optional (O) just as you did with the other tables.

STATUS	DATA ELEMENT	SAMPLE RECORD
C	OUTLET NAME	REN10101V
C	OUTLET TYPE	568A
C	NO. OF CONNECTORS/POSITIONS	4
V	CONNECTOR/POSITION TYPE	1-PAIR
O	MFGR. PART NUMBER	WJ106
C	BUILDING	REN
C	FLOOR	1E
C	ROOM	105E
O	COMMENTS	

Table 14.7: Proposed list of data elements that describe information outlets or jacks.

STATUS	DATA ELEMENT	SAMPLE RECORD
C	EQUIPMENT NAME	TH-REN1E01
C	EQUIPMENT TYPE	HUB
C	MODEL	BAY-3030
O	SERIAL NUMBER	723514
C	NO. OF PORTS/CONNECTORS:	
V	FRONT SIDE	4
V	BACK SIDE	NONE
V	CONNECTOR/PORT TYPES	SLOT
C	NETWORK ADDRESS	01102123
O	DEPARTMENT	ENG
O	INSTALLED DATE	4/5/92
O	DRAWING REFERENCE	REN-03
C	BUILDING	REN
C	FLOOR	3W
C	ROOM	3W ER
O	COMMENTS	ALL PORTS ON CARDS IN USE

Table 14.8: Proposed list of data elements that describe controlling equipment.

When you evaluate a CMS, be sure to understand what rules (if any) will be applied for associating a cable, cable pair, or fiber strand record to an outlet or termination position record. Ideally, a CMS will prohibit you from terminating a fiber strand on a 66 block, but such rules may require comprehensive tracking at this level of detail.

Information about controlling equipment must now be evaluated in similar fashion. You can simply indicate a connected Port ID as part of the cable record or as part of the individual cable pair record. If more information about the controlling equipment is required, then make a list similar to Table 14.9 of potential data elements for control devices such as type, number of ports, type of ports, etc. and rank them as Critical (C), Valuable (V), or Optional (O). As with the relationship of cable to pair, each port or connection point must become its own record, perhaps in its own table. The various ways that device connection points can be described, referred to as the configuration, are discussed in Chapter 6 and again in Chapter 21. Device configurations are usually more complex than cable configurations, and are more likely to change over time.

The final area which must be evaluated is whether a separate record is required for tracking of desktop equipment. If there is already an asset tracking system or some other method for tracking user devices, it may not be required in the CMS records. However, you may determine it is useful to track the desktop device as a beginning or ending point of a circuit. If all the information required about the desktop device is contained in another table such as a Circuit ID or User table, a separate desktop equipment record would not be required. Some CMS products may require a piece of equipment be located at each end of the circuit, while other CMS products use a Circuit ID or User ID as the beginning point of a circuit in lieu of tracking a specific device. Identifying your preference with regards to tracking of desktop equipment now will help you in evaluating the different CMS products later. If you desire to track more information about the desktop equipment, then make a list similar to Table 14.9 of potential data elements and then rank them as Critical (C), Valuable (V), or Optional (O). You may determine that it is possible for both desktop equipment and controlling equipment to be combined and tracked in a single table. Each

STATUS	DATA ELEMENT	SAMPLE RECORD
C	DEVICE NAME	PC1000
C	DEVICE TYPE	PC
V	MODEL	PS2/90
V	SERIAL NUMBER	123541
V	DEPARTMENT	ENG
C	EXPENSE CODE	163-051
V	SOFTWARE COMPONENTS	WINDOWS, EXCEL, WORD
V	ADJUNCT EQUIPMENT	SVGA MONITOR
		V.36 EXTERNAL MODEM
V	INSTALLED DATE	7/12/93
V	PURCHASE COST	$4,295.00
V	MAINTENANCE CONTRACT	MC-P1349
V	USER	ADAMS, J. L.
C	BUILDING	REN
C	FLOOR	2E
C	FLOOR	214E
C	STATUS	USED

Table 14.9: Proposed list of data elements that describe desktop equipment.

device could still be identified by type, and would be tracked by individual location.

Some information on both the cable and circuit descriptions will be common to all elements, such as building and floor designations, department codes, and expense codes. In many CMS products, such lists are maintained in separate "master" tables. An entry in any element field will be validated or checked against this list to maintain consistency for these fields. For convenience, you can usually call up these lists with a hot key or function key and select an entry from the list. Other validation lists will include cable and device types. You will want to have lists of all these designations as developed in Chapter 21.

Data Structure and CMS

As you develop these lists, you must make judgments about the merit of tracking each piece of information shown. In addition, you should begin to think about how each piece of information might be tracked in a CMS, since you will need to consider this when evaluating CMS candidates. If tracking a particular piece of information is appropriate only in certain cases, you will have to decide whether this must be an optional feature, whether you can go along with tracking this piece of information for all cases, whether the particular case will require or be forced into a unique record, or whether there is an available field which can be adapted for this purpose.

Yet another possibility is encoding the information into the "name" or "type" designation for the particular item. This becomes part of the process of developing a catalog. The major drawback to this approach is the difficulty in remembering all the built-in designations. Ultimately, developing a catalog is very dependent on the specific CMS, and cannot be addressed in a general fashion here.

For example, if you wish to track the type of metal sheath or shield on outside plant cable, you can require an optional field in the cable record for this purpose. A candidate CMS could provide this optional field for all cable types (built-in or user-defined), provide a separate record format for outside plant cable which includes this field, or simply provide a comments field where this information could be included when appropriate. Otherwise, an indication of sheath type could be included as part of the cable type designation. This is a simple example, and much more complex situations may arise.

In a similar fashion, certain items of information may require their own records rather than be a single piece of information in another record. If this is the case, you will want to understand how the CMS relates these two records - specifically, what is the functionality that allows for associating, querying, and reporting on these two records simultaneously. A simple example is whether a pathway can be tracked as a field in the cable record, or whether the pathway becomes its own record. A separate pathway record allows for additional information about the pathway to be included, and this is the general benefit of the separate record. In this case, you would want to understand how the CMS ties these two records together and how updates to one record will affect the other.

DATASET TABLES	COMMENTS
CABLES - HORIZONTAL CABLES - BACKBONE	Two separate tables are preferred to allow for tracking of the outlet as part of the horizontal cable. However, one table is acceptable.
CIRCUITS	A table is required to track Individual Circuit IDs by Service type as a means for identifying circuits without specific equipment.
PATHWAYS	A table is required to allow for tracking of additional elements regarding size and contents of the pathway, in addition to physical routing to be shown on drawings.
TERMINATION HARDWARE	A single termination hardware table is required for tracking of blocks, patch panels, and outlets (if outlets are not tracked in the horizontal cable record).
CONTROL EQUIPMENT	A separate table for control equipment is required to allow for tracking of individual port configurations and port types.
DESKTOP DEVICES	A separate table is preferred to allow for tracking of software, adjunct equipment, and maintenance contracts. However, a single table for all equipment could be utilized if all desired fields for both types of equipment were contained in the single table.

Table 14.10: Sample list of final datasets (tables) for proposed CMS recordkeeping.

As you can see, the more information that you wish to track, the more variation and complexity in the data structures that will arise. As a general rule, it is best to try for the simplest and most straight-forward data structure which meets your basic requirements. In other words, if you can live with one record type for all cable, rather than two or three or four different types, the better off you will be.

As a final step, and to begin preparation of the RFP and a proposed catalog of cable and equipment configurations, make a final table listing the major datasets you have described in Tables 14.1 through 14.9, similar to the types of datasets shown in Table 14.10. These datasets, covering circuit descriptions, cables, terminations, pathways, desktop devices and controlling equipment, provide a basis for your recordkeeping requirements and become part of your Project Scope. They can be formalized when preparing a Request For Proposal, and be included as individual tables which the Vendor can respond to on a element by element basis for each table. If performing a less formal evaluation of CMS software without the use of an RFP, the tables can be used as simple checklists where each system is compared against the list of fields to see which best matches your proposed list.. You will need to begin defining individual cable and device types within this framework as discussed in Chapter 21, and then ultimately create individual cable and equipment items with unique names or identifiers. The example shown in Table 14.10 covers essentially all elements, and you may not require all of them.

On each list, you should have indicated which information is Critical, Valuable, and Optional. As each of the CMS products is compared against your checklist of items, the prior identification as to the importance of each field will make it easier to identify which of the systems meet your most critical needs. Information that will be tracked exclusively on drawings (such as pathways, manholes, etc.) should also be indicated. The final lists should represent a consensus of the Project Team and cover all functional areas necessary for successful use and maintenance. Management and staff must make the commitment to use and support this information on a routine basis.

CMS ACCESS AND PLATFORM

15

This chapter will address the issues to be considered when looking at where the CMS records will be stored and how they will be accessed and utilized. Various CMS products are designed to function on many different platforms, with choices of operating system, database software, and communications protocols. Your goal should be to determine the most desirable combination of access and platform, while understanding how you might deal with some of the alternatives. This may be the case if you select the CMS which best meets your functional requirements but which cannot be supported in the desired fashion.

It will be critical to determine at an early stage if you will need a Single-User System or a Multi-User System. Some CMS products are specifically designed for a single-user environment, while other vendors have different product versions (and prices) for different numbers of users. Some CMS products have different levels of pricing for a specific number of users, such as a 4-User or 8-User License, or unlimited Multi-User LAN version. The total number of users should be determined and indicated in the Request For Proposal (RFP) to the vendors. This assures the price quoted is correct for supporting the expected number of users.

Single-user systems tend to be designed for a personal computer, either DOS-based, Windows-based, or for the Apple Macintosh. The DOS and Windows systems typically specify a 386 processor and four to eight megabytes of RAM as a minimum, and 10 to 100 megabytes of hard disk storage space available. A better processor and more RAM will always improve the response time of the system, especially when querying a large table of records, preparing reports, or displaying CAD drawings. There are several single-user systems available which can operate on a UNIX-based workstation, but these tend to be CAD-oriented systems.

Multi-user systems also take different forms, including minicomputer and mainframe-based software which might be typical of call accounting systems with CMS functionality. But the biggest distinction is between what we will call "traditional" LAN-based software and the client-server systems. With traditional LAN software, the CMS application can be accessed from anywhere on the network. Individual access may be limited by password, and the maximum number of simultaneous users may be fixed. On the other hand, client-server systems require that the "client" software be loaded on specific machines, and only those machines can access the CMS. Therefore access is now limited to specific users and locations, even though this is still referred to as a multi-user system. More recent developments of client-server applications are making the client software available on the network and blurring the distinction between these two approaches. How and where you plan to use the CMS information will guide your choice between these alternatives.

Determining Access Requirements

It should be recognized up front that at least one individual must assume the role of system administrator. In an enterprise-based project, it must be determined which group or organization will provide this Administrator, and the nature of their role must be defined. If there will only be one key person who will need to access the database and who will be responsible for all updates/changes to database records, a single-user system may be sufficient. To be sure of selecting the right system, you must evaluate the need for other individuals in the organization to access the CMS records. This, of course, will depend on the procedures and workflows that you develop for cable and connectivity administration.

As you review and develop CMS procedures, make a list of the individuals which will be affected by the information to be maintained in the CMS database. If they will be making changes which affect the database, it is important that they be included in the CMS project, and be given the opportunity to have access to look at records in the database for informational purposes and to determine if someone should be notified of the changes they are making. These other individuals or groups might include the Network Support, Employee Directory, Facilities Management, Help Desk, Employee Relations, etc. You may want to add managers or supervisors who may need to access the CMS records for reports or information on equipment or connections.

Based on the extent of the records, the rate of trouble calls, and the level of Move, Add and Change (MAC) activity, list the individuals who will be required to enter Service Orders and Trouble Tickets. This may be the same number of individuals currently assigned to the Help Desk or Customer Service Support group. Include those technicians who may ultimately update and monitor the CMS records. This will provide the number of individuals who may be considered for direct access to the CMS records. Some users will be able to access the CMS at the same location, and so do not require separate access. From this list and your anticipated workflow, determine the minimum number of simultaneous users and the minimum number of access locations. For the traditional LAN system, the required number of users for licensing will be the minimum number of simultaneous users, while for client-server systems it will be the minimum number of access locations.

Besides knowing who needs to access the system, you may wish to determine the different levels of access required for each user or user group. These levels may include Read Only, Add/Change, Delete, User Profile Access, and Reports, depending upon the particular CMS. Some CMS software offers these levels on a module by module basis; in other words, you can control access to the Service Order and/or Trouble Ticket functions. Access by level and module is usually controlled by the log-on ID or user password, and is managed by the system administrator. By controlling access level, you can create some security and protection against unexpected modifications or deletions, whether these actions are intentional or unintentional.

Divide your list of users into their respective groups such as administrator, technician, Help Desk, Customer Service, and supervision. List these across the top of a table as shown in Table 15.1. For your list, use common sense and your proposed workflows to determine

LEVEL	ADMIN	TECH	CUSTOMER SERVICE	HELP DESK	SUPER-VISION
ADD	X	X	X		
CHANGE	X	X	X	X	
DELETE	X	X	X		
USER PROFILE	X				
SERVICE ORDER	X		X		X
TROUBLE TICKET	X			X	X
REPORTS	X	X	X	X	X

Table 15.1: Typical CMS access levels for various functions.

what groups are appropriate. Down the left side, list the access levels shown. Based on the tasks of each group, indicate which access levels each group will require. Table 15.1 is filled in with typical requirements, but they will depend on your specific procedures and environment.

You should consider whether individual users will require their own User IDs and passwords in order to control access to the database. In some organizations, this can be controlled through LAN administration, where a LAN ID will restrict the user to certain applications on the network. If this is not controlled by LAN administration, the system you select must have the capability to assign individual User IDs and passwords to insure the integrity of the CMS records. A CMS that allows each user to change their own password is preferred.

Platform Options

There are many options regarding the possible platform for your CMS. By platform, we mean the particular computer or computers required, the operating system or systems required, the database software required, and the particular network or communications protocols required or supported. Some general aspects of the chosen platform as it relates to access were discussed in the previous section. Overall, your platform decisions will depend on what is available, what provides proper access, what works together efficiently, and what might be guidelined or mandated in your organization. Platform may also affect security, availability, backup procedures, maintenance costs, and even printing capabilities. Consider these issues and determine what the combined platform must provide.

Since the combined platform can have a significant impact, and because CMS software is available on such a broad variety of platforms, it pays to consider many different approaches rather than a single preference. By considering the variations in platform components you will be able to identify the preferred methods, but will also see where alternatives are acceptable. That way, you will be able to select from a larger pool of candidate CMS software and are likely get a better fit for your desired functionality. It would be unfortunate to think that the "perfect" CMS is not supported by your network services, so some tradeoffs will likely be required. As you consider less desirable alternatives, it becomes more likely that costs will increase for additional hardware, software, and maintenance.

While it is good to keep your options open, existing internal support can be a key factor in the viability of a CMS. If there is internal support for a particular database, then a CMS using this database is preferred. A CMS that uses the Open Database Connectivity (ODBC) standard may be able to operate with your existing database software, rather than requiring a specific or proprietary database.

To help your evaluation of platform alternatives, make a chart similar to the one shown in Table 15.2 showing the advantages and disadvantages for each of the different platform elements. For each element, list as many options as you can, and then consider their suitability according to the points just raised.

- Computer - Include all the current computer systems available or obtainable, including mainframe, VAX, minicomputer, server, workstation, or PC. List the advantages and disadvantages of each, as well as noting which ones are guidelined, preferred, or not allowed.

- Operating System - Include all the current operating systems available or obtainable by your organization, such as VMS, TSO, UNIX, DOS, OS/2, Macintosh, or Windows. List specific versions, such as Ultrix or DOS 5.0. List the advantages and disadvantages of each operating system, as well as noting which ones are guidelined, preferred, or not allowed. Consider the impact and frequency of the inevitable upgrades to new versions.

- Relational Database (DBMS) - Include all the current database software which is available or obtainable such as Oracle, Sybase, Informix, etc. List the advantages and disadvantages of each database, as well as noting which ones are guidelined, preferred, or not allowed. Here, too, consider the question of upgrades, and whether you must stay with current versions to receive vendor support. You may wish to

ELEMENT	CHOICE	ADVANTAGES	DISADVANTAGES	GUIDE LINE	PREFER	NO
CPU	VAX		SLOW RESPONSE	YES	YES	
	MAINFRAME		PHASING OUT	YES	NO	
	SERVER	BEST RESPONSE		NO	YES	
	SINGLE PC		SINGLE POINT	NO	NO	X
OPER'G SYSTEM	VMS		LIMITING	YES	NO	
	TSO		LIMITING	YES	NO	
	UNIX		NO CURRENT EXPER.	NO	YES	
	DOS	MOST APPLIC'S		YES	YES	
	OS/2		NOT APPLIED YET	NO	NO	X
DBMS	ORACLE		LACK OF SUPPORT	YES	NO	
	SYBASE	NEW DIRECTION		YES	YES	
	INFORMIX		NO EXISTING APPL'S AND NO EXPERIENCE	NO	NO	X
	INGRES		NO EXISTING APPL'S AND NO EXPERIENCE	NO	NO	X
NOS	NOVELL	CURRENT LAN		YES	YES	
	BANYAN		NO CURRENT USE	NO	NO	X
	LOCALTALK		NO CURRENT USE	NO	NO	X
COMM PROT'L	TCP/IP	USED TODAY		YES	YES	
	IPX/SPX		NOT TESTED ON LAN	NO	NO	
	PATHWORKS		PROBLEMS ON LAN	NO	NO	
	LAT		NOT TESTED ON LAN	NO	NO	

Table 15.2: Sample evaluation of platform elements for CMS planning.

consider the availability of SQL tools to do customized queries and reports. Some databases will require licensing (and fees) separately from the CMS.

Some database software such as Oracle and SyBase resides on a database server. Other systems such as DBase and FoxPro reside on the client PC, and the server is used as a file server. A database server will usually be quicker and more efficient than a file server.

o Network Operating System (NOS) and Communications Protocol - If you are considering a LAN-based system, then include all the current LAN operating systems currently in use, such as a Novell Netware, Banyan Vines, LocalTalk, etc. Add to this list the available or obtainable communications protocols such as IP, IPX, TCP/IP, Pathworks, LAT, etc. List the advantages and disadvantages of each operating system and protocol, as well as noting which ones are guidelined, preferred or not allowed by your company or organization. Consider any special requirements such as driver files, network modules, conversion software, or any restrictions between operating system and communications protocol.

Be sure to consult with the proper support groups within your organization for guidance and agreement in each of these areas.

USING EXISTING RECORDS

16

Using existing records of the cable plant, equipment, and their connections can help you with project planning, defining new recordkeeping requirements, and selecting the CMS itself. In some cases, you can reduce the time and expense of conducting a full inventory and entering the data into the new CMS tool. This chapter takes a look at how to evaluate existing records and get maximum benefit from their use.

Acquiring a CMS is a good opportunity to look at consolidation of current recordkeeping systems. Some CMS software has the ability to manage the Employee Directory or to function as an Asset Management system. Consolidation can eliminate duplicate recordkeeping between different groups and maintain this information in a more timely manner, especially if the information in question is tightly coupled to the Service Order and Trouble Ticket functions. For example, it might be easier to update the Employee Directory immediately upon closing a Service Order, rather than passing a paper document to a different group for updating at a later time.

If you desire to consolidate recordkeeping, then the current methods become a template for defining requirements in the CMS, perhaps with minor modifications. Obviously the group in charge of the Directory, in this example, must participate in the CMS planning and be assured that their needs are being met. It is recommended that you get the new CMS records up and running in parallel with the older system before you cut over, in order to assure a smooth transition. In other words, don't scrap an existing system until the new system has been tested and debugged.

Of course, existing records can be an excellent resource when you are defining the scope of a CMS project, as discussed in Chapter 13. Existing notes, spreadsheets, and drawings can give you a general picture of your facilities and cable plant without having to walk around and perform a preliminary survey. They also can provide clues as to where to look for certain things, or can indirectly show you where some inventory or survey work is required.

Limitations

In the ideal case, current electronic records can be input directly into a CMS, reducing the time and expense for inventory and data entry. We will address this point more in the next section. Unfortunately, it must be pointed out that this would be the unusual case - more often than not, current electronic records are not adequate for directly populating the CMS, and for several reasons.

First, current records (especially cable records) are typically not as accurate and/or comprehensive as you need for the new CMS. Just looking at the records, you may have little sense as to how well they have been maintained, which parts are correct or incorrect, and where significant omissions may exist. For example, cable records from the telephone company may not have been updated for several years. Other groups may have installed cable or equipment which you know nothing about. Paper records are usually complete and accurate at the time they were created, but are not necessarily updated after that point. Even connection records in the wiring closets, the proverbial "cable books", will usually have errors and omissions, particularly on disconnections.

Unless your organization and its recordkeeping have been extremely disciplined, the records are suspect. And if you choose to use this information anyway, you may quickly get frustrated with the new CMS when it spits out incorrect data. Making corrections as you go will be like chasing your tail. If you don't know the accuracy of particular records, then you should field check a portion of the information. If a sample of ten to twenty percent are found to be fairly accurate, then you can assume this accuracy for the remaining records. If a small sample is found to be inaccurate, then you should audit a larger sample. If the large percentage of errors or omissions continues, then it is likely you will have to abandon the data.

The second potential problem has to do with suitability - are the field formats, field lengths, and nomenclature exactly and consistently what you need for the new CMS? Duplicating existing information in the new CMS may be your goal, in which case there may be no problem. This is reasonable where records are up-to-date and consistent from building to building and site to site. But if you are looking for new capabilities, more consistent naming conventions, or a different way of describing components, then you may have conflicts between old and new data formats. Defining CMS requirements with a clean slate will likely lead to a more capable, more functional system.

Finally, the structure of your current records may be significantly different than the records structure in the CMS, even when they are tracking similar fields of data. For example, two pieces of data on your table can be kept in two different tables in the CMS. Or ten pieces of data on your table can be kept in four different tables in the CMS. Even worse, an entry on your table can be a relationship between two other things in the CMS, and these two things may or may not exist as such on your table. These examples may sound extreme, but they do happen. It is best to take nothing for granted about importing existing records into a new CMS.

Take the case of termination information. Your existing electronic table shows a termination position for each cable pair, with the termination

	EMPLOYEE DIRECTORY	ASSET MGMT	CABLE
ELEMENTS:	USER NAME PHONE NUMBER LOCATION DEPARTMENT	TAG NUMBER SERIAL NUMBER P.O. NUMBER LOCATION COST	PAIR COUNTS CIRCUIT NO. X-CONNECT PAIR STATUS
RESPONSIBILITY:	HUMAN RES.	PURCHASING	INFO. TECH.
FORMAT:	EXCEL	LOTUS	PAPER
ACCURACY:	95%	60%	70%
COMPLETENESS:	95%	95%	75%
EASILY AUDITED:	YES	NO	NO
CAN IMPORT TO CMS:	YES	EXCEPT LOC'N	NO

Table 16.1: Summary and evaluation of existing recordkeeping systems.

device identifier encoded into the termination name. A particular CMS, on the other hand, tracks the cable pair in one table, the termination device in another table, and must have a third table defined in order to list the termination positions for that device. The cable pair and termination position relationship is then tracked in a fourth table. Other required fields in these tables must be populated before accepting the cable pair-to-termination position entry. Simply trying to read the termination position field from one set of records to another won't work. It will require a conversion program which must be written, tested, and debugged before use. In some cases it can be done, for a price, but it is not a straightforward task. Only when records are otherwise complete, accurate, and suitably formatted should this option be considered.

One of your evaluation criteria for a CMS could be how will it is suited for accepting some of your existing records. Note that it is only after a CMS has been selected that a final determination on importing can be made.

Evaluating and Using Existing Records

At the beginning of a CMS project, you will need to identify what records currently exist and whether they have been maintained and are up-to-date. Even if the records are not being maintained, the format and types of information may be helpful in structuring what needs to be tracked in the new CMS. You will need to make a judgment on how far you can go with each set of records, based on the limitations discussed previously.

Make a table listing all of the current records available today, with as much information about their accuracy and format as possible. Current records might include cable records, desktop/asset inventory, employee directory, help desk profiles, network management system, etc. Use the sample in Table 16.1 as a starting point.

For each recordkeeping system, determine if it should be included in the CMS database, and if so, would the CMS database then replace the existing record keeping system. If not, remember to address how these other records will be maintained and updated in concurrence with the CMS. This should be clearly indicated in the new workflows and procedures which are developed for maintaining and supporting the CMS, and be done while considering the total requirements of the proposed CMS.

If the CMS will be replacing a current recordkeeping system, a plan should be developed for phasing out the old database and transitioning

DESKTOP/ASSET INVENTORY RECORDS - SPREADSHEET

DATA ELEMENT	EXAMPLE	TRACKED IN CMS?	FIELD LENGTH
TYPE	PC	YES	8
SERIAL NUMBER	113715	YES	12
PURCHASE ORDER	B4-1675	NO	12
MAINT. CONTRACT	MW-4096	NO	12
SITE	LRP	YES	8
BUILDING	REN	YES	8
FLOOR	1E	YES	4
ROOM	138E	YES	8
COST	$2,499.00	NO	10
DATE PURCHASED	4/13/94	YES	8
DEPARTMENT	ENG	YES	8
CHARGE CODE	163-041	NO	10

Table 16.2: Example of table information for existing asset management records.

to the new CMS. You should allow for a brief period of testing and fine-tuning of the CMS after it has been fully loaded and operational, just to be sure that all required information is available and properly presented. Set a target date for when the original system will be discontinued and removed from service. Access to the old system must be eliminated so that individuals will begin using the new CMS.

Obtain copies of sample pages or screens from each of the existing recordkeeping systems. You should have a complete list of all the data elements included, and should have some sense of how each system is structured. This provides you with all the current information tracked, as well as the current names or IDs being used. Go through each page to determine what information needs to be tracked in the CMS, the minimum number of characters for each field, and any additional information which you want to include in the CMS database which is not currently being tracked.

For each set of records, you may want to prepare a chart similar to the example in Table 16.2 from a hypothetical inventory spreadsheet. If a recordkeeping system has a number of individual tables, then make a chart for each table. You can add notes about field formats, required fields, case sensitivity, etc. as you see fit. If you are choosing to duplicate this table in the CMS, then be as complete as you can. Again, your evaluation must consider the workflows and procedures which are developed for maintaining and supporting the CMS, and should be done while considering the total recordkeeping requirements.

If you have determined that certain records are suitable for importing into the CMS, you may want to consider taking the existing records and conducting an audit to verify and update the information. Once the information has been verified and is accurate, it can be used to populate the new CMS database. This can be done while the CMS is being purchased and installed, so your records are ready for entry into the CMS as soon as your new system is operational.

Be sure to discuss the format of your electronic records with the selected CMS vendor to identify the best way of utilizing those records to populate the new CMS. As discussed, this may require creating special batch files to download the information from the existing database or require using certain file commands. The field titles and lengths in the existing system may need to be revised to match the appropriate field names and lengths in the CMS database. Since the vendor has the most experience with their own product, it is usually better to let the vendor do this rather than attempt it in-house.

For those situations where information must be manually transferred from system to system, it is helpful to have the intermediate paper copy in as useful and friendly a format as possible. Use whatever report-generating capabilities are available to create specific report formats which closely match the order in which the records are entered into the CMS database. For example, if the CMS lists fields A, B, and C in that order of entry on the screen, then create a report from your original records which also list the fields A, B, and C in that order. Whatever method is utilized, the most important thing is to make sure the data entry is done as accurately as possible.

As a final use of your existing records, review each set of records again for completeness and accuracy. On a separate table, list the physical areas which are missing, incomplete, or out-of-date for each recordkeeping system. For example, you may have no asset information on equipment in two particular buildings at a certain site. This information will need to be obtained through some sort of inventory process, no matter how you transfer the available records to the database. These inventory steps may be in addition to any general inventory work that you do for the CMS. Use the existing records tables as models to develop inventory forms to gather the specific information required. If the missing information is of no consequence, then simply ignore it.

CAD/GRAPHICS

17

Certain benefits can be realized through the use of a CMS that includes graphic capabilities, but there will also be additional effort and requirements to implement and maintain the graphics portion of cable and connectivity recordkeeping. Some infrastructure elements are naturally better represented with a "picture" rather than a simple text description. For example, the routing of a particular backbone cable or pathway is best represented on a drawing. The physical location of each outlet in a work area is another example.

CAD systems can also merge the details of cable layout with information about space planning and facilities management. Location and routing of other building services can be included, such as plumbing, HVAC, and electrical. Even information about who resides in which office, and how much floor area belongs to a particular department, can be shared in such a system. Of course, users who currently support CAD-based plan drawings will have the easier time implementing a CAD-based CMS than users who don't.

Special Benefits

A CMS with CAD functionality offers the ability to track the physical arrangements of campus, riser and horizontal/station cables, as well as the physical location of wiring closets, information outlets, controlling equipment, and desktop equipment. Examples are given in Figures 17.1, 17.2, and 17.3. The visual layout maximizes the amount of information on the screen at any one time and minimizes the amount of nomenclature that must be learned to navigate around the database. Graphics systems also provide a means for locating and tracking elements where there are no specific room numbers, since the item can be displayed on the drawing without requiring additional textual information to describe the physical location.

CAD drawings provide an excellent tool for instructing cable installers and equipment vendors as to where new cables or equipment should be installed, including the routing path and termination points. This helps to reduce the time required to locate specific rooms, junction boxes, and specific points within a room. Elevation drawings of equipment rooms, frames, racks, and the equipment itself can also be beneficial, but will take a substantial amount of upkeep and should therefore be considered carefully. Communications with or between third-party contractors can be made clearer and more concise, requiring less supervision of outside vendors and better coordination among groups during a major project.

CMS-based CAD drawings can also be used to show the location of network equipment. When tied in to a network management system, as can be done with some CMS software, the CMS quickly provides the location of a troublesome or broken device. Some systems are applied for this purpose more on a logical basis, with a minimum of cable and connection information included. While this approach has benefit, it is perhaps a limited form of cable management. If the same CMS has been fully implemented with cable and connection information, then the circuits into and out of the bad equipment can also be identified and managed. Applying the CMS for network and

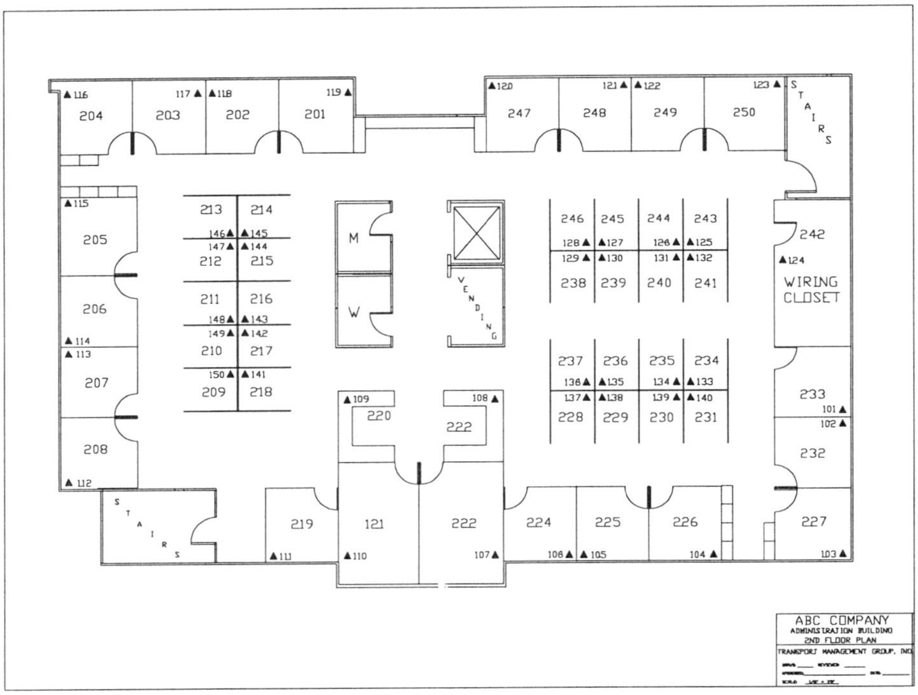

Figure 17.1: CAD-based floor plan drawing.

facility management is discussed further in Chapter 8.

Floor plan drawings alone, without a database, can be used to track your cable records. Textual information on the drawing can indicate attributes of pathways, cables, and terminations, as well as the locations of jacks and user devices. However, connection information on cable-to-cable and cable-to-port connections is difficult to track in this fashion. The usefulness of this method would depend on the availability of existing floor plan drawings, and the size and complexity of the cable and connectivity records in question.

Determining the Need for CAD/Graphics

There are several items to consider when making the decision to use a CMS system which has CAD/Graphics, including certain requirements that must be addressed in order to support CAD drawings. These considerations are listed in Table 17.1. Assuming you will have established the need for CAD drawings, does your organization or company have an existing CAD system which is used to maintain floor plans? If there is an existing system in place, this provides a source for the drawings as well as establishing criteria for the CAD/Graphics portion of the CMS. It would be beneficial if the CMS selected utilizes the same CAD system which is currently in use by your organization or company. If you do not currently

- ESTABLISHED NEED FOR CAD ?
- HAVE EXISTING CAD SYSTEM ?
- HAVE PLAN DRAWINGS AVAILABLE ?
- CABLE AND PATHWAY INFORMATION AVAILABLE ?
- WILLING TO INVENTORY ?

Table 17.1: Considerations for using CAD records with a CMS.

have a CAD system, then you must acquire one in addition to the CMS or else get a CMS with native CAD functionality. If you simply use the floor plan drawings without a CMS to document your cable and connectivity records, existing CAD drawings will provide an excellent starting base.

Assuming you have addressed the need for basic CAD capabilities, the second concern is whether floor plan drawings exist, are up-to-date, and are available. If they exist on paper, you will have to digitize or scan them into the CAD system. If they don't exist at all, you will have to start from scratch. If they currently exist in a CAD system, are they in the proper format for use in the CMS? Most CMS products with CAD/Graphics capability can utilize existing floor plan drawings if they can be put into DXF format. As you evaluate potential CMS software, be sure to check the allowed formats. The cost to generate floor plan drawings can be substantial if they do not already exist, a potential penalty for the CAD-based CMS products that require drawings before any elements can be created in the database.

The final requirement is that your inventory process must gather the appropriate information to accurately locate spaces, pathways, cables, and jacks. Your project plan must allow for the time and budget to acquire this data. This is easiest to capture during the design phase of a cabling project, rather than as an afterthought. In some cases, approximations of physical location are adequate provided this information is not used architecturally - that is, you are not relying on this information to know where it is safe to drill a hole in the wall, or lay a new conduit.

It should be noted that the TIA/EIA 606 Standard requires certain information to be tracked in a drawing format (paper or CAD). This information includes the identifier, location, and size of all pathways and spaces, the identifier and location for all backbone cables and terminations, the identifier and location of all telecommunications outlets, the identifiers and location of the grounding electrode and all

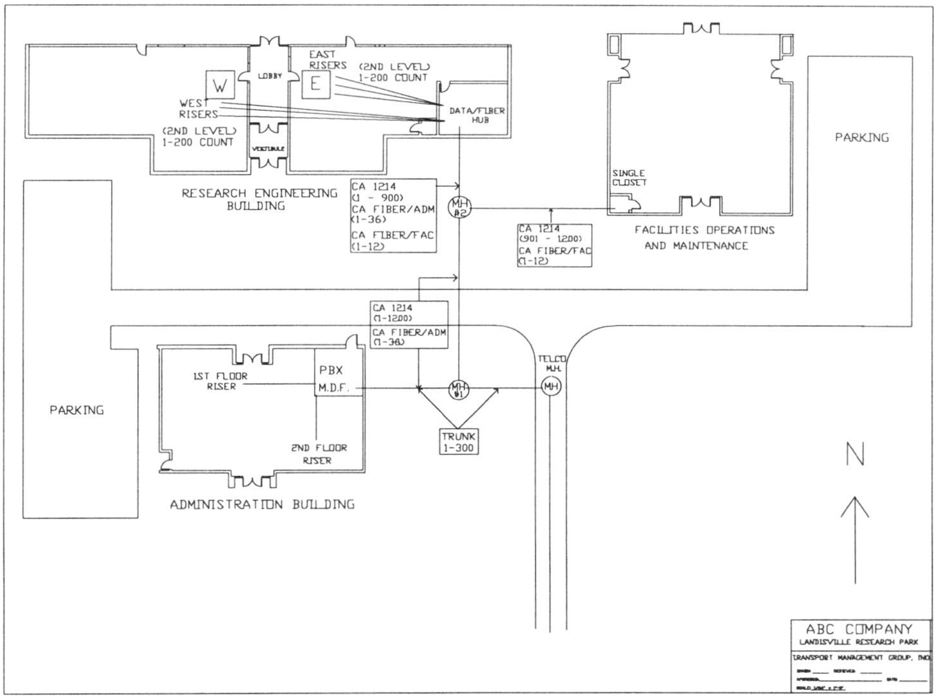

Figure 17.2: CAD-based site plan drawing.

grounding busbars, and the routing of the grounding electrode conductor from the grounding electrode to the Telecommunications Main Grounding Busbar (TMGB). See Chapter 4 for more discussion of the TIA/EIA 606 Standard.

As has been stated, the benefits of CAD-based information are obviously related to physical arrangement - location and routing. To select this recordkeeping method means to have an individual or group that is actually and routinely going to take advantage of this information and keep it current. Industrial environments, shared tenant facilities, and extensive campus arrangements are likely candidates for benefit. Other locations that place a premium on pathway capacity and location will also benefit, as will those facilities where third-party contractors are used to pull additional cable.

Facilities with cable raceway systems and underfloor cellular systems are also good candidates for CAD records, since the tracking of these systems would be very complex and difficult to follow on a purely textual basis. Knowing the routing to a particular outlet location and the intermediate access points and junction boxes can save appreciable time and disruption when installing or troubleshooting cable. Areas with moveable partition furniture often have outlet boxes located on a grid pattern throughout the floor, so that outlet access can be rearranged as the partition furniture is rearranged. Here, outlets are tracked versus a floor location rather than an actual office number, and a graphic record offers the best representation of their arrangement.

On the other hand, certain facilities will not necessarily benefit from CAD-based recordkeeping. Smaller buildings with limited backbone cable facilities are an example. Here, the closest thing to a physical pathway may be riser sleeves, which could be reasonably tracked in textual format. Other locations where there is a consistency of arrangement from floor to floor are also less likely candidates. For example, if a basic floor plan is identical throughout all office areas,

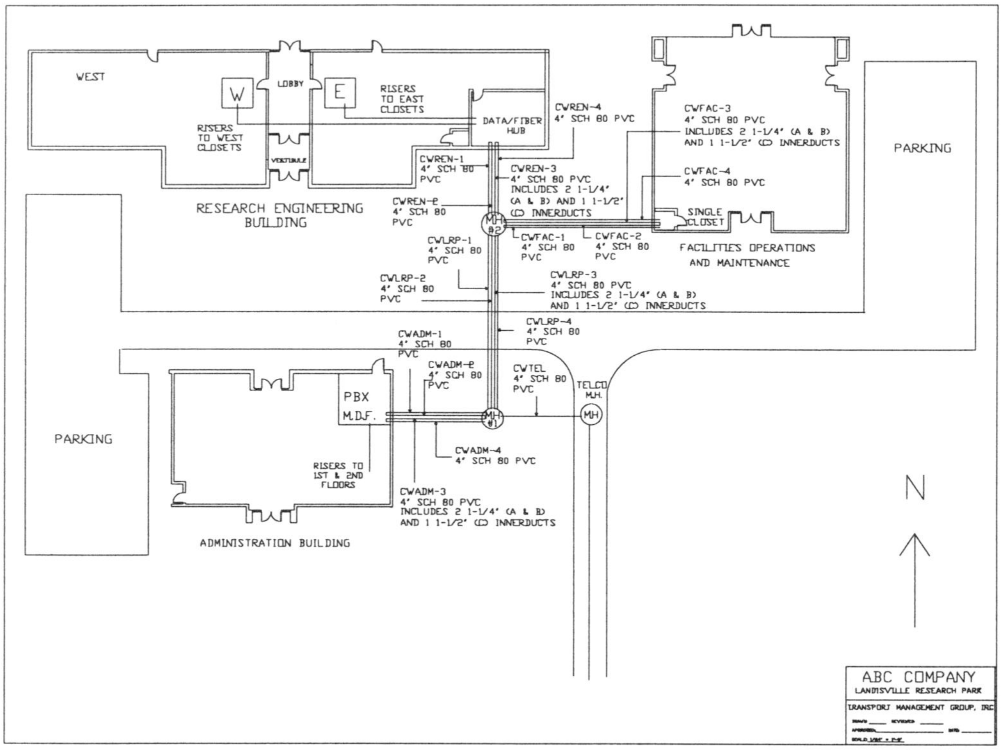

Figure 17.3: CAD-based site plan drawing showing pathway information.

then wiring closets, cable routings, information outlets, etc. are located in approximately the same location on each floor. The more consistent the location and naming of items, the easier it is to navigate without drawings since locating specific cables, equipment rooms, or office outlets is not difficult. And as with all recordkeeping, short-term leased space may not be around long enough to get benefits from extensive CAD records.

For managing connections, the CAD-based records work best when simply adding, moving, or removing an icon. This applies primarily to desktop equipment which moves from office to office and jack to jack, and to adding new horizontal cable. Connection changes in the wiring closet are still generally handled through the database or textual portion of the records. This suggests that CAD systems work best where the majority of changes are to horizontal wiring and office connections, with a minimum of changes to backbone connections. Buildings with a very consistent, permanent wiring arrangement, locations where dedicated backbone wiring is used, and certain LAN environments where only the horizontal run (port-outlet-device) is significant are good candidates for CAD-based records. Dedicated backbone wiring means jumpers or patch cords between horizontal and backbone cable is fixed, and backbone cable pairs are "dedicated" to each outlet.

In an enterprise CMS system, you will likely have a mixture of facilities where CAD will be more or less valuable. If you choose a CAD-based CMS, you will probably have to support CAD drawings for areas where they will have marginal benefit. On the other hand, if you select a database CMS with CAD interface, you will have the flexibility to utilize the CAD only in those areas where it has value.

Maintaining CAD Records

If the answers to the above questions indicate a CMS with CAD/Graphics is desired, the next step is to determine if the value received from a CAD/Graphics system is worth the added requirements to implement and maintain the graphics portion on an ongoing basis. Adding graphics to your CMS records involves some additional requirements in order to take advantage of graphical information, with a partial list given in Table 17.2.

First a plan for easy and consistent naming of drawings needs to be established. Such a plan should consider systematically naming the drawing files by building and floor, for easier recall. Then, database elements need to be located on the floor plan drawings in their appropriate location. For some systems, this requires a separate step in addition to creating the item record in the database.

During routine operation, additional time is required to load each drawing in the system. Loading drawings over a network will also take additional time. Displaying and moving around a drawing to view specific areas can require additional time for drawing regeneration each time it is recreated on the screen. A 17" monitor is preferred over the standard 14" or 15" models. CAD/Graphics systems often utilize more hard disk drive space and Random-Access Memory (RAM), a Super-VGA display, and may require a faster and more capable processor.

Using a CAD/Graphics system requires floor plan drawings to be updated to reflect actual physical changes. Each time a floor plan is revised, the drawing file must be revised in the

- PLAN FOR NAMING DRAWINGS
- ADDITIONAL PREPARATION, LOADING, AND TRAINING
- POSSIBLE UPGRADES FOR PC OR WORKSTATION
- PRINTING AND PLOTTING CAPABILITY

Table 17.2: Some additional requirements for CAD maintenance.

CMS database. Information outlets and other items which have been relocated during renovations need to be relocated on the updated drawing. If architectural revisions on the drawings are done by a different group, arrangements must be made to get updated drawings in a timely fashion.

For organizations which track a large amount of information such as voice, data, video, security, alarms, etc., the drawings can become very crowded and difficult to work with when making changes and performing queries. Consideration needs to be given to how much information should be stored on the drawings, and in which layers. This also requires separate icons for each type of equipment which is represented on the drawings, and the user may be responsible for creating and maintaining these icons. Some systems allow selection of the icon for a particular item, while others are fixed and only include a minimum number based on the structure of database tables. For example, if the database has a single table for termination hardware records, then there may only by one icon provided to indicate all types of termination hardware. Representing many items in a small area can be troublesome, such as when using icons for each termination device in the Telecommunications Closet or Equipment Room.

CAD/Graphics systems often require additional training, since users must learn some of the CAD commands in order to navigate around the floor plan drawings, zoom in and pan around the drawing, and move items from one location to another, especially if moving between drawings. Mastering additional system commands will be necessary for printing and plotting functions.

If the primary drawing files are maintained in the CMS system, the CMS must be capable of handling modifications to the base drawing, such as moving or adding walls and partitions, changing room numbers, etc. This we have referred to as "native" CAD functionality. Such a system avoids the need to export the drawing to a CAD system where drawings are revised, and then re-importing the drawing back into the CMS. If all drawing maintenance is handled by the facilities or building management department, you may not need the capability to maintain the base drawings in your CMS, since that will be handled for you.

To truly benefit from CAD drawings, they must be available as paper copies. Be sure to determine what printing and plotting requirements are needed. A CMS database should have the capability to print copies of floor plan drawings or circuit paths on a drawing, and should be able to print these at a local printer, a network printer, or a full-size plotter. Drawings are most readable when printed in color, with different colors for different drawing elements.

Some laser jet printers can be configured to emulate a plotter using an add-on cartridge or a special printer driver file, but this will still be limited to black-and-white printing on letter size paper. A full E-size drawing at this print size can be difficult to read depending on its level of detail. Using a network printer in this fashion is usually not practical. The capability to perform screen prints also varies from system to system, particularly with Windows-based software, and may require the use of the DOS Graphics command prior to loading the CMS.

Be sure to have the CMS vendor identify any specific printing equipment or software required in order to print floor plan drawings or cable routing information from the CAD/Graphics module. Determine the flexibility to direct printing to different devices on the network, and whether this can be done from within the CMS.

REPORTS

18

Each organization's process for connectivity management will require that different types of reports be developed as one way of accessing the CMS records. Reporting requirements for your CMS are significant, since the records are only valuable if they can be retrieved in a manner that is useful and can provide all the information required. Documenting specific reporting requirements will assist your planning and ultimately the evaluation of CMS software.

A textual report can be defined as a formatted set of fields of information presented on the screen or in a paper document. The fields of information could be from a particular table of records, or be fields combined from records from different tables. The report could include all records from a particular table, or those selected by a specific query. The report should have column headings and be easy to read. If the report is prepared for printing, then the CMS must include appropriate printer setup strings so the report is properly formatted for the specific printer. Graphical reports may also be available, as we'll discuss in a moment.

As described in Chapter 6, CMS software usually comes with a set of canned reports with fixed formats and based on specific queries. Some

- SUPPORT TABLES
  - BUILDINGS
  - FLOORS
  - USERS
  - DEPARTMENTS
  - CONNECTORS OR CONFIGURATIONS
  - DRAWINGS
- CABLE TYPES BY NAME
- EQUIPMENT TYPES BY NAME
- PATHWAY CAPACITY AND UTILIZATION
- CABLES BY ID
- EQUIPMENT ITEMS BY ID
- CABLE-TO-CABLE CONNECTIONS
- CABLE-TO-EQUIPMENT CONNECTIONS
- CIRCUIT COMPONENTS BY CIRCUIT

Table 18.1: List of minimum recommended report formats for a CMS.

systems come with a report generator capable of preparing ad hoc or custom reports and formats. Other systems will require purchase of the Structured Query Language (SQL) software for the specific database. The SQL tools are based on specific commands in batch files which access the database tables directly to create custom reports. You should evaluate CMS candidates to determine the built-in reporting capabilities and whether additional tools or software is needed. You will have to develop some additional expertise to use the report generator or SQL software for the CMS. In some cases, the CMS Vendor can provide customization of report formats or screens.

We suggest looking for a standard, minimum set of report formats like those listed in Table 18.1. Every significant table such as cable and equipment types, cable and equipment items, and all support tables should be available as a report. Further, you should be able to query and select certain records for the report from each of these tables based on matching field entries. If different types of cables are maintained in different tables, then there should be a report available for each table. There should be capacity reports available whereby you can list the usage status of every pair in a cable for example, or every port/connection on a device. This information may also be presented on a higher level, so to speak, with the percentage used, bad, and available given for each cable.

There must also be some way to report on circuit connections. This is most easily done by listing the connections in a table, such as cable pair to cable pair or cable pair to equipment port. From these reports, circuit information can be derived. Again at a higher level, actual circuits can be displayed listing all components in circuit sequence. This may be available one circuit at a time, or less often as a list of a number of circuits. The latter approach, as a printed report, is a very involved query for the CMS and is difficult to format, and hence is not commonly found.

Your report requirements should include formats that are similar to those currently available, as well as new reports which will save time and effort in the future. Make a list of the reports currently used today. Obtain a copy of at least one page of each report currently being used in managing MACs, or otherwise used in managing connections on cables and controlling equipment. Be sure to specify each of these reports in an RFP or CMS evaluation form so that each of the requirements can be met. It may even be helpful to provide a sample copy of each report with the RFP to clearly indicate the specific reporting requirements.

Make a list of the new reports which will be required from the new CMS database. Be sure to include all the fields which are to be included in the report, noting the desired format or order of information. Consider reports which will be useful in auditing the records in the database, not only to verify the information entered through the data entry process but also for use in auditing the actual connections in the wiring closets against the records contained in the database. This provides for an easy method to measure the accuracy of the CMS records and procedures to be sure all changes are being captured. A number of generic, sample report formats are included at the end of this chapter based on the ABC Company's LRP site (Tables 18.3 through 18.8). The actual CMS reports will depend on the table and record structure of the specific database.

If your interest is in recordkeeping according to the TIA/EIA 606 Standard for administration, the standard offers suggestions for reports on an optional basis. The standard recommends a circuit report showing end-to-end connectivity including user code, cables, and termination positions. Remember that desktop and control equipment is not considered. The standard also recommends summary reports for pathways, spaces, cables, grounding and bonding, and cross-connections as shown in Table 18.2.

If the CMS database is to include a Service Order or Trouble Ticket section, be sure to specify your printing requirements for Service Orders and Trouble Tickets and what information must be contained in the printed ticket. Some systems even provide the option for the user to specify whether or not the circuit information is to be included with the printed ticket. It is also valuable to consider other types of reports which may be required, such as all Open Orders or Tickets, Orders or Tickets by assigned technician, Trouble Tickets of a particular type, or by location, etc. There is a wide variety of ways to look at this information. Be sure to specify your requirements for obtaining and printing these records.

- PATHWAY SUMMARY REPORT - FOR EACH PATHWAY:
  IDENTIFIER, TYPE, PERCENT FILL, AND PERCENT LOADING
- SPACE SUMMARY REPORT - FOR EACH SPACE:
  IDENTIFIER, TYPE AND LOCATION
- CABLE REPORT - FOR EACH CABLE:
  IDENTIFIER, TYPE, TERM. HDWE. IDENTIFIER FOR EACH END
- CROSS-CONNECT REPORT - FOR EACH SPACE:
  TERMINATION POSITION TO TERMINATION POSITION
- GROUNDING/BONDING SUMMARY REPORT - FOR EACH TGB:
  TMGB IDENTIFIER, BONDING COND. IDENTIFIER, TGB IDENTIFIER
-
- CIRCUIT REPORT - FOR EACH CIRCUIT:
  USER CODE, CABLES, AND TERMINATION POSITIONS

Table 18.2: Recommended reports for TIA/EIA 606 recordkeeping.

Check with managers and supervisors of the groups which will be affected by the CMS recordkeeping to determine the reports and information they want to be able to retrieve from the CMS records. This is very important in gaining their support and commitment and to make sure their goals for the project are met.

If you have selected software with CAD capabilities, you should consider the types of reports which might be displayed on a floorplan drawing. A typical example is to provide the location of each item of a certain type of equipment, such as showing all PCs of a certain model on a given floor. Another useful report is to show the locations of all LAN printers on a floor with their network addresses indicated. This could help determine where a particular print job should be sent, or help users identify different printers when requesting copies of reports or documents. A report showing the routing of a particular circuit (through specific cables and terminations) is very useful when troubleshooting or relocating that circuit. Examples of graphic reports are given in Figures 18.1 and 18.2 at the end of this chapter.

It is important to list all the printing options which will be required for the CMS reports. With some CMS products, the printing is controlled by the local LAN software when a user profile is set up, while other CMS products specify a designated printer or printer address when the system is initially set up, such as LPT1.

Consider all possible printing options which are available with the various software systems and that will be used on a routine basis - print screen capability, printing to local printer, and printing remotely across the LAN. Specify these as printing requirements which must be available with the CMS. Determine if reports need to be downloaded or created in ASCII format for inclusion in other files or reports, or for import to a spreadsheet. Another requirement might be the capability to link the report to an Electronic Mail System so that reports can be sent to multiple users.

CABLE UTILIZATION REPORT

Cable Name	Originating End Bldg.	Originating End Floor	Originating End Room	Terminating End Bldg.	Terminating End Floor	Terminating End Room	No. of Pairs	Pairs In Use	% Utilization
CC-FAC1	FAC	1	125	ADM	B	MER	400	89	22%
CC-RENB	REN	B	105E	ADM	B	MER	800	563	70%
RC-ADM1	ADM	1	1100	ADM	B	MER	200	77	39%
RC-ADM2	ADM	2	2100	ADM	B	MER	200	96	48%
RC-REN1E	REN	1E	130E	ADM	B	MER	200	73	37%
RC-REN1W	REN	1W	145W	REN	B	105E	200	67	34%
RC-REN2E	REN	2E	230E	REN	B	105E	200	104	52%
RC-REN2W	REN	2W	245W	REN	B	105E	200	108	54%
RC-REN3E	REN	3E	330E	REN	B	105E	200	95	48%
RC-REN3W	REN	3W	345W	REN	B	105E	200	116	58%

Table 18.3: Sample cable report, listing cables by name.

EQUIPMENT REPORT - BY CATEGORY

Equipment Name	Category Class	Type Model	Bldg.	Floor	Room	Dept.	Installed Date
PC0001	PC	PS2/70-386	REN	1E	122E	ENG	1/12/94
PC0002	PC	PS2/70-486	ADM	1	1104	ADM	5/4/94
PC0003	PC	Compaq/50M	REN	2E	2221E	ENG	9/15/94
PC0004	PC	Compaq/66M	FAC	1	116	OPR	6/8/94
PC0005	PC	DEL-33D	ADM	2	2110	SLS	3/14/94
PC0006	PC	PS2/90	REN	3W	231W	ENG	8/24/94
PC0007	PC	MAC SE	ADM	1	1142	MFG	4/20/94
PC0008	PC	GW/4DX2-66	FAC	1	108	OPR	12/6/94

Table 18.4: Sample equipment report, listing devices by category.

CABLE CONNECTIONS - BY CABLE NAME

Cable Name	Cable End	Pair/ Conn.	Equipment/ Cable Name	End or Side	Port or Pair
RC-REN1	1	1	REN1E101	2	1
RC-REN1	1	2	REN1E101	2	2
RC-REN1	1	3	REN1E102	2	1
RC-REN1	1	4	REN1E102	2	2
RC-REN1	1	7	REN1E103	2	1
RC-REN1	1	9	REN1E104	2	1
RC-REN1	1	10	REN1E117	2	1
RC-REN1	1	11	REN1E106	2	1
RC-REN1	1	12	REN1E106	2	2
RC-REN1	1	14	REN1E114	2	1
RC-REN1	1	16	REN1E116	2	1
RC-REN1	1	17	REN1E124	2	1
RC-REN1	1	18	REN1E124	2	2
RC-REN1	2	1	CC-RENB	1	101
RC-REN1	2	2	CC-RENB	1	102
RC-REN1	2	3	CC-RENB	1	103
RC-REN1	2	4	CC-RENB	1	104
RC-REN1	2	7	CC-RENB	1	112
RC-REN1	2	9	CC-RENB	1	123
RC-REN1	2	10	CC-RENB	1	115
RC-REN1	2	11	CC-RENB	1	126
RC-REN1	2	12	CC-RENB	1	127
RC-REN1	2	14	CC-RENB	1	114
RC-REN1	2	16	CC-RENB	1	133
RC-REN1	2	17	CC-RENB	1	141
RC-REN1	2	18	CC-RENB	1	142

Table 18.5: Sample (partial) cable connection report by cable pair number.

DRAWING REPORT - BY NAME

Drawing Name	Drawing Description	Bldg.	Floor	Last Rev. Date
ADM-01	Administration Building - 1st Floor	ADM	1	3/14/94
ADM-02	Administration Building - 2nd Floor	ADM	2	8/24/94
ADM-BA	Administration Building - Basement	ADM	B	4/20/94
FAC-01	Facilities Ops & Maintenance Building - 1st Floor	FAC	1	12/6/94
FAC-02	Facilities Ops & Maintenance Building - 2nd Floor	FAC	2	12/6/94
REN-01	Research Engineering Building - 1st Floor	REN	1	1/12/94
REN-02	Research Engineering Building - 2nd Floor	REN	2	5/4/94
REN-03	Research Engineering Building - 3rd Floor	REN	3	9/15/94
REN-BA	Research Engineering Building - Basement	REN	B	6/8/94

Table 18.6: Sample drawing report by drawing name.

OPEN SERVICE ORDER REPORT - BY S.O. NUMBER
1/24/94

S.O. Number	Requested By	Date Received	Due Date	Technician Assigned	Order Type/Description
94010301	Johnson, L .R.	1/3/94	1/24/94	Hutchins	Move Phone & PC to Rm 123E
94010302	Allison, B. R.	1/3/94	1/24/94	Sweeney	Add Modem Line for PC
94010402	Matthews, T. R.	1/4/94	1/26/94	Sweeney	Add LAN Connection for PC
94010407	Haley, M E.	1/4/94	1/26/94	Sweeney	Add User to Shared Printer
94010610	Roberts, A. F.	1/6/94	1/26/94	Hutchins	Move Phone & PC to Rm 247W
94010613	Alexander, J. R.	1/6/94	1/29/94	Sweeney	Needs Answering Machine
94010701	Williams, T. A.	1/7/94	1/30/94	Robinson	Add Bridged Extension in Shop
94010702	Craig, J. A.	1/7/94	1/30/94	Hutchins	Change Phone to Digital Set

Table 18.7: Sample Service Order report by open Service Order number.

OPEN TROUBLE TICKET REPORT - BY TICKET NUMBER
1/15/94

Ticket No.	Reported By	Date Rec'd	Equipment I.D.	Technician Assigned	Problem Description
94011101	Alexander, J. R.	1/3/94	PC1014	Hutchins	PC Doesn't Display Main Menu
94011102	Haley, M E.	1/3/94	123-1243	Sweeney	No Dial Tone
94011202	Roberts, A. F.	1/4/94	123-4325	Sweeney	Static On Line
94011207	Matthews, T. R.	1/4/94	PC0135	Sweeney	Cannot Connect to Network
94011210	Allison, B. R.	1/6/94	PR0045	Hutchins	Not Printing through Network
94011213	Johnson, L .R.	1/6/94	123-1021	Sweeney	Dead Phone

Table 18.8: Sample Trouble Ticket report by open Ticket number.

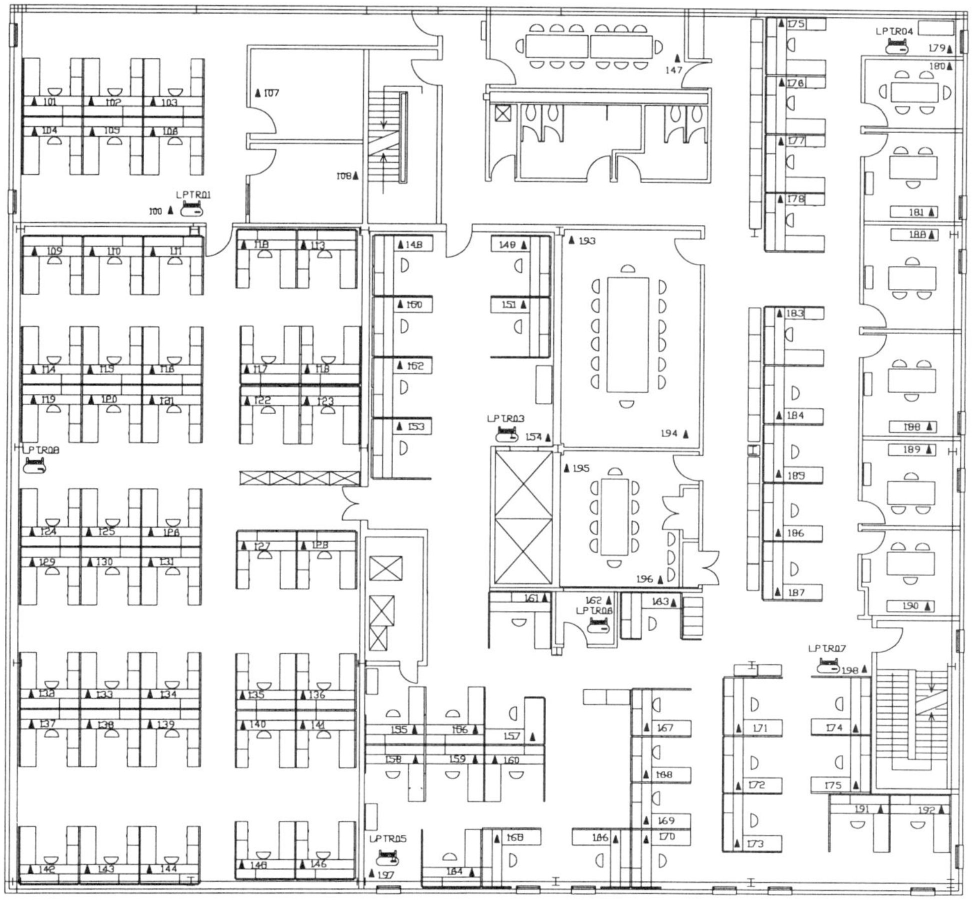

READ BUILDING - 2ND FLOOR

Figure 18.1: Graphic report showing all LAN printers on a specific floor.

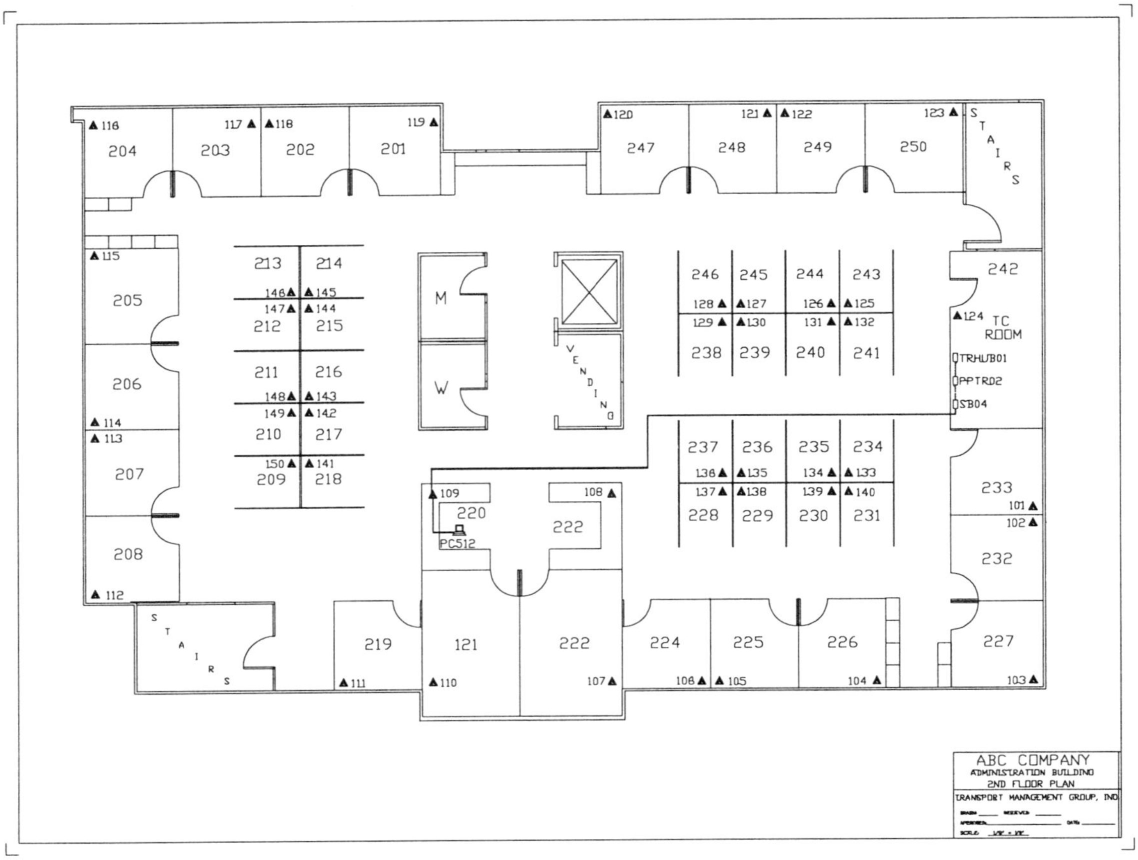

Figure 18.2: Graphic report showing the circuit path for troubleshooting a problem on PC512.

DEFINING NEW WORKFLOWS

19

It has been emphasized throughout the Handbook that cable and connectivity management is a process, and that records must be used and maintained in order to provide benefits. This chapter looks at the administrative procedures and workflows that define your interaction with the CMS and which are required to use and support the CMS records. These workflows must be integrated with other management functions such as LAN and PBX administration, and they must be mandatory and enforced. The integration must consider normal Moves, Adds, and Changes (MACs), network reconfigurations, and any changes made while resolving problems during troubleshooting.

The need for defined procedures has been clearly demonstrated early in the development of connectivity recordkeeping. Where it was assumed that records would naturally be used and maintained by individuals, without clear procedures or accountability, there was little demonstrated success with these systems. Group responsibility usually results in no one taking responsibility at all, and the recordkeeping becomes ineffective. A management process that uses Service Order and Trouble Ticket documents to convey information and assure timely updates will be more effective than one that does not. Some organizations use workflow software, such as Lotus Notes, to assign tasks and provide information about the changes requested. It is beneficial to have the workflow software interface directly with the CMS Service Order so that MAC information can be distributed to the proper individuals.

While your current procedures and workflows may be essentially adequate, it is likely that some degree of adaptation will be required. When there are several individuals involved, several organizations involved, or several steps in the process, it is reasonable (even required) that the workflows be documented and agreement reached on any required changes. Roles, responsibilities, and timing should be clear to each individual so that the work proceeds smoothly and records are properly updated. Team members and supervision must reach consensus on any changes. Additional considerations for CMS administration are given in Chapter 7.

Considering Current Workflows

A good starting point for developing new workflows is to take a look at the current processes for handling trouble calls and routine service requests. This may require that the current flows be documented in a block diagram, which helps to identify what information is provided with a request and how that information eventually gets to the technician. It is also useful to document where the current workflow ends, and whether the current process calls for the technician to report back that the request was completed. This is critical in determining where holes exist for information to get lost, so these areas can be addressed in the new workflows.

As you document the current process, be sure to address each of the different types of requests which might be handled including trouble calls, routine MAC requests, simple requests for name changes, or relocation of a phone or computer within an office. Depending on how much information is tracked in the CMS database, each of these changes may impact the accuracy of the CMS records.

The team should discuss other ways in which changes can be made, such as user groups or departments that have control over their data connections, or the times when a technician is grabbed by a user and asked to make a change right then without going through the normal channels. These are actions which will negatively impact your CMS records, and must be addressed in the new workflows. If it is anticipated that the CMS become a tool for Customer Service or Help Desk functions, their roles and interactions with the CMS must be considered as well.

Creating New Workflows

In order to determine what the new workflows will look like, you will need to consider all the steps in the process of working with your new records: From the initial user request, to how the information is recorded and transferred to the technician, to the completion of the circuit modification, to the updating of the CMS database, and to the confirmation with the user that the work has been completed. The final confirmation with the user on completion of the request is optional; however, it is a step which improves the quality of service as perceived by the user. This evaluation must consider *every* type and piece of information that you would consider for your recordkeeping and how it will be maintained.

As you first begin to look at answering the questions regarding the flow of information, consider several different approaches. It is best to document each of the ideas, then massage the information into a couple of different options. You should address different workflows, one for routine MACs, another for major relocations and network reconfiguration, and another for trouble calls and problem resolution. It is possible, depending on the organizational responsibilities, to have separate workflows for voice and data, and even individual workflows for specific data networks or activities where outside contractors are used. Use the following questions as a basis for determining workflow requirements for your new CMS records.

- What format will be used by individuals to submit trouble and change requests (paper forms, electronic mail, fax, customer service desk, etc.)
- What information will the user be required to provide?
- How will these requests be tracked and transmitted to the technicians?
- Will the CMS include a Service Order/Work Order module which will provide the technician with the instructions for making the circuit changes?
- Will the Service Order/Work Order be created in the CMS by a clerical administrator?
- Will the new circuit connections be pre-assigned by the clerical administrator, or created in the field by the technician?
- Will the Service Order/Work Order be closed out by the technician or a clerical administrator?
- If the new circuit connections are determined and created by the technician, who will be responsible for updating the CMS records with the new circuit information? How will information flow from the technician to the administrator?
- Will the CMS include a Trouble Ticket module which will provide the technician with the circuit information for problem calls?
- Will the Trouble Ticket be created in the CMS by a clerical administrator?
- How will changes resulting from Trouble Calls be updated in the CMS? Will the technician update the CMS records, or record the changes on the Trouble Ticket so updates can be done by a clerical administrator?

As you start to chart out the workflows, you must also address the roles and responsibilities of specific individuals and organizations during each step of the process. This includes the responsibilities of technicians when making circuit changes or finding a connection which was not previously in the database. It is their responsibility to report that connection back to the clerical administrator or update the database themselves by entering the connection in the CMS records themselves. These roles and responsibilities need to be clearly spelled out and agreed to by the appropriate organizations.

You also need to take into consideration whether the existing staff can handle all of these responsibilities, whether additional personnel will be required, or whether a new role must be filled by adjusting current assignments. This is a key to making sure there is sufficient staff with the proper training available to handle the tasks associated with the new CMS recordkeeping. If personnel cannot keep up with the changes, updates will fall behind and errors are created in the circuit records. Overall, efficiencies should be reached which will maintain or reduce current staffing levels.

Once you have discussed the flow of information and activities, it will be necessary to document the process in a flow chart. Some sample workflows are provided at the end of this chapter for your reference. You may even want to use these as a starting base, and then modify and expand them into greater detail to meet the requirements and organizational structure for your environment. We must emphasize that every organization's current structure and management objectives are different, and there is no one universal approach, at least not yet. Your evaluation must consider where you are today, where you want to be, and how you might get there. Therefore, the specific workflows will look somewhat different for each organization.

After complete workflows have been defined, it is a good idea to test them by walking through each step with someone familiar with each specific task. Try to look for weak areas and places where the procedures break down or don't cover all cases, especially with regard to the exchange of information between parties. This process will help to narrow down the best alternatives and provide some fine-tuning, and gives a means for making sure the process has as few areas as possible where changes could be made without being captured in the CMS database. Remember when trying to find possible holes and weak areas, remind team members to keep an open mind and that judgments are not being made against them. This process is necessary to make sure the workflows are as fool-proof as possible.

At this point, it is time to finalize the workflows in a formal way and communicate them to the individuals and organizations whose daily routines will be affected by the new process. It is important to review each of the steps to be sure individuals are clear as to their respective roles and responsibilities. It takes the commitment and cooperation of all individuals to make the CMS records provide the value and benefit desired. It only takes one individual making undocumented changes to make the information in the CMS database unreliable and inaccurate.

We recommend that you make final flowcharts of all procedures and workflows prior to your selection of a CMS. For each flowchart, write a brief description of the process and why it is justified or required to use and support the CMS as part of your connectivity management process. This set of documents, considered to be an Administration Plan, can be used as part of the request for final project authorization and for evaluation of candidate CMS systems, particularly when a Request For Proposal (RFP) is used.

Applying New Workflows

You should make it a point to implement the new processes as soon as possible for those areas where the CMS recordkeeping has been implemented. Start using the new process so that requests are funneled through the appropriate channels. If a new position of CMS Administrator is required, have that individual in place and trained as early as possible, before any inventory has begun. Do not wait until your records are complete and the database is fully loaded to begin administration, since inventory will take a fair period of time and the records will become quickly obsolete and inaccurate as changes are made which are not captured in the database.

Even if the CMS is not quite ready, it may still be valuable to handle the requests in the proposed manner. Circuit information may not be available, but it will at least serve as a test and practice of the process. It's easier to introduce new procedures and monitor the progress when a minimal amount of damage can be done, and only a few records or areas are affected. This allows for modifications to the procedures to be made early in the project. It also helps to identify problem areas quickly, so that changes can be made (such as in cataloging) which will prevent major problems or inaccuracies in the records later in the project. It also helps to get individuals familiar with the project in the initial stages, when ideas for changes and modifications can be accommodated with the least amount of effort.

Since the administration of the CMS is the real key to success or failure, new rules and procedures are best put in place quickly. Resources and commitment must be sufficient from the beginning. Unexpected problems or needs can be addressed as they arise.

It is important to periodically review the workflows to be sure they are still appropriate, useful, and provide a means for capturing all the changes in your cable and connectivity records. Team meetings with appropriate organizational representatives should continue on a periodic basis to make sure all needs are being addressed, and to surface any concerns or issues which need to be resolved. This can be done on a monthly or quarterly basis. In the early stages of the project, this should be done on a monthly basis to keep organizations informed of which buildings or office areas have been completed and which areas are next in the implementation plan.

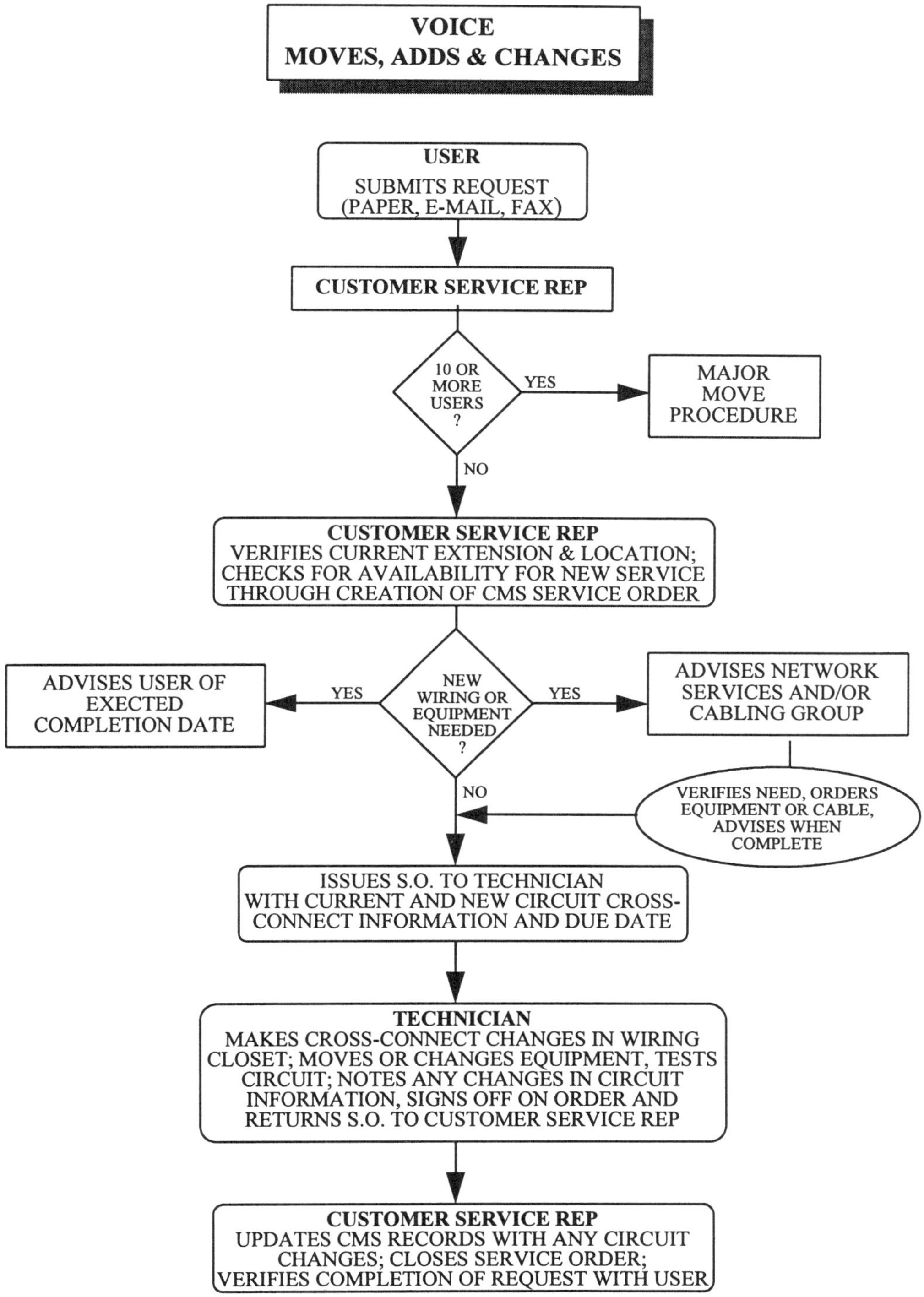

Figure 19.1: Sample workflow diagram for Voice Moves and Changes.

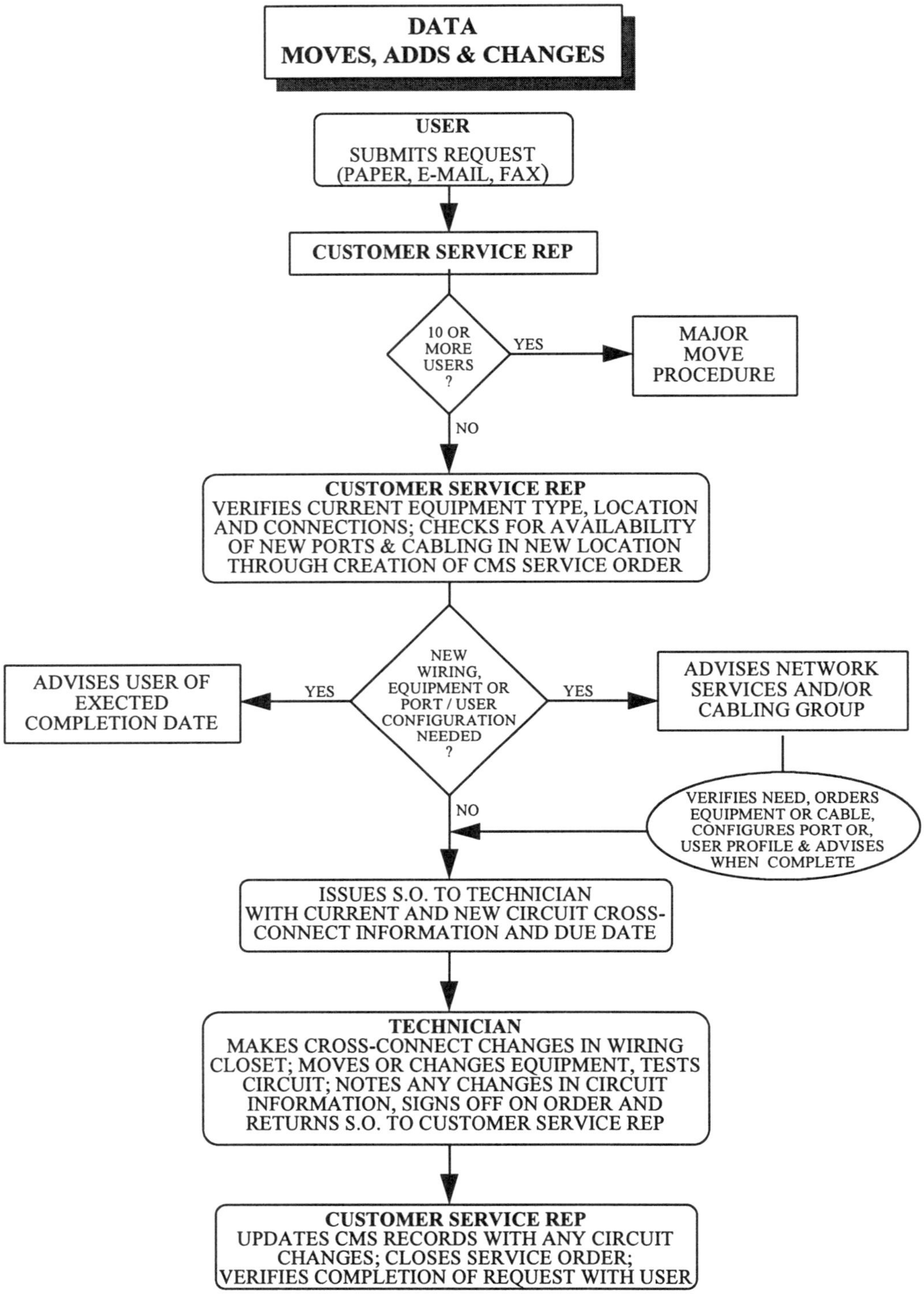

Figure 19.2: Sample workflow diagram for Data Moves and Changes.

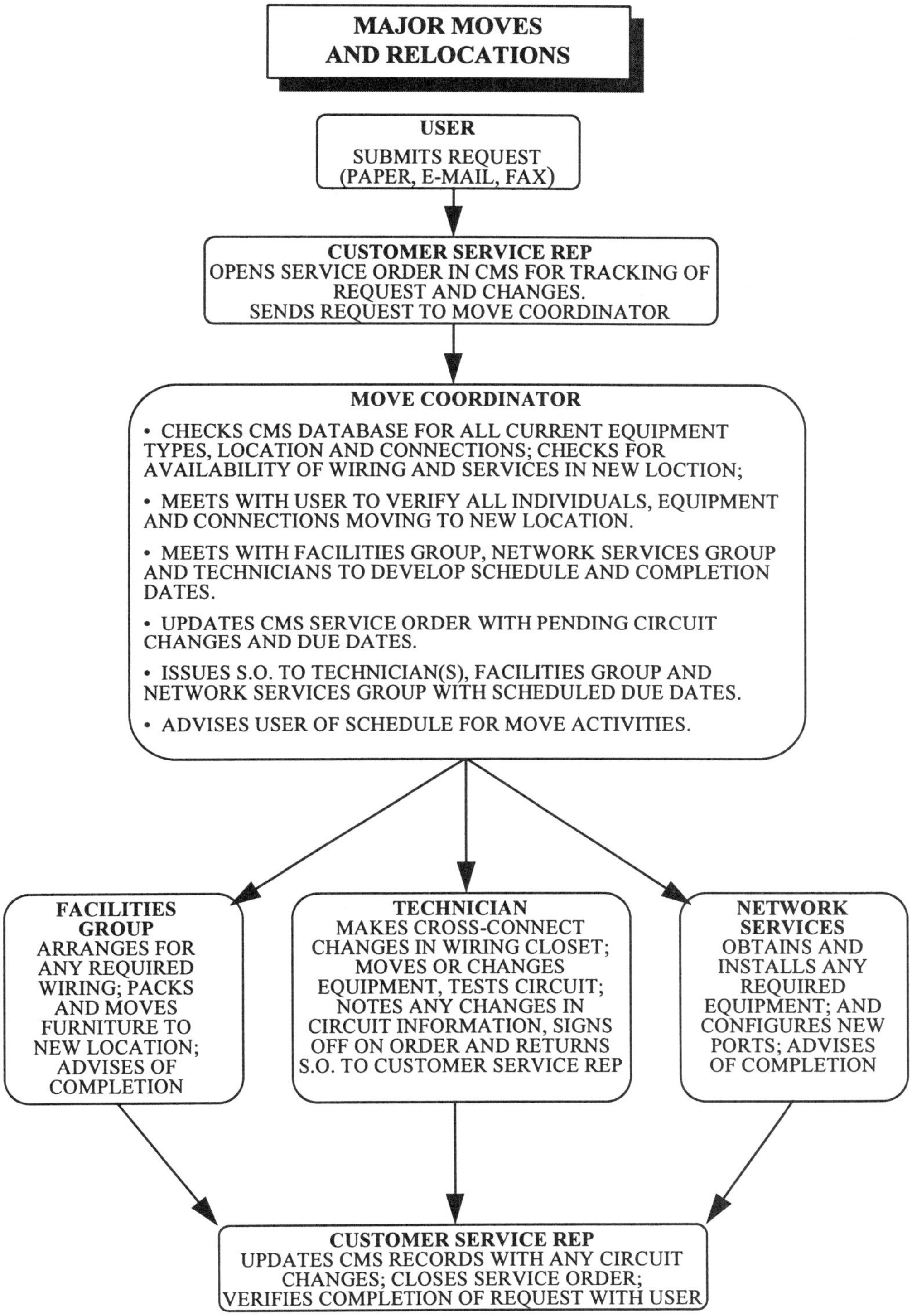

Figure 19.3: Sample workflow diagram for major Moves and Changes.

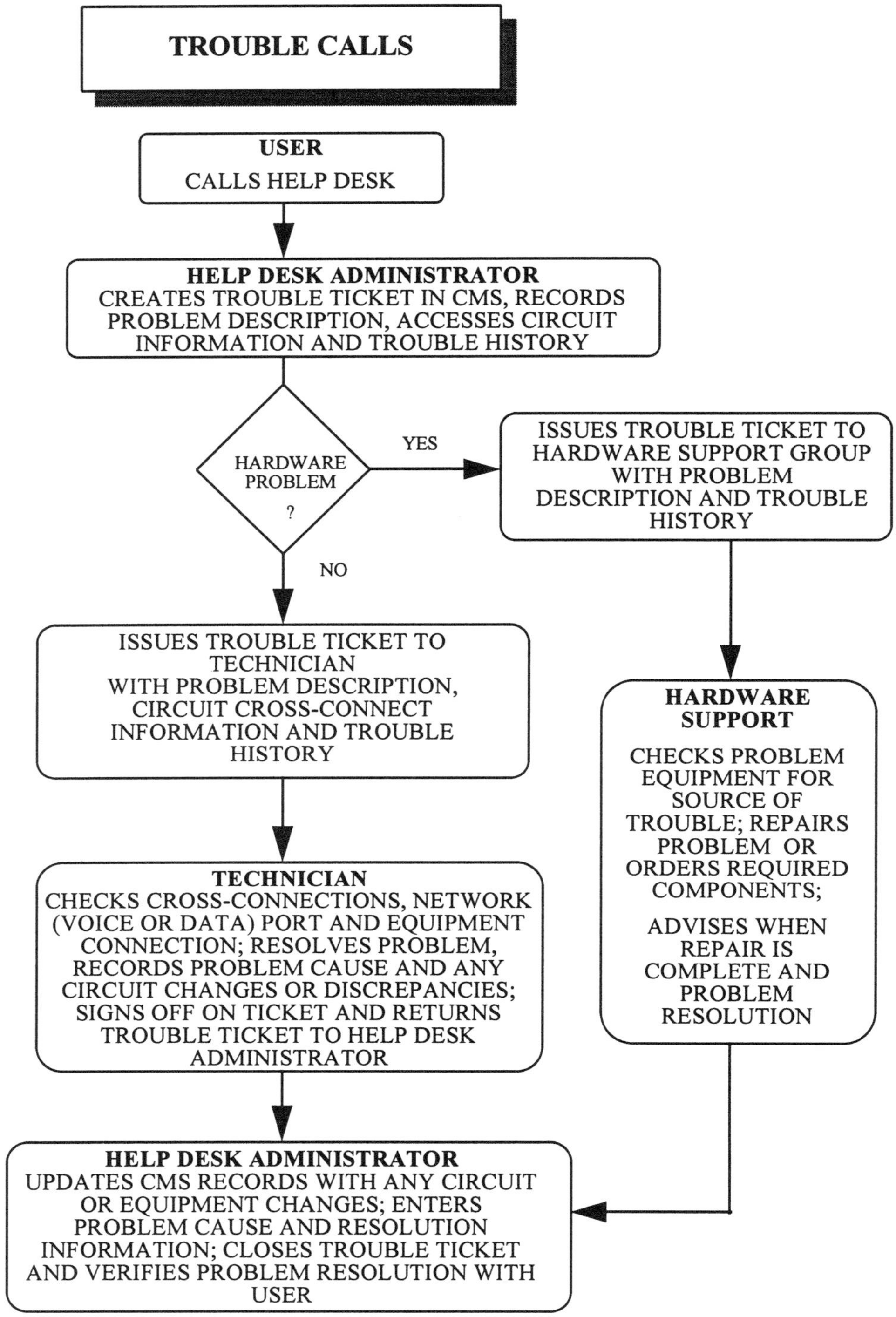

Figure 19.4: Sample workflow diagram of the Trouble Ticket process.

PLANNING AND AUTHORIZATION

20

At some point, you will likely be required to obtain the organizational commitment and authorization necessary to implement the CMS project, and to ensure continued support and resource allocation on an ongoing basis. In addition, you must now get some sense of the tasks before you and create a basic project plan. Each organization will have different requirements for project proposals and documentation, from purely verbal to highly developed. The following discussion provides a reasonable basis for documenting the important aspects of an enterprise-based project, and can serve as a template for your requirements. Even if your formal documentation will be limited, review this chapter to be sure you can answer any questions that may arise.

- STATEMENT OF GOALS AND OBJECTIVES
- PROJECT SCOPE
- ADMINISTRATION PLAN
- ANTICIPATED BENEFITS
- PROJECT PLAN AND TIMELINE
- ESTIMATED COSTS

Table 20.1: Documentation to consider for an Authorization Request.

In setting out to develop a cable management plan, you likely had some specific goals in mind. As you have performed evaluations discussed in previous chapters, you have been thinking about the direct impact of your records structure - what tasks will benefit, and what new effort is required. Your goals and objectives may now be modified somewhat but they should be very clear, and based on your proposed administration plan (workflows). Similarly, you have considered the Project Scope and the records structure of the information you want to track, each now also consistent with your proposed administration plan. All of the above information can then be used to estimate the value of expected benefits and to prepare an overall plan and timeline for the project. With the plan and timeline, you can also estimate both the upfront project costs and the ongoing costs for support, even estimating a payback period if you desire. All of these items listed, summarized in Table 20.1, are now appropriate to include in your authorization request.

The Authorization Request

The overall goals for a connectivity management plan are discussed in Chapter 1. You may recall that meeting many of your goals for cable management involved the management of people and their tasks. Therefore the statement of your goals and objectives will be a summary of administrative benefits as well as reduced expenditures for the cable plant and network hardware. This statement should be brief, two or three paragraphs at the most, and address issues in terms that your management wants to see. The following list offers some different ways to consider and state your project objectives:

- o Reduce technician time for tracing each connection for routine trouble calls and routine Moves, Adds & Changes (MACs).
- o Automate the Service Order process to facilitate a faster turn-around time on trouble calls and routine service requests.
- o Reduce network down-time by resolving problems more quickly with available network connection information.
- o Provide CMS information to the Help Desk to improve response time and improve customer perception.
- o Expedite the Planning/Engineering functions with available information on network equipment, cabling, outlet locations, and connections to provide for faster relocations and reorganizations.
- o Provide documentation for use when communicating with cable and equipment vendors. Provide cable path and routing information for use in designing or planning new circuits.
- o Provide Management reports on cable and network equipment utilization, asset inventory, and for space planning.
- o Provide a method for tracking utilization of cable plant for planning, circuit routing and expansion purposes, without conducting a cable survey.
- o Provide a method for tracking network controlling equipment, equipment locations, and port utilization.
- o Provide an inventory of desktop equipment by type, model, and department to assist in troubleshooting, and in budget planning for upgrades and replacement of older equipment.
- o Provide a means for identifying problem-prone equipment by make/model, purchase date, and location.

As you can see, there is overlap among some items only to underscore that goals must be defined in terms specific and meaningful to each organization. They represent a commitment you are making to yourself and management as to what you intend to make happen. They will not happen automatically; it will take your focused efforts to make them a reality.

After the statement of project goals and objectives, you should include the Project Scope which you developed in Chapter 13. This is one end result of your team evaluations, briefly describing the extent of the project. This should indicate to Management the extent of what needs to be done.

Next, your documentation should include a copy of the Administration Plan which you developed in Chapter 19, including the workflows required to use and support the CMS for MACs and troubleshooting procedures. If significant changes to current workflows are being proposed, it is assumed that Management has been a participant in the planning process, or at least has been kept informed along the way. Some essential procedural changes may indicate new roles and responsibilities, different manpower requirements, and a new system of accountability which Management must embrace and uphold.

The fourth item to consider for your Authorization Request is your best estimate of benefits and savings from the cable management plan. Understanding these benefits is covered in some detail in Chapter 5. As discussed there, some benefits can be quantified while others are more subjective. However, even these "soft" benefits can be approximated and can be shown to have merit towards meeting overall Management objectives. Estimating the payback for each organization is different, and often difficult to determine. This requires an understanding of what your costs are today for an average trouble call or Moves, Adds, and Changes, and they are often difficult figures to determine. Many organizations do not have a method for tracking the actual current costs for routine services, so estimating the savings which will be achieved through recordkeeping is not easily identified.

Estimating Anticipated Benefits

You may use or adapt the following formulas to determine the minimum payback from CMS record keeping. These numbers may be higher or lower, depending on your actual costs and the anticipated time savings for trouble calls and Moves, Adds, and Changes (MACs). The

• REDUCE LABOR COSTS
• IMPROVE LEVELS OF SERVICE
• FASTER RESPONSE TIME ON TROUBLE CALLS
• TRACKING OF TROUBLE HISTORY
• CAPTURE NEW CABLE PLANT RECORDS
• AUTOMATIC CIRCUIT ROUTING
• CABLE CAPACITY TRACKING
• PLANNING TOOL FOR LARGE MOVES
• IDENTIFY UNUSED CIRCUITS
• LESS TECHNICAL EXPERIENCE/TRAINING
• MANAGEMENT OF WORK ASSIGNMENTS
• DISASTER RECOVERY
• TRACKING OF CABLE PERFORMANCE CHARACTERISTICS

Table 20.2: Benefits of a good connectivity management plan.

time savings used here are based on experience with fully implemented CMS record-keeping systems. You should use only those benefits that are intended to result from, and are built into, your connectivity management plan. Possible areas for savings are summarized again here in Table 20.2.

The major savings elements which can be reasonably estimated are developed individually and can then be put in a summary format. These elements include time savings for MACs and troubleshooting, recovery of spare inventory, and avoidance of repeated wiring surveys. Whether you present these numbers as actual savings or the ability to accommodate future growth in these services is up to you. The factors that influence these numbers are discussed in Chapter 5. Additional savings or cost avoidance items can be included in the summary.

Anticipated savings for routine MACs, are based upon estimating the average time saved per MAC by having a complete record of the current and proposed circuits. The project team should reach a consensus on this number, and then multiply by the appropriate number of annual MACs and the hourly cost (with overheads) for this work. In the example in Figure 20.1, an organization with 2000 MACs per year (75% involving connections), an estimated time savings of one-half hour per MAC, and a labor cost of $40 per hour could save up to $30,000 annually.

A similar approach is taken with trouble calls, whereby the project team must estimate the average time saved per trouble call involving cables, ports, or connections (connectivity). Multiply this estimate by the appropriate number of trouble calls per year and the hourly cost (with

2000 MACs per Year
75% Involve Connections
1/2 Hour Saved per MAC
Labor at $40 per Hour

$$2000 \text{ MACs} \times 0.75 \times \frac{0.5 \text{ Hours}}{\text{MAC}} \times \$40/\text{Hour} = \$30{,}000$$

Figure 20.1: Estimated savings for Moves and Changes.

3000 Trouble Calls per Year
40% Involve Connectivity
1 Hour Saved per Trouble Call
Labor at $40 per Hour

$$3000 \text{ Troubles} \times 0.40 \times \frac{1 \text{ Hour}}{\text{Trouble}} \times \$40/\text{Hour} = \$48{,}000$$

Figure 20.2: Estimated savings for trouble calls.

overheads) for this work. In the example in Figure 20.2, an organization with 3000 trouble calls per year (40% related to connectivity), an estimated time savings of one hour per trouble call, and a labor cost of $40 per hour could save up to $48,000 annually.

If the system is tied in to a network management system, significant additional benefits are obtained for tracking down hardware problems on the network. In this case, faster response time may avoid substantial costs for lost productivity, especially when a group or department's functions are completely shut down during the network outage. Based on your experience, this benefit may be quantified and included in the savings summary.

During the process of conducting an inventory of active circuits, a percentage of circuits are often found which are still active but are no longer being used. These circuits can represent a savings in a number of ways. Freeing up ports on a voice switch or data controlling device can represent cost avoidance in purchasing additional capacity, or redistributing the capacity to other areas. If voice lines are obtained from an outside provider such as the local telephone company, this represents an annual out-of-pocket dollar savings. The average number of unused lines which are located during the inventory process is approximately 5 - 10% of the active lines. To calculate your savings in disconnecting unused circuits, take 5% of your estimated active voice and data connections times the annual line cost (or use an approximate average cost of $200 to $400 per connection.).

In the example in Figure 20.3, a typical site of 2500 users will have approximately 5000 individual voice and data circuits. Assuming 5% recovered capacity (250 circuits) at $300 per connection gives a savings of $75,000.

By capturing and maintaining circuit information in your CMS records, you have available an accurate, up-to-date account of your cable pair and port utilization. This avoids the recurring surveys of wiring closets or station outlets to determine spare capacity of cables or controlling equipment. These wiring surveys of cable plant, controlling equipment, and patch panels can be quite costly and only provide a temporary benefit. Without maintaining the

5000 Active Voice and Data Circuit
5% Active but Unused
$300 Cost per Voice or Data Circuit

$$5000 \text{ Circuits} \times 0.05 \times \$300 \text{ per Circuit} = \$75{,}000$$

Figure 20.3: Estimated savings for recovered active but unused circuits.

10,000 Backbone Cable Pairs
20% Inventory per Year
2 Frames per Cable
$1,500 per Frame per 1000 Cable Pairs

$$10{,}000 \text{ Pairs} \times 0.20 \times 2 \times \frac{\$1{,}500}{1000 \text{ Pairs}} = \$6{,}000$$

5,000 Horizontal Cables
20% Inventory per Year
$5 per Horizontal Cable (Frame and Office)

$$5{,}000 \text{ Cables} \times 0.20 \times \$5 \text{ per Cable} = \$5{,}000$$

Figure 20.4: Estimated savings for avoidance of recurring cable and outlet surveys.

information captured during the survey, the information quickly becomes outdated. The cost for these surveys will vary depending on the size of the facilities, and the number of connections. This work may be done on a project basis by outside contractors, or by in-house staff during the routine process of handling major moves or reorganizations. You may have actual dollar figures available for wiring audits and surveys which have been completed in recent years.

It costs between $1000 and $2500 per thousand cable pairs to inventory a single frame, and of course each cable has two ends which must be inventoried. In the example in Figure 20.4, a typical site of 2500 users may have 10,000 backbone cable pairs or more, as well as 5000 to 10,000 outlets. Assuming that 20% must be inventoried at each end in a given year, and using a cost of $1,500 per thousand cable pairs provides a savings of $6,000 per year. Inventory of 5,000 outlets or horizontal cables is also included at $5 per outlet for frame and office inventory at an annual rate of 20%, yielding annual savings of $5,000.

An additional savings factor could be added by avoiding costs for unnecessary rewiring. Horizontal wiring costs between $200 and $1000 per drop. Avoiding this rewiring at 1% of the total (50 lines) could save an additional $10,000 to $50,000 per year.

Yet another consideration is the growth in network services that requires adding additional staff. If proper information and instructions are available through a CMS, persons with less experience can be used for many routine tasks. These individuals can be obtained at lower wage rates, often at a savings of $5 to $10 per hour over highly trained technicians. The skilled personnel can be used for more technical, higher-level network support and operations.

As discussed in Chapter 5, many benefits of a CMS emerge which are not so easily quantified, but yet may be a driving force behind the project. Do your best to have the project team put a dollar value on these strategic benefits, and include them in the savings summary. For example, you can estimate the improvement in organizational productivity through an improvement in Level of Service, where faster delivery of services helps the organization achieve better business performance. Cost avoidance during a disaster situation can also be considered, based on reducing lost productivity following the loss of service.

Your specific list of anticipated benefits can be summarized as shown in Figure 20.5.

Summary Statement of Anticipated CMS Implementation Savings
ABC Company

I.	Labor Savings	
	Reduction in MAC Costs	$ XX,XXX
	Reduction in Troubleshooting Costs	XX,XXX
	Elimination of Cable/Outlet Surveys	XX,XXX
	Labor Total:	$ XXX,XXX
II.	Cable and Equipment Savings	
	Recovery of Active but Unused Circuits	$ XX,XXX
	Avoid Unnecessary Cabling	XX,XXX
	Cable and Equipment Total:	$ XXX,XXX
III.	Indirect Benefits and Savings	
	Level of Service	?
	Billing Platform	?
	Improved Design Capability	?
	Disaster Recovery	?
	Indirect Total:	?
	PROJECT TOTAL SAVINGS:	$ XXX,XXX

Figure 20.5: Summary statement of anticipated CMS implementation savings.

Project Plan and Timeline

The steps involved in implementing your CMS and the approximate timing of these steps is another important part of your planning. You should establish a Project Schedule similar to Table 20.3 which will anticipate certain phases or activities. At this point, the plan should be general and will be modified and finalized after you have selected a CMS and actual inventory requirements are defined. Beginning from the point of project authorization, the project will require time for CMS evaluation and selection, CMS installation, final system preparation, and then the actual inventory and data entry process. The complete implementation process is discussed and flow-charted in Chapter 1.

You should allow plenty of time for the selection and installation of your CMS database, unless you have already determined paper records will provide the necessary requirements for your record keeping. The best option for evaluating

PHASE	WEEKS
I. CMS EVALUATION AND SELECTION	
Prepare Request For Proposal (RFP)	2
RFP Responses	4 - 6
Final Selection	2
II. CMS INSTALLATION	
Hardware	0 - 2
Software	4
III. FINAL PREPARATION	
Catalog And Nomenclature	4
Set-up And Testing	1
Training	0 - 2
IV. INVENTORY AND DATA ENTRY	(See Table 20.4, Item 2)

Table 20.3: Sample project implementation schedule.

and selecting the best CMS software is to prepare and issue a Request For Proposal (RFP). This process provides the best method for comparing the different systems on a common basis, provides a firm pricing quote, and documents the functionality and performance of each system.

The process of writing an RFP, issuing it to the CMS vendors, evaluating the responses and making a final selection takes approximately eight weeks. This includes two weeks for RFP preparation, four weeks to receive Vendor responses, and two weeks for evaluation and system selection. While this does add to the length of the project, the value received through the RFP process is worth the extra time to be sure the best system is selected.

The usual length of time for system installation, following final CMS selection, is a minimum of 4 weeks. This allows sufficient time for the vendor to schedule the installation, and for the necessary purchase orders to be issued. If hardware must also be purchased, and is not being supplied by the CMS vendor, allow an additional 2 weeks for the delivery and installation of the hardware. It is always best to have the hardware installed and tested, especially with a network system, before installing the CMS software.

Once the CMS has been loaded locally, the project administrator or project team can receive initial training from the Vendor. There will likely be a period of four to six weeks for final development and testing of the catalog and nomenclature, and preparation of specific inventory forms and procedures. Software customization and initial training can also be completed during this period.

Based on the number of buildings, floors, information outlets, and number of circuits, you should be able to provide a rough estimate of the time required to conduct any necessary inventories, to load the information into the database, and to test the performance of the CMS. The easiest (but not necessarily the most accurate) method for estimating inventory timing is based on an average time per office or per outlet. You may want to use the formulas below to come up with an estimated timing for your CMS project. These time estimates are based on using individuals who are trained in CMS inventory techniques and are assigned full time to the CMS project.

There are generally three distinct tasks or roles during the inventory process, discussed in detail in Chapter 5. The first is for a project manager, administrator or coordinator who is responsible for project management and technical development. This individual will usually be the best candidate to handle the inventory of backbone cable and equipment as well. The time for backbone inventory must be considered in the project schedule, but other technical responsibilities of the administrator will run concurrently with the remaining inventory tasks. Therefore, the basic formula assumes an average of 2 days per wiring closet to inventory and label appropriate cables and cable counts, network controlling equipment and ports, and appropriate termination hardware. This will vary depending on the number of closets and the amount of equipment contained in an average closet. This does not include the inventory of each cable pair or port connection on the network equipment.

The second inventory role is for the inventory of office equipment and cross-connections of the cable and hardware, often referred to as frame inventory. The combined time for office and frame inventory ranges from 10 to 30 minutes per office for one person. Teams of two could perform at a little more than twice this rate. The actual pace of the inventory will depend on the overall inventory manpower and the other responsibilities that may have, so they will likely not be this efficient over a long period of time. Other natural events will cause delays, such as locked doors, rechecks, meetings, training, etc. The flow of office and frame inventory is the major factor in the overall inventory duration, and should be used for setting the project schedule.

The third task during inventory is the actual data entry. While data entry consumes a certain amount of time, it happens concurrently with the inventory and does not add much to the overall length of the project, provided that it can keep pace with inventory. A typical rate for data entry, based on a fairly comprehensive recordkeeping system, ranges from 10 to 20 minutes per office. This number includes entry of desktop equipment, horizontal/workstation cables, backbone cables, controlling equipment and ports, and circuit cross-connects, and is based on using experienced data entry personnel.

You can prepare a chart similar to Figure 20.6 showing the estimated manpower and duration for the physical inventory tasks. Note how the duration of office and frame inventory is adjusted for total manpower, efficiency and other

ESTIMATE OF PHYSICAL INVENTORY

I. INVENTORY OF BACKBONE CABLE AND EQUIPMENT

20 Closets X 2 Days per Closet = 40 Man-Days (Non-consecutive)

II. OFFICE AND FRAME INVENTORY

$$2500 \text{ Offices} \times \frac{20 \text{ Min.}}{\text{Office}} \times \frac{1 \text{ Hour}}{60 \text{ Min.}} \times \frac{1 \text{ Day}}{7 \text{ Hr.}} \times \frac{1}{0.8} = 150 \text{ Man-Days}$$

Duration Based on Two Inventory Persons:

$$\frac{150 \text{ Man-Days}}{2 \text{ People}} \times \frac{1 \text{ Week}}{5 \text{ Man-Days}} = 15 \text{ Weeks}$$

III. DATA ENTRY

$$2500 \text{ Offices} \times \frac{10 \text{ Minutes}}{\text{Office}} \times \frac{1 \text{ Hour}}{60 \text{ Minutes}} \times \frac{1 \text{ Day}}{7 \text{ Hours}} = 60 \text{ Man-Days}$$

Figure 20.6: Estimate of timing for physical inventory tasks.

factors. Use the entry for office and frame inventory duration as the basis for Item 4 in Table 20.3 to complete your estimate of overall project timing.

Estimated Costs

The final document for your Authorization Request must be an estimate of the initial implementation costs and anticipated ongoing costs. The initial costs include the hardware and software necessary to support the CMS and the labor costs for implementation. Each organization will have a different approach to evaluating costs, depending upon internal accounting methods and whether internal manpower costs will be considered. The discussion that follows treats all costs as out-of-pocket, so that you can see all the potential cost elements.

Refer to Chapter 5 for a complete discussion of the project cost elements and some typical ranges of hardware and software costs. A summary statement of project costs is given in Figure 20.8 at the end of this chapter, although at this point you may have to make some assumptions about specific items. The estimates given here assume a multi-user, enterprise system

The cost of the hardware to support the CMS software can range anywhere from $3,000 for a simple PC with enough memory, speed and disk space to handle your records up to $50,000 for a server, additional PCs or workstations, and supporting equipment. This depends on the complexity of the CMS package selected, and the number of users on the network who will need access to the records. The cost may include items to upgrade existing equipment, or provide a plotter, printer, modems, network cards, backup systems, etc. These costs will vary depending on whether existing hardware can be utilized for storing the CMS software and records. If you need a ball-park estimate, $50,000 for hardware will usually support several dedicated users on a fully-functional, networked system with all support equipment considered.

The cost of the CMS software itself can range anywhere from $2,000 to over $100,000, depending on the CMS chosen, the number of modules selected, and the number of concurrent users. A basic, non-CAD product with excellent features can be priced in the $15,000 to $20,000 range, while more sophisticated systems with CAD, call accounting, billing, service orders and trouble tickets, plus links to Network Management can be priced as high as $200,000. If you are not sure what your needs may require, use an estimated cost of $50,000 for multi-user software to cover any anticipated functional and access requirements. You may have to consider additional costs for database software or licensure, CAD software, communications software, and network and CPU operating systems.

In the previous section there are models given for manpower and cost requirements. The factors used in these models are representative, and you may substitute your own as appropriate. The larger the scope of the implementation, the more inventory time that will be required. If only a portion of the information needs to be gathered and recorded, such as only the office inventory, or only the wiring closet connections, then base your number on the percentage which needs to be

QUALIFIED CONTRACTOR RATES:		
• PROJECT MANAGEMENT	$45 - $75/HR	$360 - $600/DAY
• INVENTORY PERSONNEL	$20 - $35/HR	$160 - $280/DAY
• DATA ENTRY	$15 - $20/HR	$120 - $160/DAY

Table 20.4: Cost ranges for outside project services.

completed. Finally, use labor rates which are representative of your industry and geographic location.

Clearly, out-of-pocket costs will increase the more you rely on outside vendors or contractors to perform certain tasks. But in terms of real costs, or overall project timing, there may be benefits to this approach. Using in-house personnel will reduce the amount of dollars actually spent outside the company, but the cost is still there nonetheless, since the cost for in-house personnel still exists whether the inventory work is conducted during working hours, or after hours on an overtime basis. In-house personnel are more familiar with your facilities, and get the opportunity to familiarize themselves with the CMS records and process. On the other hand, they may have limited time available, lack formal training in inventory work, and may cost significantly more on an overtime basis. Overall implementation time may increase.

Outside personnel may be specifically trained in inventory methods and can devote full time to the inventory. If using outside contractor personnel, refer to the rates in Table 20.4 or use actual rates paid by your company for the various levels of temporary help to establish cost estimates.

The estimated cost for each of the physical inventory functions should be based on your estimates in Figure 20.6 using your own labor rates or qualified contractor rates. An example is given in Figure 20.7. Note that the overall cost for project administration includes amounts for management (including technical development) at 2 days per 5 days of inventory, as well as the estimate for backbone inventory at 2 days per wiring closet.

Besides the actual project costs, there will be some costs for ongoing support and maintenance of the system. Such things as staff training and CAD drawing maintenance will likely

PHYSICAL INVENTORY COSTS

I. PROJECT ADMINISTRATOR

MGMT. AND TECHNICAL DEVELOPMENT

$$15 \text{ Weeks} \times \frac{5 \text{ Days}}{\text{Week}} \times \frac{2 \text{ Days}}{5 \text{ Days}} \times \$500/\text{Day} = \$15{,}000$$

BACKBONE INVENTORY

20 Closets X 2 Days X $500/Day = $20,000

II. OFFICE AND FRAME INVENTORY

150 Man-Days X $250/Day = $37,500

III. DATA ENTRY

60 Man-Days X $145/Day = $8,700

Figure 20.7: Summary of anticipated physical inventory costs.

be covered by normal staff overheads. But there are three areas that deserve early consideration.

First is the possibility of staffing changes or new staff requirements. In a large organization with a high volume of MACs, there will typically be a clerical administrator(s) for creating Service Orders and Trouble Tickets, and there is a need for someone to serve ongoing as CMS administrator. If it is anticipated that the current customer support or help desk staff can handle these tasks, there may not be any additional cost for this administration. If additional staffing resources are required, you want to be sure to get clear commitment that these resources will be supported ongoing.

Another cost which must be included is for ongoing maintenance and support of the CMS software from the Vendor. This annual cost is normally 15 to 20% of the initial purchase price, and usually includes enhancements, upgrades, and system support for user questions and problems.

The third item is miscellaneous costs for software support, including upgrades to new versions of the database, CAD, or operating system software. Such upgrades seem to naturally have some minor bugs or conflicts which must be diagnosed and resolved. Additional software customization or reporting functions from the CMS Vendor or in-house staff should also be considered.

Summary of the Authorization Request

You should now have a good base of information regarding your project so that you can get the commitment and authorization to proceed. Documentation that you have prepared for the Authorization Request began with a brief Statement of Goals and Objectives, so the overall intent of the project is conveyed. You can include the Project Scope document prepared in Chapter 13, and the Administration Plan covered in Chapter 19. Finally, as you have prepared in this chapter, you can include summary statements of anticipated savings and anticipated costs, along with a Project Schedule outlining the flow and timing of the general project tasks.

This information will provide management with a reasonable picture of the project and help build confidence in the commitment and ability of the organization to pursue a structured connectivity management process.

Summary Statement of CMS Implementation Project Costs
ABC Company

I.	Hardware	
	Database Server with Tape Backup	$ XX,XXX
	Computers and/or Workstations w/ DOS and Windows	XX,XXX
	Network Interface Cards	XXX
	Modems	XXX
	Printer	XXX
	Plotter	X,XXX
	Hardware Total:	$ XX,XXX
II.	Software	
	Server Operating System	$ XXX
	Network Operating System and Modules	XXX
	Database User License	X,XXX
	Database SQL Tools	X,XXX
	CAD Software	X,XXX
	Communications Software	XXX
	CMS Software	XX,XXX
	Software Total:	$ XX,XXX
III.	Project Services	
	Inventory	$ XX,XXX
	Data Entry	XX,XXX
	Project Administration	XX,XXX
	Project Services Total:	$ XX,XXX
	PROJECT TOTAL:	$ XXX,XXX

Figure 20.8: Summary statement of project implementation costs.

BEGINNING YOUR CATALOG

21

In Chapter 14, you laid out in a general fashion the lists of information that you wanted to track about cables and circuits. You went on to consider which items on each list might require their own tables, and how these items would then be related together to describe a circuit. Now you must go into serious detail about the various components to be tracked in your CMS database, how each of these items will appear, what categories they will be organized under, what will be tracked about each, how they will be connected to other components in the database, what they will be called, and how they will be labeled. This is a very important and sometimes difficult task, but one which will greatly impact how the database is set up and how records will be managed from a connection and reporting standpoint. We will refer to many of the charts and lists discussed in Chapters 13 and 14, so it may be helpful to have those lists handy as you work through this Chapter. Your catalog may require some adjustment once the final system is selected, but it will provide an excellent tool for evaluating the performance characteristics of each of the different CMS systems.

The first step in creating your catalog involves setting up the "buckets" or categories into which you can put different components or records. These categories should be based on the items in your list of data elements that describe circuits, similar to Table 14.2, and should generally reflect items for which you desire usage and availability information rather than just trivial circuit detail. These categories might be reference tables such as building codes or abbreviations, departments or divisions, expense codes, location codes, as well as equipment and cable types. After setting up the categories, it can be determined how each of the items will be managed with respect to connections. When preparing these lists, it will be beneficial to put them into some type of electronic spreadsheet, since it may be possible to load the tables directly into the CMS database through a batch file transfer. We will cover each of these areas individually to help you better understand the impact of creating a CMS catalog.

Creating Support Tables

Any recordkeeping will likely require the development of tables of supporting information such as building names, department codes, etc. that will be applied many times to the different elements of your connection records. A typical list is shown in Table 21.1. As you set up your database, you must consider the names of locations for each of the components, such as where controlling equipment or backbone cables

- SITES
- BUILDINGS
- FLOORS
- ROOM NUMBERS
- DEPARTMENT CODES
- EXPENSE CODES
- USER INFORMATION

Table 21.1: List of supporting tables for a CMS.

are located. A structured approach to location names will be beneficial if a CMS is selected which automatically creates a location code through a combination of the building, floor and room. Use the planning tables which you prepared in Chapter 13 as a basis.

When managing multiple sites, you will need to differentiate between cables and devices at the different sites. If you can establish a separate database for each site, then site codes are not required. If multiple sites will be included in a single CMS database, it will be valuable to set up a table of site codes or site abbreviations. These codes can then be used in a "Site" field for each element record, or else be built into the naming conventions for cables and equipment to create unique IDs which differentiate between cables and equipment at different sites. It is important to be consistent in setting up your site codes. All site codes should be the same number of characters. Two or three character site codes are the easiest to work with. If the CMS does not have a site code field, you may consider using the site code as part of the building code.

Next, develop a table of building designations using either building abbreviations or building numbers which can be utilized in the database for locations of offices and wiring closets, or in setting up floor plan drawings. If you are planning to use abbreviations for the buildings rather than requiring the full building name to be entered for each record, be sure to stay consistent in the number of characters used for each building. An example would be that all building abbreviations would be 4 characters. This would mean ALL buildings would be 4 characters, not some with 3 and some with 4. If a building number will be used, be sure to keep the number to the same number of characters, such as Building 01, 02, ... 10, 11, 12, etc. This helps to ensure all records are created consistently as the information is loaded into the database.

Use the same table created for your building designations and add a column for the floors in each building. List all the floors or floor/wing combinations (such as 1N, 1S or 1A, 1B, etc.), depending on how the room numbers are divided and referenced. Setting specific numbers to be used consistently for floors will be important to insure records are entered properly, especially if the CMS does not validate the floor. One record might be entered as 1, while another might be 1N, or 01. This may not present a problem initially, but a later search may not find all the records if they were entered differently. The same format should be followed when using a floor in the name of a cable or piece of equipment. If there are more than 9 floors, a good rule is to always use two

SITE: LANDISVILLE RESEARCH PARK

BUILDING CODE	BUILDING DESCRIPTION	FLOOR	ROOM NUMBER
ADM	ADMINISTRATION BUILDING	B	B01-B99
		1	101-199
		2	201-299
FAC	FACILITIES OPERATIONS & MAINT.	1	G101-G140
		2	NONE
REN	RESEARCH ENGINEERING BUILDING	BE/BW	B01-B99 E/W
		1E/1E	101-199 E/W
		2E/2W	201-299 E/W
		3E/3W	301-399 E/W

Table 21.2: Building, floor, and room designation codes for the LRP site.

characters, such as 01, 02, ... 10, 12, etc.

It is also valuable to establish a standard number of characters for room numbers as well as a standard format. Some office areas are divided into cubicles where each cubicle has a letter designation, such as cubicles A, B, C and D in Room 100. These workstation location formats must be standardized as well. If both 100-A and 100A are used, it will make it difficult to perform queries and create reports since the database will consider 100-A and 100A to be two different locations, and a query for equipment in Room 101A will not find equipment entered in Room 101-A. A completed table showing building, floor, and room number codes for the LRP site described in Chapter 4 is given in Table 21.2.

Some of the CMS products validate against a pre-defined list of room numbers to avoid inconsistency in how room numbers are created, but for those systems that don't a little pre-planning is well worth the effort. If the CMS requires the creation of all room locations prior to loading any records, it may mean you will need to create a list of room numbers for each floor. It may be difficult or impossible to revise the room numbers at a later date, since some systems do not permit changes in the location fields for cables and equipment once an item has been connected to another component in the database. Be sure you are consistent in the number or name used for wiring closets, communications rooms, and equipment rooms, etc. If you use the actual room number in one record and TC Room, Comm Room, or MER on other records, some databases will consider these to be two different rooms and will not permit connections between items which it considers to be in two different locations.

The next set of tables to consider are lists of departments or divisions which you may want to track in your CMS records. Many CMS products have fields for tracking the department or division to which a piece of equipment, cable, or an employee belongs, as well as the expense or cost code for these items. If you plan to track this information in the database, you will need to develop the appropriate lists so they can be loaded into the CMS database prior to adding cables and equipment. Generally, entries for these fields are usually validated against a table, therefore the tables must exist prior to adding a record which refers to the table.

As with the site and building abbreviations, it is helpful if a consistent number of characters is used in setting up abbreviations. This may not always be possible since many organizations have already set up abbreviations for their departments or divisions, and sticking with existing designations where possible will make the database easier to use. Developing a standard list of abbreviations is especially important if the selected CMS does not have a validation table for departments or divisions. This will help to avoid the same department being entered several different ways, depending on who created the record. An example of this might be the Employee Relations Department, which could be designated as ER, Emp Rel, or EMPREL, E-R, etc.

Another table which may be utilized in the CMS might contain the expense codes or cost codes for each department or division. This code might be used for charging a particular department for connections to the network, or for special circuits, or even for service requests. Many of the CMS products have a Service Order module which provides a field for the expense code which should be charged for the services provided on the work order. A list of valid codes might be obtained through the finance department or billing section.

A decision must be made on whether to track employee/user information in the database. This usually generates a lot of discussion, since employee information is normally maintained by the Personnel Department in a separate database. It is often difficult and time-consuming to maintain this information in the CMS database, since the Information Systems groups don't always get involved in personnel changes.

User information can be helpful when troubleshooting or handling routine requests for MACs. The Help Desk or Customer Service Center can be more responsive when they have access to information about the user's voice connections and network services. If a user calls with a request, the Help Desk or Customer Service Center can look at the CMS records for that individual user and verify what service is active and needs to be repaired or relocated. Some of the commercial CMS products have a well-developed user/employee directory module which can handle directory inquiries as well as print employee directories for distribution among users.

If user/employee information is maintained in a separate database, there are several different approaches to dealing with the existing system and the CMS that are covered in Chapter 8. Considering the following questions will help determine whether user/employee information should be maintained in the CMS database:

- Is there a current database for Directory information?
- What information is tracked in this database?
- Who maintains the Directory?
- Is the Directory connected with the Call Accounting System?
- Could the Directory functions be replaced by the CMS?
- How much effort would be required to maintain accurate directory information in the CMS database?
- Would this be a duplication of effort?
- Which database would be kept most accurately?
- How much value is provided?
- Can the existing Directory handle file transfers from other databases?
- Can employee information from the current Directory be periodically downloaded to another database?

If user/employee names and information will be tracked in the CMS, it is also an area where consistency will save time and frustration later on. If the information exists in a personnel or directory file which can be electronically transferred to your CMS database, this simplifies the process. However, if this information will be entered as it is recorded during the inventory process, it is important to establish a policy for how the information is to be recorded and entered. The CMS products vary in how they track user information. Some systems have one field for user name, while others provide separate fields for first and last name. Again, these fields can often be completed inconsistently from person to person. Consider the following examples for how one name might appear:

William L. Matthews
W. L. Matthews
Bill Matthews
William Matthews
W. Matthews
Matthews, W. L.

These examples should provide a good indication of how complicated it might be to search for a listing of all the equipment or connections for Bill Matthews. If the CMS has separate fields for Last and First Name, it would make your job a little easier, but still it points out the need to establish a specific method for entering user information. Any combination of first and last name will work, so it really depends on your organizational preference and from where the information is obtained.

Desktop User Devices

A decision on tracking desktop equipment is affected by whether there is a current recordkeeping system for desktop inventory, and whether this information needs to be tracked in the CMS database as a part of the circuit information or as a means of identifying the circuit. Items such

PHONES	PCs	TERMINALS	MODEMS	PRINTERS
ANALOG	386SX	VT	EXTERNAL	DOT MATRIX
DIGITAL	386	SNA	INTERNAL	INK JET
FAX	486	SUN	V.35	LASER
ANS. MACH.	MACINTOSH	ANSI	V.22BIS	PLOTTER
CENTREX	PENTIUM			

Table 21.3: Initial list of basic desktop equipment and categories.

as phones, computers, printers, fax machines, and answering machines usually fall into the category of desktop equipment and represent the user equipment connected to outlets or horizontal cables at the office location.

The desktop equipment can be a convenient starting or ending point for a circuit, and can also provide an alternate way to determine the availability/status of outlets in a given office location. If no inventory tracking system exists, a CMS can provide a means for tracking desktop inventory with information on manufacturer, model, serial number, department, installed date, cost, and purchase order or maintenance information, as well as location and connections to various services. If movement of all desktop equipment is controlled through the communications group, tracking of these devices in the CMS records will be valuable and beneficial. If users or individual departments can swap or relocate the equipment, then the desktop inventory information in the CMS database will quickly become outdated and obsolete. Tracking of desktop equipment in the CMS database requires that each equipment item be uniquely identified and labeled, and changes in connections and locations must be maintained.

Start by making a list of all the types of desktop equipment which might be found in each of the office areas. At this point, let's just consider office equipment such as phones, computers, terminals, modems, fax machines, answering machines, printers, etc. Add as many items to the list as the team can think of. Looking at the list of types, make separate categories such as phones, computers, printers, etc., and then list the types which fall into each category. An example would be for Phones - Analog, Digital, Wall, or you may be more specific by identifying the make or model such as AT&T/7406, Panasonic, etc. Your initial list of types might look something like the sample in Table 21.3.

Once you have sorted your desktop devices into categories, you need to determine how detailed your catalog will be. Discuss the different options with the team as to whether the catalog should be very specific or somewhat broad. The sample in Table 21.3 shows a broad list of sub-categories which can be set up under each major category. However, you can be much more specific in defining your catalog. This will really depend on why you are tracking desktop equipment in your CMS records. If the purpose is asset tracking, you may need to build the catalog at the level of the specific make and model, as shown in Table 21.4. If the primary reason for tracking desktop equipment is to determine the type of connection, a generic description may be sufficient. If an IBM PS/50 can be replaced by a Compaq/50M without your involvement and knowledge, it's better to call it a PC.

Building your catalog to include specific models or types can provide some benefits. It can give you the ability to identify how many computers of a specific type exist, which might be useful if a specific model-related problem develops or if an upgrade can be made by installing a new card. This would be an easy way to determine how many computers of an older model or type exist, and where they are located, so they can be considered for upgrades. It can save technician time if they know the exact type of equipment so they can bring the appropriate replacement parts with them on the trouble call.

The next step in building your catalog is to define the connection configuration for each

PHONES	PCs	TERMINALS	MODEMS	PRINTERS
2500-SET	COMPAQ/50M	IBM-3270	HAYES-2400	DESKJET
ATT-712	IBM-PS/50	IBM-6286	HAYES-9600	HP-III
ATT-7406	IBM-PS/90	VT-100	HAYES-V.36	HP-IV
PANASONIC	DELL-386	VT-340	U.S.-ROBITICS	LA-100
	AST-486			XEROX-4030

Table 21.4: Detailed list of desktop equipment and categories.

DEVICE TYPE	NUMBER OF CONNECTORS	CONNECTOR TYPES
PHONE	1	RJ
ANALOG	1	RJ11
DIGITAL	1	RJ45

Table 21.5: Possible connector configurations for a telephone.

type of desktop device. Defining the configuration means identifying the number of connections which can be attached to each device and what type of connector is used in each case. Some CMS systems determine their own set of configurations based on the equipment category, such as computer, phone, printer, etc. The best way to convey what is meant by the configuration is to use some specific examples of desktop devices and their connections. We will start with a phone, since this is the most familiar type of desktop device, and then consider terminals, fax machines, printers, PCs, modems, answering machines, etc.

When you think about installing a phone, it requires connecting a mounting cord into the back of the phone. The phone cord has a plug on the end which fits into a jack or connector on the phone. The phone can be defined as a device with a connector on the back or bottom side. The next step is to determine how many connectors are available on the phone and what to call them. There is just one connector and it is an RJ11 or an RJ45, depending on the model of the phone. You can be detailed or general in your configuration. To illustrate this, see the example in Table 21.5.

This example shows three different configurations which can be created for phones. Taking the most general approach would be to treat all phones as having one connector, called an RJ, Modular, or whatever you wish. A generic configuration for all phones may be the easiest from a database set-up point, and also in creating connections in the database. The other option is to create a configuration for each type of phone, depending on the type of connector. Using different configurations requires the administrator to remember the different types of connectors on different phone types.

When you think about installing a terminal, it requires connecting a cord into the back of the terminal. The cord has a plug or connector which fits into a jack or connector on the back of the terminal. The next step is to determine how many connectors are available on the terminal and what to call them. There may be one or more connectors, such as DB25, RJ11, RJ45 or Modular, depending on the model of the terminal. You can be specific in your configuration detail or general, as shown in Table 21.6.

This example shows several different configurations which can be created for terminals.

DEVICE TYPE	NUMBER OF CONNECTORS	CONNECTOR TYPES
TERMINAL-A	1	RJ
TERMINAL-B	2	RJ & DB25
VT-100	1	RJ11
VT-340	2	1 RJ45 & 1 DB25
IBM-6286	3	2 RJ45s & 1 DB25

Table 21.6: Possible connector configurations for a terminal.

DEVICE TYPES	NUMBER OF CONNECTORS	CONNECTOR TYPES
RICOH-1000	1	RJ11
SHARP/UX-181	2	2 RJ11s

Table 21.7: Possible connector configurations for a facsimile machine.

Taking the most general approach would be to treat all terminals as having two connectors, one RJ and one DB25. However, some of the newer terminals have multiple RJ45 connectors, so an option might be to create 3 connectors - two RJ45s and one DB25. Another option would be to create a configuration for each type of terminal make/model, depending on the number and types of connectors. Connectors can even be numbered 1, 2, 3, with each connection taking the next available connector. How you name/number the connector types depends on what information is valuable to the technician working with the connections. The key to remember is to keep your database as simple as possible. A general configuration for all terminals may be the easiest from a database set-up point, and also in creating connections in the database.

When you think about installing a facsimile machine, it usually requires connecting a phone cord into the back of the fax machine. As with the phone, the cord has a plug which fits into a jack or connector on the fax machine. Depending on the make and model of the fax machine, it may have a separate connector which then connects to the phone. Newer models usually have a phone and keypad built in, and only have a single outlet. The fax machine can therefore be defined as a device with at least one outlet, and possibly two. Both outlets are usually RJ11 type. You can be specific in your configuration detail, or general. See the example in Table 21.7.

This example shows two different configurations which can be created for fax machines. Taking the most general approach would be to treat all fax machines as having two connectors of type RJ11, as shown in the diagram in Figure 21.1. Even though some fax machines only have one connector, creating a configuration with the two connectors allows you to utilize a single configuration for all fax machines.

Answering machines are very similar to fax machines. Installing an answering machine requires connecting a phone cord into the back of the answering machine. As with the phone, the cord has a plug which fits into a jack or connector on the answering machine. Depending on the make and model of the answering machine, it may have a separate connector which then connects to the phone. Newer models usually have a phone built into the answering machine, and only have the single outlet. You can define an answering machine as a device with at least one connector, and possibly two. Both outlets are usually RJ11

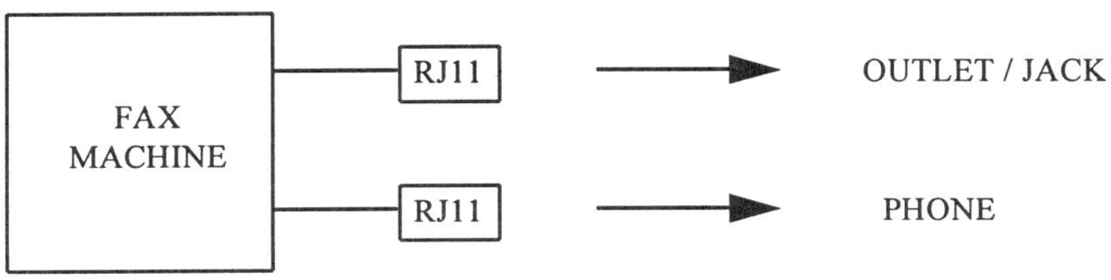

Figure 21.1: Diagram showing possible connections to a facsimile machine.

DEVICE TYPES	NUMBER OF CONNECTORS	CONNECTOR TYPES
PANASONIC	2	2 RJ11s
PHONE MATE	1	RJ11

Table 21.8: Possible connector configurations for an answering machine.

type. You can be specific in your configuration detail or general, as shown in Table 21.8.

This example shows two different configurations which can be created for answering machines. The most general approach would be to treat all answering machines as having two connectors of type RJ11. Even though some answering machines only have one connector, creating a configuration with the two connectors allows you to utilize a single configuration for all answering machines.

When you think about installing a printer, there are a number of options. One option is to connect it to the local PC or terminal through the use of a DB25 or Centronics connector, while another approach could use multiple connections to a single printer, such as a shared printer or a network printer. There is more than one type of connector possible, and different printer models will have varying numbers and types of connectors. As with your previous equipment, you can define one general configuration for all printers, or you can be specific depending on the type of printer. A printer can be defined as a device with multiple connectors of different types, including DB25 and Centronics. The examples are given in Table 21.9.

This example shows three different configurations which can be created for printers. The most general approach would be to treat all printers as having two connectors, one DB25 and one Centronics, which could be connected to either a local PC or to a wall outlet. Even for shared printers where multiple PCs are connected to the printer, this can be done through an adapter which is connected to the parallel port. The other option is to look at the specific configuration of connectors on the back of each type of printer and creating a configuration for each printer type. Again, this will take additional set-up time, and requires that the CMS administrator or technician remember which connectors exist on the different printers and what type of connector is used for what type of connection. You should begin to see that as you increase the number of connectors and connector types for each of type of equipment, the configuration becomes a little more complex.

When you think about installing a PC, there are several options on how the PC could be set up. It can be a stand-alone PC with no connections to the communications network, or it could have multiple connections to different network applications such as 3270 or LAN. The PC could also have a connection to an external modem, and even another connection to a local printer. While it may seem simple to just connect the PC to a horizontal cable or outlet, many of the CMS systems want to know where on the PC to

DEVICE TYPES	NUMBER OF CONNECTORS	CONNECTOR TYPES
PRINTER	2	1 SERIAL, 1 PARALLEL
HP-111	2	1 CENTRONICS, 1 DB25F
PR-625	2	1 CENTRONICS, 1 25-PIN

Table 21.9: Possible connector configurations for a printer.

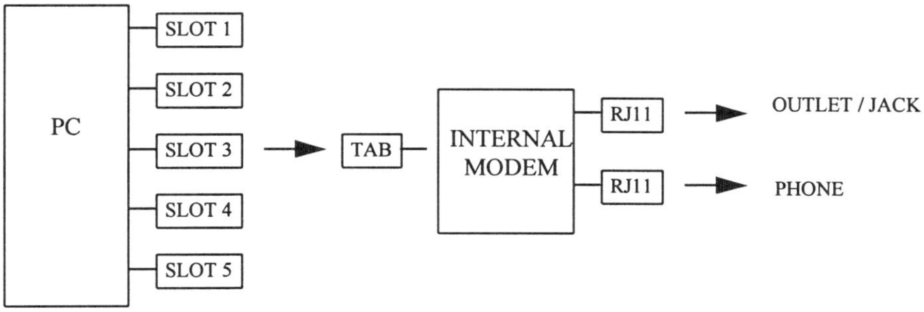

Figure 21.2: Configuration for a PC with an internal modem.

connect to. Some systems only allow one connection per port which then requires the PC to have multiple ports or points for connection. You need to determine a way to connect each of these other components to the PC, and need to have a point at which to connect them. Each of these points of connection needs to have a name and type, so each connection can be differentiated from the others. Further, these connections can be determined by add-on boards within the PC, such as a Network Interface Card (NIC). Table 21.10 illustrates several ways to configure the PC.

This table shows different configurations which can be created for PCs. Taking the most general approach would be to treat all PCs alike and give them generic connectors numbered one through five, in order to handle a connection to the printer, two potential network connections, a modem, and any other connection which might be required. As each connection is added, it would be connected to the next available connector (1 to 5). In this case, the term Slot has been applied to indicate this could be an add-on board as well. Another method is to use a set of defined connectors as shown in the second example. This more closely represents the types of connectors one might find on the back of the PC, but it requires the administrator to know which port or connector is actually used for which connection. A third option is to create connectors for the different types of network connections, such as EN (Ethernet), SN (SNA/3270-Mainframe), TR (Token Ring), AT (AppleTalk), SL (Slot - for Internal Modem/LAN Board), or RS (RS-232). You can approach this in many ways, but try to create a configuration that will not have to be changed or modified later.

You must consider how to show an internal modem which is actually installed into a slot in the PC. The modem could take on the identity (name) of the PC, but will have its own circuit ID (the phone extension) to as well. This really requires the modem to be created as a separate component which would be attached to the PC. When handling a request to move the PC, the modem and extension would be considered connections to the PC that must also be moved. Refer to Figure 21.2 for this configuration.

DEVICE TYPE	NUMBER OF CONNECTORS	CONNECTOR TYPES
PC	5	SLOTS (SEQ. NUMBERED 1 TO 5)
IBM-PS/50	5	2 DB25, 1 SLOT, 1 BNC, 1 DB9
IBM-PS/90	6	1 EN, 1SN, 1 TR, 1 AT, 1 SL, 1 RS

Table 21.10: Possible connector configurations for a PC.

DEVICE TYPE	NUMBER OF CONNECTORS	CONNECTOR TYPES
INTERNAL MODEM	3	2 RJ11s, 1 TAB
EXTERNAL MODEM	3	2 RJ11s, 1 DB25F
MODEM (GENERAL)	4	2 RJ11s, 1 TAB, 1 DB25F

Table 21.11: Possible connector configurations for a modem.

If you are planning to use the CMS to track desktop devices as assets, it may require tracking additional information about the PC such as the Manufacturer, Make/Model, and Serial Number. You may also want to track the monitor type and the cards installed in the PC, such as a LAN board (NIC), 3270 board, memory board, etc. Each of the CMS products has a different method for handling the components of a PC. Be sure to describe your preferences to the Vendors so they can properly respond as to whether their system can meet your needs. The different components of a PC such as keyboard, monitor, mouse, and internal boards are normally considered attributes of the PC rather than as individual records with unique identifiers.

When you think about installing a modem, it requires connecting a phone cord into the back of the modem. As with the phone, the cord has a plug which fits into a jack or connector on the modem. Depending on the make and model of the modem, it may have a separate jack which can then be connected to a phone. Modems also have a connection to a PC or some other type of equipment. This connection normally uses a DB9 or DB25 type connector. Modems which are designed as a card and fit into a PC have a tab-like section which fits into a slot in the PC. A modem can be defined as a device with two RJ11 connectors, one RS-232 connector (DB9 or DB25), and possibly a tab, depending on whether it's an internal or external modem. You can be specific in your configuration detail or general. Examples of modem configurations are given in Table 21.11.

This example shows three different configurations which can be created for modems. The most general approach would be to treat all modems as having four connectors, two RJ11's, one DB25, and one Tab (even though the tab only applies to internal modems).

This brings us to a new feature in setting up configurations for equipment -- sides of the equipment. The two RJ11 jacks are on the back of the internal modem, while the tab is toward the front or bottom of the modem. Figure 21.3 demonstrates the concept of sides of a device, using front and back connections. Does it matter where the connectors and connections are made? The answer may depend on the CMS software which is chosen. Refer to Chapter 3 for consideration of the logical "Front" and "Back" of

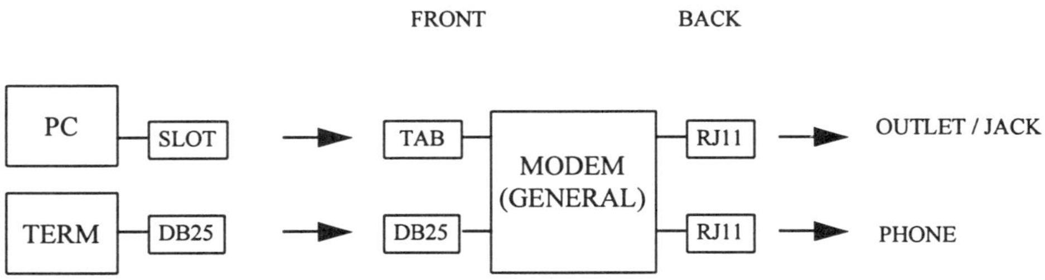

Figure 21.3: Configuration with connectors on front and back.

a device. Generally, you should assign connectors to front and back according to the flow of the circuit connections, rather than the physical front or back of the device itself.

The next step in developing your desktop equipment catalog is to come up with a naming scheme for each of the desktop items. Make a chart similar to the one in Table 21.12 and list each of the categories with the specific device types found in each of those categories. Beside each device type, note the naming scheme that will be used. In the final column, list the maximum number of characters in the device name.

The easiest approach for voice equipment is to use the Telephone Extension as the Name, and to be sure the Extension appears on the piece of equipment. This minimizes the need for developing new naming conventions. In some cases, modems or network equipment may take on a network address or special Circuit ID rather than a voice extension number.

The remaining types of desktop equipment will require naming conventions. There are several options on how this might be done:

- o Some organizations attach a tag number to all equipment when it is installed. This may be the easiest identifier to use as the equipment name, since it is most likely already a unique number which is located in a similar place on each item.
- o Another option is to assign a sequential number to each device as it is inventoried and identified. This not only provides a unique identity, but also keeps track of the number of devices inventoried and tracked in the CMS.
- o You can track each device with a prefix to describe the device type, followed by a sequential number. This approach might refer to PCs with a label such as PC1000, PC1001, etc., while Printers could be PR1000, PR1001, etc., and Terminals could be TM1000, TM1001, etc.
- o You can also use the Serial Number as the name provided the serial number is still attached to the equipment and is accessible. You will need to have a method for creating a fictitious serial number where one does not exist or is not accessible and label the device accordingly.

Some CMS products automatically assign a sequential number to each device as it is created in the database. Be sure to discuss this feature with the Vendor. If you select a CMS with this feature you may not be able to assign your own Device Names, and labeling must be done after each device is created in the CMS database.

EQUIPMENT TYPE	NAMING CONVENTION	FIELD LENGTH
PHONE	808-123-4456	12
FAX	808-123-4452	12
MODEM	808-123-4459	12
ANSWERING MACHINE	808-123-4457 OR	12
	ANS-123-4459	12
PC	TAG #, SERIAL NUMBER, PC1000, PC1001, ETC.	12
TERMINAL	TAG #, SERIAL NUMBER, TM1000, TM1001, ETC.	12
PRINTER	TAG #, SERIAL NUMBER, PR1000, PR1001, ETC.	12

Table 21.12: Sample naming conventions for desktop equipment.

Otherwise, names could be developed and used for labeling, and then entered into a secondary field on the device record for cross-reference purposes.

The final step in developing your desktop equipment catalog is to come up with a method for labeling each of the desktop items. If you are using existing tag numbers, phone numbers, or serial numbers, it may not be necessary to label the equipment. Labels should be placed where they are visible to the user. Refer to Chapter 3 for more specific requirements on device labeling.

Network Control Equipment

Having covered different methods for describing desktop equipment and how connections will be made, you need to consider network controlling equipment such as LAN hubs, 3270 controllers, network servers, bridges, routers, etc. Controlling equipment can be an important part of a CMS since a port on a control device is usually the ending or beginning point of a circuit. A CMS can assist in tracking utilization of ports on these devices, and the port ID identifies the type of circuit. Most controlling equipment is made up of a set of ports which are connected to horizontal cables, and some controlling equipment may contain cards which provide a number of ports. A CMS database can track additional information about the type of controlling device, manufacturer, model, serial number, date installed, port types, and availability.

Deciding to include control equipment will depend on whether the organization has control over the connections made to the equipment. Take the example of a PBX - if the ports are reconfigured or changed other than as a result of a specific work request, accurate tracking of port assignments may not be possible. Another situation might be where connections to an SNA Controller or LAN Hub are handled by a local departmental coordinator, and control over port assignments cannot be maintained by the CMS organization.

With some CMS products, a port on a controlling device is required in order to complete the circuit. It is important to have the Vendor indicate if their product requires specific components in order to complete a circuit. As noted earlier, if automatic circuit creation or path routing is a desired requirement of the CMS, be sure to have the vendor specify what components are required in order to generate a new circuit. Also, it is important to determine if port naming can be specified by the user rather than always following a numeric sequence beginning with 1. An example of this is a 3270 controller with ports 0 - 31, or ports that are referred to as an 8-character Terminal ID or log-on address.

As with the desktop equipment, start by making a list of all the types of equipment which might be found in each of the wiring closets, computer rooms, data centers, etc. Add as many items to the list as the team can think of. Looking at the list of types, make separate categories such as hubs, servers, controllers, routers, repeaters, muxes, etc. and then list specific equipment types

CONTROLLERS	HUBS	SERVERS	ROUTERS	BRIDGES
AT&T	IBM	DEC	CISCO	AT&T
IBM	CABLETRON	HP	MAC	IBM
	AT&T	SUN	KINETICS	
	BAYNTWKS	ZYPLEX	SHIVA	
	HP	3-COM		
8-PORT	MAU-8	FILE	ETHERNET	ETHERNET
16-PORT	HUB-4	DATABASE	TOKEN RING	TOKEN RING
32-PORT	HUB-12			

Table 21.13: General list of network equipment and categories.

CONTROLLERS	HUBS	SERVERS	ROUTERS	BRIDGES
3174-11R	IBM-8228	DEC-500	SHIVA-FP	IBM-L150
3274-31C	BAY-3030	COMPAQ-486	MAC-II	ATT-2000
3274-61C	BAY-3000	XYPLEX-1600	KINETICS-2C	
3708	ATT-STARLAN	HP-720	CISCO-6000	

Table 21.14: Detailed list of network equipment and categories.

which fall into each category. You may wish to differentiate similar equipment from different manufacturers. Your categories might look something like the sample in Table 21.13.

For network controlling equipment, it is more important to define equipment down to the make or model, since different models can take on a different number of ports or have different port configurations. Take the Bay Networks 3030 Hub as compared to the 3000 models - the 3030 only has 4 slots, while the 3000 has 12. This will make a difference how each of these items is defined in the CMS. An alternative is to treat the equipment purely in terms of ports or connections rather than as a specific model, such as setting up a 32-port Controller as the equipment type. This way all 32-port controllers would be set up the same, regardless of the model number. The same approach could be taken with network hubs. The types might be HUB-12 for a 12-slot hub and HUB-4 for a 4-slot hub regardless of the manufacturer and model.

Some systems require you to set up the configuration for each type of equipment. Then each time you add another piece of equipment of that type, the database automatically creates the designated number of ports. Other systems require the user to specify the number and type of ports each time a new device is added to the inventory. No matter which type of system is selected, it still requires defining the configuration for each equipment type prior to loading the database.

You must determine at what level of detail to define and name the types of control equipment. Discuss the different options of whether to make your catalog very specific, or

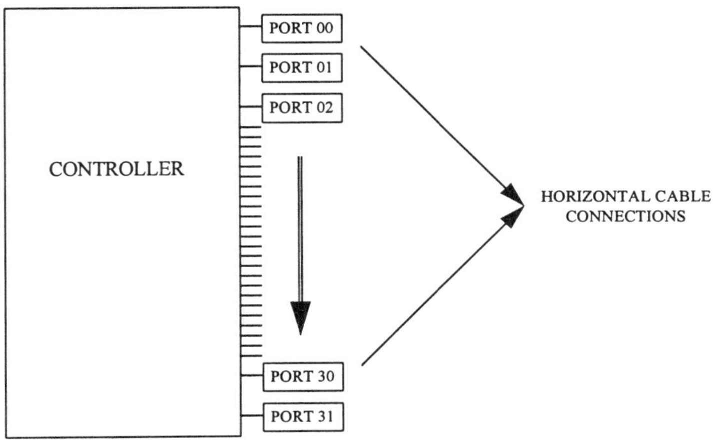

Figure 21.4: Ports located directly on the front of the controller.

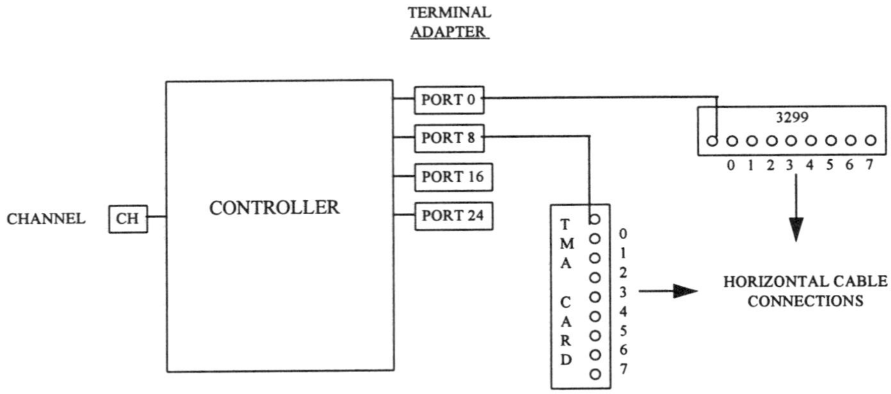

Figure 21.5: Controller ports connected to a 3299 device or TMA controller cards.

somewhat general. The example shown in Table 21.13 is a broad list of sub-categories which can be set up under each major category. You can be more specific, depending on the system selected and how many different models of network equipment exist in your organization. If network equipment is similar at all locations, tracking your equipment by manufacturer and model number may not be necessary. If there is another source of device information, such as a network management system, tracking the general configuration may be sufficient. Some CMS products have additional fields for tracking the manufacturer and model number of specific devices if desired.

If the network control equipment is being tracked for asset management, you may need to build your catalog down to the specific make and model of the equipment, as shown in Table 21.14. Other benefits of tracking specific models are similar to those given for desktop equipment in the previous section.

Once you have identified the types of network control equipment, the next step is to define the configuration of each type of network device. We will describe the process for defining your own configurations so that you can make a choice as to which method best meets your specific environment. You must look at the number of connections which can be attached to each of the devices, and what type of connector is used. The best way to convey what is meant by the configuration is to start with some specific examples of network devices and their connections. We will start with a 3270 controller.

When you think about installing a controller, there are usually two types of connections. There is a circuit of some type (fiber, copper) which connects the controller to the network or Front-End Processor (FEP). This circuit can be direct (channel-attached) or via modem. There are also a specific number of ports or connectors to which connections are made to the individual users on the network. These connections can be done a number of ways. If it is an older 3270-type controller, there are 8 to 32 BNC connectors directly on the front of the controller which are available for individual connections. The connections could be set up as shown in Figures 21.4, 21.5, and 21.6.

These examples show different configurations which can be created for controllers. There are likely others as well. The most general approach, shown in Figure 21.4, represents the ports located on the front of the controller with all connections made directly to the controller device. This reduces the number of records and items which must be tracked in the database. If you create cards or patch panels, these are usually treated as separate components with their own ID and record as shown in Figures 21.5 and 21.6. The cards or panels must then be connected to the controller through some type of cable connection such as a balun or patch cord to

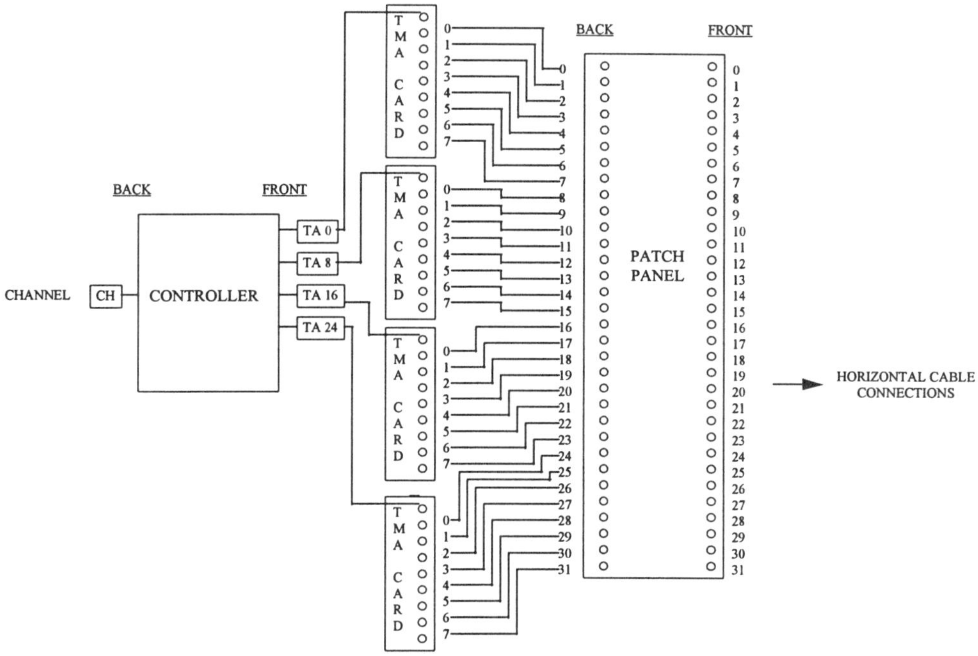

Figure 21.6: Controller ports connected to patch panel.

each port. If different configurations are used at different locations, it may be valuable to track the actual connections and configurations.

You must also look at the connection to the network. This is handled through a channel, modem, or network card which is connected to the controller. Since you have attempted to describe the ports as being connected to the front of the controller, you would want to show the logical continuation of those 32 circuits out the back of the controller to the host through the channel, token ring adapter or asynch emulation adapter, shown here as a Channel connector. This has now added another component to the configuration for your controller - a connector or set of connectors on the back of the device.

Turning our attention to LAN equipment, there are two basic styles of LAN hubs. The first is non-expandable, with a fixed number of connectors and ports. Once the device is configured, it will never change. The IBM 8228 token ring hub is a good example. These ports are connected to individual horizontal cables, and can also be connected to a server or another hub device. When you set up this type of device in the database, you need to take into consideration all these connections. You need to define the number of connections each device can have and what type of connectors are used for each.

There are actually 10 ports or connectors on the front of the IBM-8228 MAU. One port is an "Inbound" (Ring In) connector which connects to the Server or to another MAU. Another port is the "Outbound" (Ring Out) connector which can connect to another 8228 hub, with the remaining 8 ports used to attach individual users to the network. Each connector is usually referred to as an IBM Data Connector, which you could designate a DC type. The configuration could be defined as in Figure 21.7.

This again addresses "sides" of a device, and how a circuit continues through a device and onto another part of the network. This circuit begins at the Server which is connected to the "Ring In" (RI) connector on the front of MAU #1. The circuit continues out the "Ring Out" (RO) connector on the back of MAU #1 and connects to the "Ring In" (RI) connector on MAU #2. Port 5 of MAU #2 is then connected to a horizontal cable which is then connected to a PC. Note the use of the designations "RI" and "RO" for these connector types rather than DC in this example.

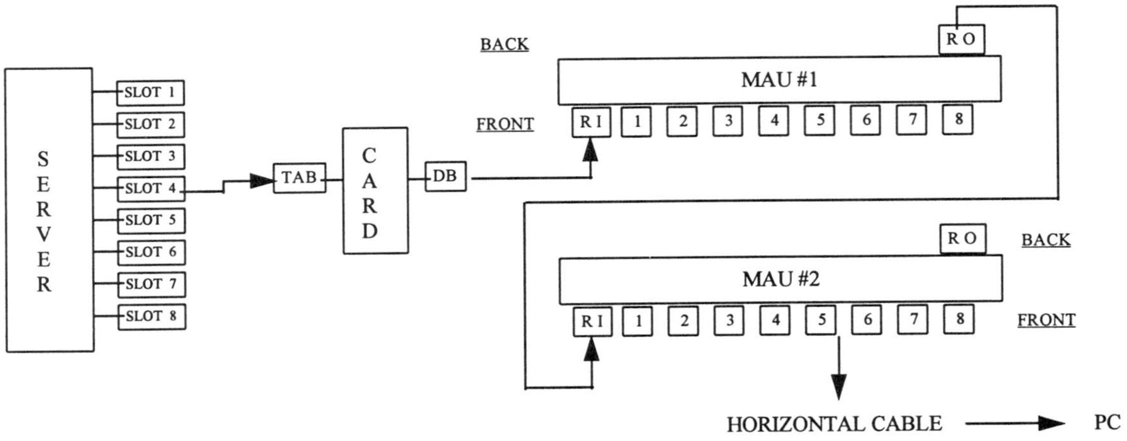

Figure 21.7: Connections between MAUs, PCs and Servers.

You will also note that the "Ring Out" (RO) connector is shown as located on the back of the MAU. While this is not the real case, it is configured this way so that the circuit can continue through the first MAU to the second MAU.

Note that some CMS products already contain a library of configurations for network equipment, such as a "Token Ring MAU", that may or may not be what you are looking for. If your desire is to track connections with actual connector types and names consistent with specific devices, discuss the use of pre-defined equipment configurations with the CMS Vendor.

Single-unit Ethernet hubs are set up very much the same way, in that they have a set of connectors (BNC type for 10Base2 and RJ45 for 10BaseT) which are for connecting users to the network. There are also connectors for carrying the circuit from the hub to the network, and for connecting several hubs together through an RJ45 or AUI connector.

The second style of LAN hub has an expandable chassis, to which cards or modules are inserted as shown in Figure 21.8. Since these cards can be added or changed, the database configuration of the device is not necessarily

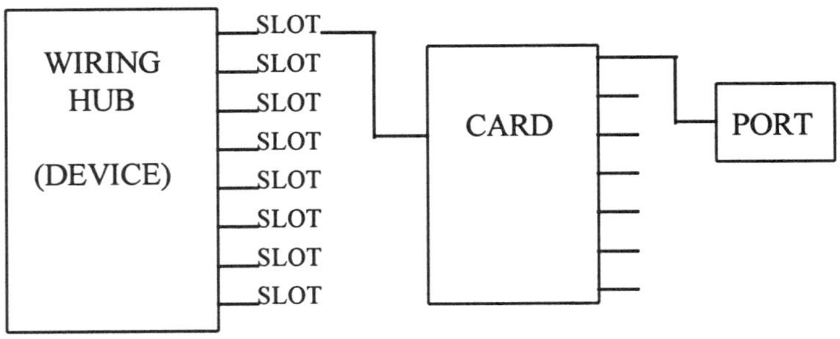

Figure 21.8: LAN hub with expandable chassis.

fixed. Such a device demands a little more consideration as to how it might be configured.

There are several different types of what are commonly referred to as wiring hubs. These hubs usually have a specific number of slots where different types of cards can be installed. Each of these cards can have a different number of connectors, and the connectors may consist of several different types. Let's take the example of a Bay Networks 3030 Hub, which has 4 slots. One of these slots is usually reserved for the Network Interface Card. Even at this level, there are several different types of Network Interface Cards, depending on how the hub is connected to the network. The other three slots are usually filled with User Interface Cards, each containing anywhere from 8 to 24 ports depending on the type of card used.

If you simply define the device to have ports according to the current population of cards, you may have a problem if cards are added or changed. In some CMS software, ports can be added to the device after it is in-use (connected). In other CMS software, the device configuration is fixed and cannot be changed. To add a card and more ports, the device must have all its connections broken and then the item must be removed from the database. A new device (with the original name) must be added with the new configuration, and all connections re-entered.

There are several options on how to represent this equipment in the database. The hub itself may have a very simple configuration, consisting of 4 Slots on the front of the hub. The cards would then take on a basic configuration whereby the back side has a Tab (in the sense of a Tab fits into a Slot). This allows the front of the card to have the actual ports to which individual users are connected through horizontal cables. If there are restrictions as to which types of cards can go in which slots, then you may have to differentiate these connections by creating more than one type of Slot or Tab.

The port types will differ from card to card, as will the connector type. If you define cards separately from the chassis, you may enter the cards as "children" of the hub or you may have to separately connect each card to the hub. Sometimes each device can automatically have a number of ports defined, in which case you can identify the card type or ID as part of the port number. In other words, you may get to define both the connector type and port number, or just a port name or number alone.

For an example, refer to Figure 21.9. This diagram shows the specific card

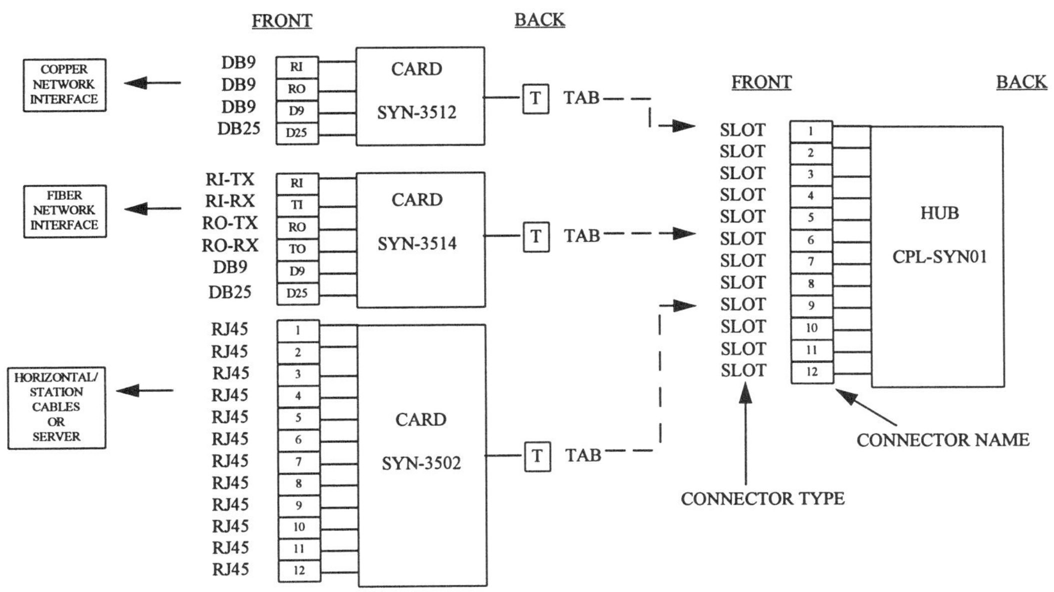

Figure 21.9: LAN hub with different cards and connector types.

configurations that are used with a Bay Networks 3000 hub. The 3502B is a User Interface Card with 12 ports (where each port is an RJ45 Connector), and the 3514F is a Fiber Interface Card with 4 ports (Each port is an ST Fiber Connector).

There are several ways to represent the cards and ports. The ports can be set up as individual entities, or as an attribute of the Card. The answer will depend on the CMS software which is selected, as well as the naming conventions for your ports. If the ports take on an ID which may be different than the Card Name, it may require the ports to take on individual identities of their own in order to indicate some reference to the actual service the user is connected to, such as the ring or server and segment. Another approach would be to indicate the Server or Ring/Segment in the Card name, so that the ports would not require individual IDs and could simply be numbered sequentially. Again, each port may require both a connector type and a name or sequence number.

A server is very similar to a PC, and often is a high-powered PC which supports files or databases which multiple users can access. They have multiple slots and connectors on the back which can be connected to hubs or other servers. The best approach to take is to look at the back of the server and determine the number of slots or connectors where other equipment can be connected, and define a configuration to match the number and type of connectors. A sample configuration which can be used for a server is shown in Figure 21.10.

Network bridges and routers will also vary in how they fit into the CMS. They may have multiple inputs and a single output or vice versa. Look at the number of slots or connectors on the bridge or router, and create a connector for each possible connection to the network. Again, certain CMS products have their own configurations and method for setting up specific types of equipment such as bridges and routers, so discuss this point with the CMS Vendor.

Tracking of bridges, routers, and their network connections may not be desired since changes to these devices are not made as often as voice and data services are changed at the user level. However, tracking of these devices may be necessary to represent how multiple circuits connect at a single point or to show how a circuit path may have multiple routes. Tracking of these items provides additional information when

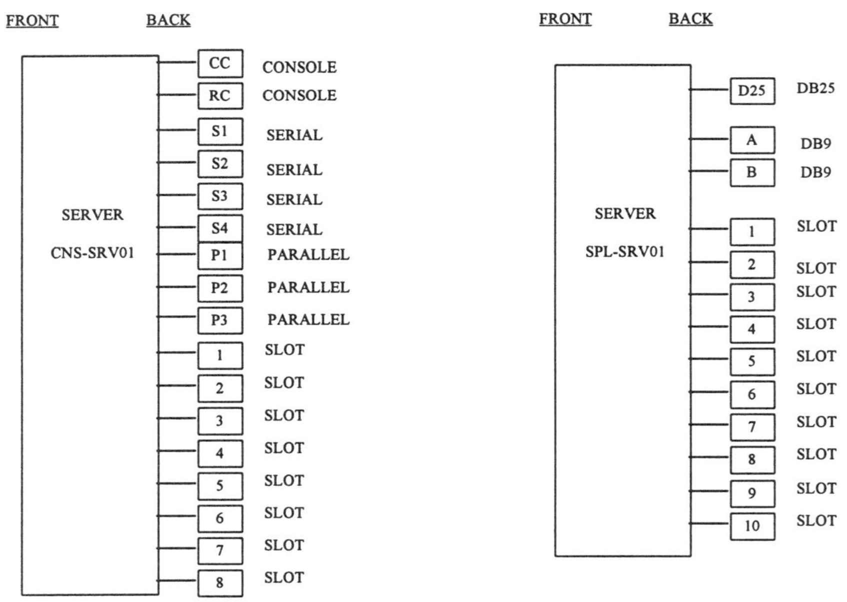

Figure 21.10: Sample configurations for network servers.

troubleshooting network problems, and is useful in a CMS interfacing with a network management system. These devices can provide information on traffic flows but can contribute to traffic problems.

The next step in developing your network equipment catalog is to come up with a naming scheme for each of the devices. Make a chart similar to Table 21.15 and list each of the equipment categories, with the specific device types found in each of those categories. Next to each of those device types, list the naming scheme which will be used. The alternatives of physical and logical names are discussed in Chapter 3. Points raised there will become more evident now as you look at each of your specific types of equipment, and determine if they have existing network names or addresses which may suffice for their name/ID in your CMS records.

Many times servers and controllers have a network address, and these may be the best names to use in the CMS as it provides a common reference point shared by both recordkeeping systems. As noted elsewhere, this does not provide for asset tracking of devices which are spare, under repair, or otherwise not currently active in the network. There may be ways in a CMS to give the devices physical names while using the logical network name for each port.

Sometimes it is easier to use a sequential numbering system with a prefix to describe the equipment type, as shown in Table 21.15. Hubs usually do not have names or addresses in the network management system, but need some name or ID if they are to be tracked in the CMS as part of the circuit information. There are several different combinations of physical location, equipment type, and sequential numbering which you can use. It is helpful to keep the names as simple as possible, but to include some sort of identifier which is meaningful to the technician when it is referenced as part of a circuit. Other options which might be used for naming of network control equipment are as follows:

- o Some organizations attach a tag number to all equipment when it is installed. This may be the easiest identifier to use as the equipment name, since it is most likely already a unique number, and is usually located where it is visible to the technician.
- o Another option is to assign a sequential number to each device as it is inventoried and identified. This not only provides a unique identity, but also keeps track of the number of devices inventoried and tracked in the CMS. A prefix for each device class could be set up to differentiate one equipment type from another. Some examples might include HUB0001, SRV0001, BRG0001, CTL0001, etc. This provides a quick way to identify the number of hubs, servers, bridges,

EQUIPMENT TYPES	NAMING CONVENTIONS	FIELD LENGTH
HUB-BAYNTWKS 3000	LOC+HUB-01	12
HUB-BAYNTWKS 3030	LOC+BAY-01	12
HUB-CABLETRON	LOC+CAB+01	12
HUB-IBM-8228	LOC+MAU+01	12
SERVERS	LOGICAL NETWORK ADDRESS, OR - LOC+SRV+01	12
CONTROLLERS	LOGICAL NETWORK ADDRESS, OR - LOC+CTL+01	12
BRIDGES	LOGICAL NETWORK ADDRESS, OR - LOC+BRG+01	12
ROUTERS	LOGICAL NETWORK ADDRESS, OR - LOC+RTR+01	12

Table 21.15: Sample naming conventions for network control equipment.

CAMPUS	RISER	LOCAL / TIE
200-PAIR	200-PAIR (DATA)	25-PAIR
400-PAIR	300-PAIR (RISER)	50-PAIR
600-PAIR		100-PAIR
1200-PAIR		
36-STRAND	12-STRAND (LAN)	

Table 21.16: Sample backbone cable types.

controllers, etc., which were inventoried and entered into the CMS.

- You can use the Serial Number as the name provided the serial number is not too long, is still attached to the equipment, and is accessible. You may need a method for creating a fictitious serial number where one does not exist. It may also require creating a new label with the Serial Number and placing it where it will be visible to the technician.

Some CMS products automatically assign a sequential number to each device as it is created in the database. Be sure to discuss this feature with the Vendor. If you select a CMS with this feature, it means you may not be able to assign your own Device Names and labeling must be done after each device is created in the CMS database. Otherwise, names can be developed and used for labeling, and then entered into a secondary field for the device record.

The final step in developing your network equipment catalog is to come up with a method for labeling each piece of controlling equipment. If you are using existing tag numbers, it may not be necessary to label the equipment. Place labels where they will be most visible to the user. Refer to Chapter 3 for more specific requirements for device labeling.

Cabling and Wiring

After you have covered the methods for describing desktop and network control equipment, it is time to define the cables which will connect these components together to create circuits. This includes both backbone cables (campus and riser) and horizontal cables. It will be easiest to handle these separately.

To start developing the cable catalog, make a list of all the types of backbone cables which might be found in each of the wiring closets, computer rooms, data centers, etc. Add as many cable types to the list as the team can think of. Take each of the these and make categories such as copper, fiber, or coax, or campus, riser, local or tie, etc. The best method for determining categories is to work with the common names and language which is used when technicians and other personnel refer to the cable plant. Create a chart listing a sample of each of the different types of cables, including the different pair counts that might exist for backbone cables. Your categories might look something like the sample given in Table 21.16 or may be entirely different. The cable types, pair counts, and local jargon will vary from organization to organization.

For cables, it will be very important to define your cable catalog down to the pair level with as much supporting information as possible regarding the type of cable, pair count, manufacturer, part number, installation date, etc. It helps to record the type of hardware connected at each end of the cable, as shown in Table 21.17. We will discuss the options for handling termination hardware later in this chapter.

This exercise will identify whether there are different types of cables with the same number of pairs, such as two different styles for the 200-pair cables. For example, one type of cable may be used for the campus cabling between buildings, which possibly could be a gel-filled cable, while another type with different characteristics may be

CABLE NAME	CABLE TYPE	PAIR COUNT	MFGR. SPEC #	TERM. HDWE.
CAMPUS A	200-PAIR	1-200	GC-3102	66 BLOCKS
CAMPUS A	400-PAIR	201-600	GC-3104	66 BLOCKS
CAMPUS A	200-PAIR	601-800	GC-3102	66 BLOCKS
CAMPUS A	400-PAIR	801-1200	GC-3104	66 BLOCKS
FIBER-1	36-STRAND	1-36	BEL/T225181	FIBER CABINET
HC-1N	300-PAIR	1-300	MOHAWK/5300	66 BLOCKS
LOCAL-1	25-PAIR	1-25	MOHAWK/5025	110 HARDWARE
LOCAL-2	50-PAIR	1-50	MOHAWK/5050	110-HARDWARE
TIE-1	100-PAIR	1-100	MOHAWK/5100	110-HARDWARE
DATA-1N	200-PAIR	1-200	MOHAWK/5200	110-HARDWARE
F-1N	12-STRAND	1-12	SIECOR/33212	FIBER CABINET

Table 21.17: Sample backbone cable types with pair counts and termination hardware.

used for the riser cabling. While you may decide not to track the specific cable type and manufacturer, it is helpful to know whether this information is available or currently tracked for future reference when troubleshooting or installing new cables. Using the same cable type, by part number, for new installations will help to maintain consistency in the design of your cable plant.

For each different type of cable you will track in the CMS, you will need to define what that cable and its connections will look like in the database - what type of connectors (pairs, pins, strands, modular, Amphenol, etc.), the number of connectors, the construction material, and the cable usage such as voice, data, video, etc. For each different cable type listed on the chart you just created, define its configuration for how it will look in the database. Some examples are provided in Table 21.18.

Note that there are three different types of 25-pair cables defined at the bottom of the table. For the first 25-pair cable, up to 25 individual connections can be made on a pair-by-pair basis, while the second 25-pair cable has a single 50-pin Amphenol connector on each end. The third cable has a different configuration on each end, and might be used for making multiple connections to a printer. It has a single Amphenol connector on

CABLE TYPE	CONFIGURATION NAME	CONNECTOR NUMBERS	CONNECTOR END 1	CONNECTOR END 2	USE
200-PAIR	1-200C	1-200	PR	PR	CAMPUS
200-PAIR	601-800C	601-800	PR	PR	CAMPUS
200-PAIR	1-200DR	1-200	PR	PR	DATA RISER
300-PAIR	1-300VR	1-300	PR	PR	DATA VOICE
400-PAIR	201-600C	201-600	PR	PR	CAMPUS
400-PAIR	801-1200C	801-1200	PR	PR	CAMPUS
36-STRAND	1-36ST	1-36	ST	ST	DATA/CAMPUS
12-STRAND	1-12ST	1-12	ST	ST	DATA/LAN
50-PAIR	1-50PR	1-50	PR	PR	DATA/TIE
25-PAIR	1-25PR	1-25	PR	PR	DATA/TIE
25-PR/AMPH	1-25AM	1	AMPHENOL	AMPHENOL	DATA/TIE
25-PR/AM/6	1-25AM/6RJ	1 / 6	AMPHENOL	RJ45	PRINTSHARE

Table 21.18: Sample cable types and configurations.

one end and an adapter on the other end which has 6 individual RJ45 or Modular connectors. This demonstrates that there are different ways to represent cables in your recordkeeping system. You can take a general approach and set up all 25-pair cables with 25 available connection points (pairs) on each end. However this may require 25 connection records if the cable is connected to a device with an Amphenol connector, versus a single connection record of an Amphenol connector to another Amphenol connector.

Each CMS has a different approach to defining and setting up cable types. The better you have your cable catalog defined, the easier it will be to evaluate and select the CMS which best meets your needs. The more information you document now about your cable plant, the easier it will be to get your CMS records up and running.

After determining all cable types and defining their connection configurations, the next step is to develop a naming convention for each cable type. The simplest approach is to use existing naming conventions wherever possible. However, existing names may be inconsistent or may be duplicated from building to building.

Refer to the table of backbone cables

CABLE NAME	CABLE TYPE	TERM. HDWE.	CMS NAME
CABLE A	1-200PR	66 BLOCKS	CA-BLD-1
CABLE A	201-600PR	66 BLOCKS	CA-BLD-2
FIBER 1	1-36ST	FIBER CABINET	CA-FIB-1
FIBER 2	1-36ST	FIBER CABINET	CA-FIB-2
BUILDING 1			
HC-1N	1-200PR	110 BLOCKS	BLD1/HC-1N
HC-1S	201-400PR	110 BLOCKS	BLD1/HC-1S
HC-2N	1-200PR	110 BLOCKS	BLD1/HC-2N
HC-2S	201-400PR	110 BLOCKS	BLD1/HC-2S
HC-3N	1-200PR	110 BLOCKS	BLD1/HC-3N
HC-3S	201-400PR	110 BLOCKS	BLD1/HC-3S
FIBER 1N	1-12ST	FIBER CABINET	BLD1/F-1N
FIBER 1S	1-12ST	FIBER CABINET	BLD1/F-1S
FIBER 2N	1-12ST	FIBER CABINET	BLD1/F-2N
FIBER 2S	1-12ST	FIBER CABINET	BLD1/F-2S
FIBER 3N	1-12ST	FIBER CABINET	BLD1/F-3N
FIBER 3S	1-12ST	FIBER CABINET	BLD1/F-3S
BUILDING 2			
HC-1N	1-200PR	110 BLOCKS	BLD2/HC-1N
HC-1S	201-400PR	110 BLOCKS	BLD2/HC-1S
HC-2N	1-200PR	110 BLOCKS	BLD2/HC-2N
HC-2S	201-400PR	110 BLOCKS	BLD2/HC-2S
HC-3N	1-200PR	110 BLOCKS	BLD2/HC-3N
HC-3S	201-400PR	110 BLOCKS	BLD2/HC-3S
FIBER 1N	1-12ST	FIBER CABINET	BLD2/F-1N
FIBER 1S	1-12ST	FIBER CABINET	BLD2/F-1S
FIBER 2N	1-12ST	FIBER CABINET	BLD2/F-2N
FIBER 2S	1-12ST	FIBER CABINET	BLD2/F-2S
FIBER 3N	1-12ST	FIBER CABINET	BLD2/F-3N
FIBER 3S	1-12ST	FIBER CABINET	BLD2/F-3S

Table 21.19: Sample backbone cables with type and CMS naming.

developed in Chapter 13 (Table 13.6) which was used in determining the project scope. Add another column to the chart for CMS Cable Names. As you can see in Table 21.19, there may be areas where cable names are duplicated from building to building. If you are using paper records this will not be a problem, but electronic records cannot have two different cables with the exact name. There are many options for how you name your backbone cables, such as using a combination of the cable type and destination, the current name plus a cable type, the cable count and originating or terminating location, etc.

As with the equipment naming, some commercially available CMS products also utilize an automatic cable naming convention where the cable type is combined with a sequential number to create the cable name. Be sure to discuss the assigning of cable names with the Vendor. If the system creates or assigns the name, it often requires relabeling the cable after it has been created in the system, or creating some way to reference the new cable name with the current cable labeling.

Table 21.19 shows how many different types of cables might exist, and how often cable names might be duplicated as you go from building to building. In this example, Cable A is used to feed Buildings 1 and 2 but with different pair counts. While this does not create a problem when working with paper records, an electronic database would not be able to track Cable A as two different cables. The name must be changed slightly for each location as indicated in the "CMS Name" column by adding the destination building to the original cable name. Another example shown in this table is the use of the cable names HC-1N through HC-3N in each building. Again, an electronic database would need some way to distinguish HC-1N in Building 1 from HC-1N in Building 2. This is where some adaptation is required in the current naming conventions to work effectively with a new CMS database. These examples simply add a prefix or suffix with the Building Name to accomplish the creation of unique cable names.

The final step in completing your catalog of major backbone and riser cables is to come up with a method for labeling each cable. Be sure each cable is labeled on both ends with the exact Cable Name, and the Pair Counts. Sometimes this requires labeling the termination hardware, or the backboard where the cables are terminated. It is also helpful to include a label of where the other end of the cable is terminated if it is not obvious from the cable name. Refer to Chapters 3 and 4 for more specific requirements for cable labeling.

Having covered backbone cables, you can now evaluate horizontal wiring in the same fashion. Make a chart similar to Table 21.20 listing each of the different types of horizontal cabling which are utilized in your facilities, such as 4-pair UTP (Unshielded Twisted Pair), IBM Type 1, RG-62 coax, etc. You may want to refer to the list of cables which were developed in Table 13.8. Most automated CMS products have some flexibility in how you define your horizontal cables, which will impact the simplicity or complexity of making connections and tracing circuits during daily use and maintenance of the CMS records.

Once you have listed the cables, you will need to define what these cables will look like in the database, and how they will be connected to

CABLE TYPE	MFGR'S SPEC./ PART NO.	NUMBER OF PAIRS	OFFICE CONNECTOR	WIRING CLOSET CONNECTOR	NAME OR LABELING
COAX	BELDEN E73-3104	1 CONDUCTOR	BNC	BNC	OFFICE NO.
CAT. 5	BELDEN 1700A	4 PAIRS	RJ45	110 HARDWARE	101D, 102D
CAT. 4	AT&T-3162	4 PAIRS	RJ45	110 HARDWARE	101D, 102D
CAT. 3	BELDEN 1229A-8	4 PAIRS	RJ11 (2)	66 BLOCKS	101V, 102V
CAT. 2	GC-153064	2 PAIRS	RJ45	66 BLOCKS	101V, 102V
TYPE 1	MONTROSE 4716743	2 PAIRS	IBM DC	IBM DC PANEL	101T, 102T

Table 21.20: Sample types of horizontal cables.

equipment or other cables. Coax cables and IBM Type 1 cables are relatively simple in that they normally have identical connectors at each end and have only one connection per cable end. Four-pair cables can be terminated in a number of different ways, depending on the preference of the installer or organization. Some examples of how a 4-pair horizontal cable might be terminated are listed below.

- All 4 pairs terminated in a single RJ45 outlet.
- The first 3 pairs terminated in one RJ45 or RJ11 outlet, with the 4th pair terminated in a second RJ11 outlet.
- The first 2 pairs terminated in one RJ45 or RJ11 outlet, with the 3rd and 4th pairs terminated in a second RJ45 or RJ11 outlet.
- The 1st and 4th pairs terminated in one RJ45 or RJ11 outlet, with the 2nd and 3rd pairs terminated in a second RJ45 or RJ11 outlet.

In addition to the different combinations of pair terminations between outlets, there are also different wire specifications, such as Category 3, 4, and 5. The number of possible connections per outlet can also vary from organization to organization. While the preference would be to have one connection per outlet, many organizations utilize splitters or adapters in order to split the pairs from one outlet into separate connections, so you might have a phone and a modem both connected to the same outlet. This is often done where a minimum amount of cable is feeding each workstation, or where the cost to install additional cabling did not justify adding a second cable. As you attempt to define these different wiring configurations in the CMS, you need to determine how to describe these cables and the connections to them.

The most general option is to define horizontal cables on a pair basis, where all connections will be made to individual pairs. While this option does provide a true-life picture of the cable pair usage, it may require additional steps and a higher level of knowledge by the average user. We will present some different scenarios to help demonstrate these points. Tracking horizontal cable connections on a pair basis means the administrator or technician who is making connections in the database must know if the connection uses Pair 1, 2, 3, or 4, or a combination of the pairs based on the type of circuit. While this may not sound like a big deal, it will impact the number of connections which must be made for each circuit created in the database. The number of possible connection options for horizontal cable pairs shown in Table 21.21 may help to demonstrate the level of knowledge required by a CMS administrator when creating circuit connections where horizontal cables are tracked on a pair basis.

CIRCUIT TYPE	PAIRS USED (568B CONFIG.)
ANALOG PHONE	PAIR 1
SNA 3270	PAIR 2
MODEM LINE	PAIR 4
TOKEN RING	PAIRS 1 & 3
DIGITAL PHONE	PAIRS 2 & 3
TP-PMD (FDDI)	PAIRS 2 & 4

Table 21.21: Possible horizontal cable connections.

This demonstrates that the user must know which pairs will be used for the connection. Depending on the CMS selected, this could require multiple connections records, one for each cable pair used. If a standard procedure is *always* used for making connections to specific cable pairs for each circuit type, this may not present a problem. The key is the word “always”, since any time a technician makes a connection which does not follow the procedure the database will not correctly represent that circuit.

Sometimes tracking termination hardware such as 66/110 blocks and wall outlets as separate entities is required in the database. Tracking the individual pairs of horizontal cable will require four individual connections (one for each pair) at each end of the cable to the termination hardware. This would represent a total of 8 connection records for each 4-pair cable.

This also requires setting up the configuration of the termination hardware for the outlet or jack so that assignments or connections can be made at the pair level. We will cover this

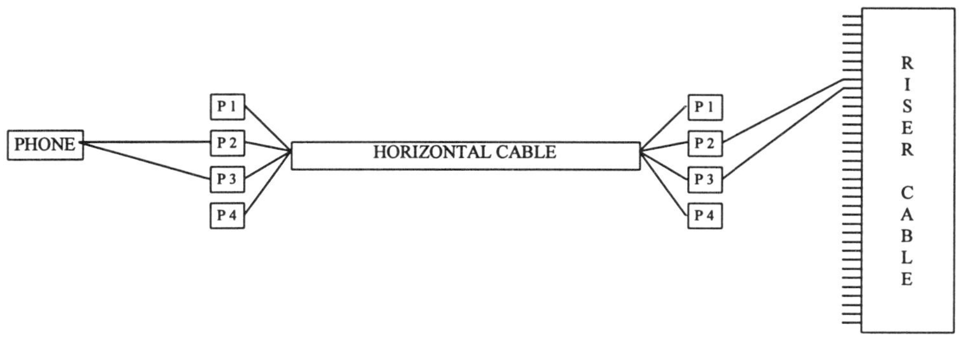

Figure 21.11: Horizontal cable connections tracked on a pair-by-pair basis.

in more detail in the following section on tracking of termination hardware. If you decide to track the horizontal cables down to the pair level, it will require making multiple connections when more than one cable pair is used in a single circuit. Compare the two diagrams shown in Figures 21.11 and 21.12 to better understand this point.

As you can see in Figure 21.11, if you connect a phone to the cable, it is easy to identify which pairs are carrying the circuit, but it requires making multiple connection in the CMS to create this circuit. However, in Figure 21.12, the cable is treated as having a single connector on each end. This option does not provide information on the actual horizontal cable pairs in use, but requires fewer connection steps. It does provide a simplified approach to making and tracking circuit connections where a minimal amount of knowledge is required by the average user.

Some other options for defining horizontal cables are shown in Figures 21.13 and 21.14. Both of these examples offer a simplified approach to tracking multiple connections to horizontal cables through the use of defining a set of connector types representing either the type of connection, or the number of connections. Figure 21.13 uses an approach where the four pairs of the horizontal cable are divided into two connection points, a 3-pair connector for multiple-pair circuits and a 1-pair connector for single-pair circuits. Figure 21.14 offers a similar approach where the 4 pairs of the horizontal cable are separated into two similar connectors, Connector 1 and Connector 2. The first circuit connected to the cable would be on Connector 1, while the second connection would utilize Connector 2.

When tracking horizontal cables on a connector basis, multiple connections can be made

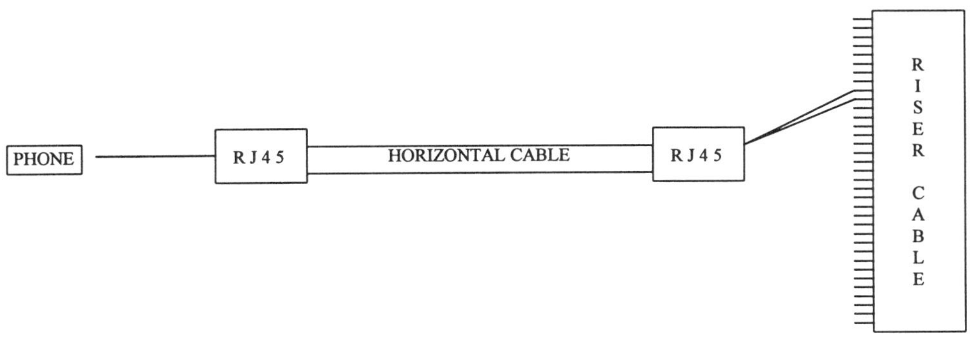

Figure 21.12: Horizontal cable connections tracked on a single connector basis.

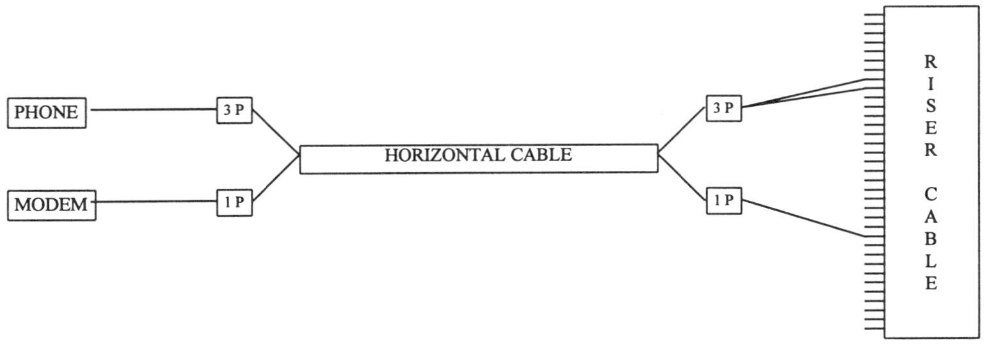

Figure 21.13: Horizontal cable connections tracked on a 3-pair and 1-pair connector basis.

to the cable without having to specify the individual pairs utilized. These different approaches demonstrate how connections can be made to the individual connectors with only a single connection record in the database, and avoids the requirement for the administrator to know which cable pair(s) are required for the individual circuit types. It makes the pre-assignment of horizontal cables easier, but puts the burden on the technician to know how each connection should be handled.

Each CMS handles connections in a slightly different manner. Some systems do not permit multiple connections to a single connector, in which case the options shown in Figures 21.12, 21.13 and 21.14 would not allow two individual riser cable pairs to be connected to the single RJ45 connector or the 3P, 1P, C1, or C2 connectors. Be sure to discuss this point with potential Vendors so they can indicate how the CMS achieves your desired approach.

After determining how to configure and track the connections to your horizontal cables, the next step is to develop a naming convention for each cable and configuration type. The ideal approach is to use existing naming conventions (usually based on outlet number) wherever possible, and adding a building/floor or building/wing prefix designation to make each cable unique.

If you are working with a new installation, where no naming or numbering convention exists, some possible numbering schemes are provided here. The simplest method is to begin with the office at the southeast most corner of the building and work your way

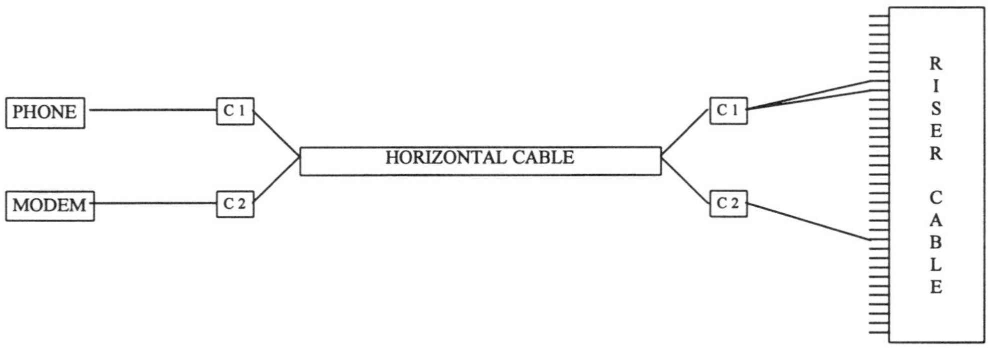

Figure 21.14: Horizontal cable connections tracked on a two-connector basis.

clockwise around each floor. Begin with a number of 101 and assign each horizontal cable an individual number. The same numbering scheme can be used for each floor, adding a building/floor prefix to the cable name to create unique IDs on each floor. Some examples would be REN1E101, FAC01101, and ADM01101.

Another approach for buildings with 800 cables or less per floor might be to use a number series starting with 1001 for the first floor, 2001 for the second, 3001 for the third, and so forth. The reason for an estimated 800 cables or less is that it allows for a growth of 25% to accommodate additional cables which might be added at a later date. Some examples would be REN1001, REN2001, and REN3001.

If there are two or more wiring closets per floor, either of the above numbering schemes could be used with a prefix or suffix of A, B, C, etc. to indicate wiring closet A, B, or C. If wiring closets are located in specific geographical locations in a building, then N, E, S, or W could be used to indicate the closet in the North, East, South or West section of the building. Some examples would be REN1A101 and REN1B101, or REN1E101, REN1W101, etc.

All of the above approaches consider the use of a sequential numbering plan so that as additional cables are added, they continue to use the next available number and position on the termination block. If cables 101 through 178 were initially installed, the next new cable automatically would become 179 and would be located next to 178 on the termination block in the wiring closet, making it easier to locate for future changes.

The above methods are preferred over using room numbers to identify cables. If during the initial installation, four cables are run to Room 101 and are labeled 101A, 101B, 101C, 101D, the next cables on the block would be 102A, 102B, 102C, 102D. In order to add any additional cables to room 101 (101E, 101F, 101G, or 101H), these new cables might be at the end of the wiring block where cables 178A, 178B, 178C, 178D might be located. As more and more cables are added to this individual wiring closet, it may be difficult for the technician to locate a cable, especially if cables are added as they are randomly needed. The wiring block might have cables in any order, such as 101E,F,G,H, 114E,F,G,H, 127E,F,G,H, 103E,F,G,H, etc. Another problem with using room numbers is when a renovation makes three rooms where two rooms used to be. Do you now call the room in between these two rooms 101A? What do you call the horizontal cable, 101AA? If you currently have horizontal cables labeled by room number, you most likely have come across this type of problem.

Some CMS products create their own cable names through the use of combining a Cable Type with a sequential number. Be sure to discuss any current naming schemes for your horizontal cables with the Vendor to make sure you do not have to use their cable names, which may mean relabeling your existing cables or outlets.

The final step in completing your catalog of horizontal cables is to come up with a method for labeling each cable. Be sure the cable is labeled on both ends with the corresponding CMS cable name. This requires labeling the termination hardware or the backboard where the cables are terminated, and on the incoming cable itself. When working with a new installation, it is a good idea to label the inside of the outlet where the horizontal cable is terminated. This will be beneficial when building renovations require faceplates to be removed.

In buildings where a building/floor prefix is used for the cables to make them unique in the CMS database, it is not necessary to label each termination position or outlet faceplate with the prefix. However, it is suggested the prefix should be indicated somewhere on the termination hardware in the wiring closet just to verify which cables are at that location. Refer to Chapters 3 and 4 for more specific requirements for cable labeling.

Termination Hardware

There are many different options for tracking termination hardware in your CMS records. While the TIA/EIA 606 Standard requires tracking the termination position of a cable, it does not specifically indicate the way in which termination hardware must be addressed. To determine what method is best for your organization, consider the following factors.

You must first consider the method in which termination hardware is currently labeled and how it is referred to. Termination blocks may be referenced by block, row and position, or by cable and pair, or by network port name. If technicians and network support groups are used to referring to termination hardware in terms of cable and pair or port name, the best approach may be to simply treat the termination hardware as an attribute of the cables or ports. On the other hand, if current practice is to refer to block, row and position, it may be better to track the termination hardware as a separate entity to which cable pairs and ports are connected.

Another factor is whether termination hardware and office outlets are of the same type for all locations and for all service types. If the same type of hardware is used in all offices and wiring closets, the simplest approach is to treat the termination hardware as an attribute of the cable or port name, rather than creating a separate entity. However, if the termination hardware used is different from location to location, it may be best to track the termination hardware as a separate entity in order to identify the specific type of hardware and connectors required for each cable or port connection. The diagrams shown in Figures 21.15 and 21.16 illustrate different approaches for configuring terminating hardware. Tracking the termination hardware as separate entities significantly increases the number of records and connections to be created.

In a general sense, if termination devices (outlet and wiring block) are tracked as separate records you will have three records for every horizontal cable you create -- the outlet, the block, and the cable itself. This requires two sets of connection records. Each of the four pairs of the horizontal cable must be connected to the outlet in the office location and to the Termination Block in the wiring closet. Each CMS will vary in how it creates these connections, with some systems requiring that termination devices be specified at the time the cable is created while other systems may require the connections to be done in a separate step, on a different screen in the database. In any case, three individual records and two connections are required.

If you plan to track horizontal cables down to the pair level, it would only make sense to define the outlet in a similar four-connection fashion otherwise the pair-by-pair usage information is lost. The left side of the diagrams in Figures 21.15 and 21.16 show how this can be done. If you take a simple approach and incorporate the termination hardware into the connectors on the ends of the cable, as shown in Figure 21.16, you reduce the number of records for each horizontal cable from three to one, and

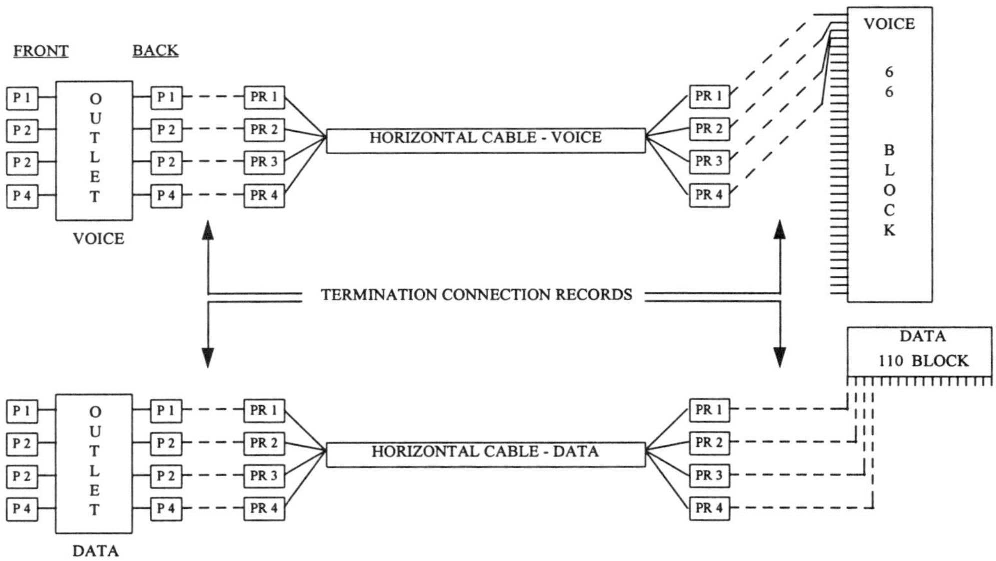

Figure 21.15: Office outlets tracked as separate entities.

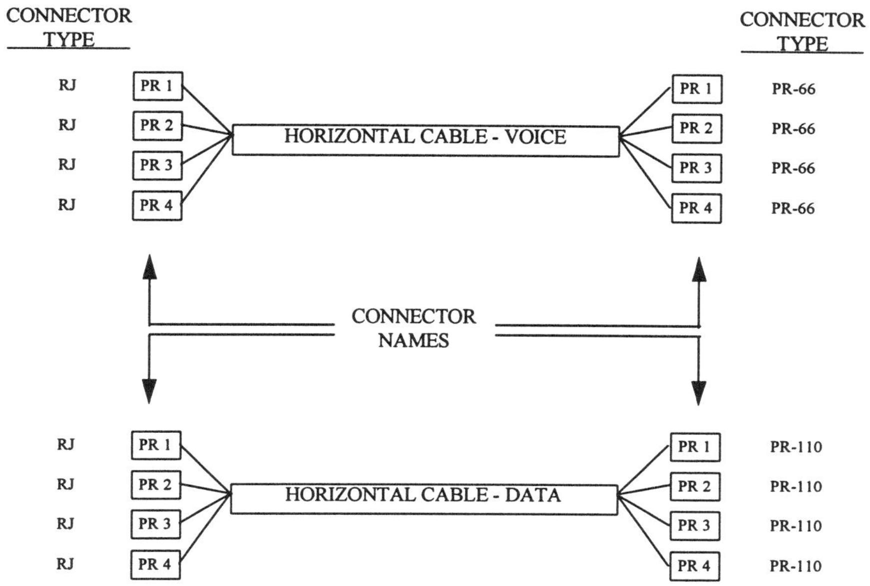

Figure 21.16: Office outlets tracked as attributes of the horizontal cable.

their are no permanent connections which must be made. This keeps the number of records in the database to a minimum.

Looking more closely at specific types of termination hardware, you must first consider whether outlets or jacks should be treated as an element of the horizontal cabling or as a separate device. This is a topic which should be given considerable discussion, since it will greatly impact the set-up, loading, and ongoing maintenance of the CMS records. All station cabling needs to have some type of connection point at the office end, defined on a pair basis or as a single entity that can be part of the cable or a separate device. The following examples provide different methods for handling outlets or jacks.

The first example shown in Figure 21.15 uses an office with two outlets, one for Voice and one for Data. If the outlets are tracked as individual components (separate from the horizontal cables), there are multiple steps for creating the horizontal cables and outlets in the database. Each outlet must first be created in each office, then each horizontal cable must be entered. Finally, each cable must be connected to a termination position in the wiring closet and to an outlet in the office.

The second example shown in Figure 21.16 also uses an office with two outlets, one for Voice and one for Data. If the outlets are tracked as attributes of the horizontal cables, only a single step is required for creating the station cables and outlets in the database. The end connection points (termination and outlet) are defined as each horizontal cable is entered.

The first example (Figure 21.15) results in four records and four connections in the database (depending the configuration of the horizontal cable) before any circuits are created for this cable. This assumes the termination hardware has already been defined. The second example (Figure 21.16) results in only two records created, with the connections included as part of each cable record. When the horizontal cable is configured in the database, it would show an outlet (RJ45) or multiple connectors at the office end, which would then be connected to phones, computers, printers, etc.

The capability to automatically generate a circuit based on beginning and ending points may require that the user specify a separate outlet or jack as the beginning or ending device in order to determine the path and connection points for the circuit. It is important to determine if automatic

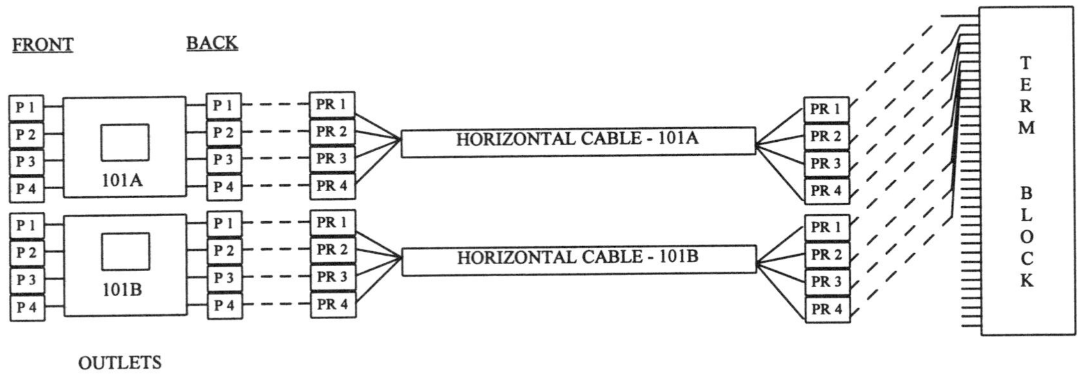

Figure 21.17: Individual office outlets set up for each connector.

circuit generation is a high priority in your list of requirements, and to have the Vendor indicate what circuit elements are required by their CMS in order to create the circuit or utilize an automatic circuit generation feature.

You should take into consideration whether office outlets have the same name or label as horizontal cables, and whether technicians typically refer to the horizontal cable or to the outlet (jack). Also consider whether the same type of outlets are used in every office or whether the type of outlet and number of connectors (jacks) per outlet varies from office to office. Review any existing records to determine whether listed connections refer to an outlet connector or specific horizontal cable pair.

An outlet can be set up as a single device for each RJ45, RJ11, BNC-type connector, or as multiple ports on a single outlet, such as might be the case with a "Quad" type outlet. If there is a need to track outlet connections on a pair or pin basis, the best approach will be to set up each outlet as an individual record with the pairs/pins specified as the connection points on the outlet, as shown in Figure 21.17. This would simplify the configuration for each outlet.

If it is not necessary to track horizontal cables down to the pair level, the simplest approach to defining multi-jack outlets such as a quad or duplex is to treat the outlet as a single device, with each jack or port on the outlet treated as an individual connector as shown in Figure 21.18.

A method of naming each outlet will be needed. The simplest approach is to use existing naming conventions wherever possible, adding a building/floor or building/wing prefix designation to make each outlet identifier unique. Often, the outlet can take on the same name as the horizontal cable since cables and termination hardware are in different tables in most CMS products.

If you are working with a new installation, where no naming or numbering convention exists, some possible numbering schemes are provided here. The simplest method is to begin with the office at the southeast most corner of the building and work your way clockwise around each floor. Begin with a number of 101 and assign each outlet an individual number. The same numbering scheme can be used for each floor by adding a building/floor prefix to the outlet name or number to create unique IDs on each floor. Some examples would be REN1E101, FAC01101, and ADM01101.

Another approach for buildings with 800 outlets or less per floor might be to use a number series starting with 1001 for the first floor, 2001 for the second, 3001 for the third, Etc. The reason for an estimated 800 outlets or less is that it allows for a growth of 25% to accommodate additional cables/outlets which might be added at a later date. Some examples would be REN1001, REN2001, REN3001, etc.

If there are two or more wiring closets per floor, either of the above numbering schemes could be used with a prefix or suffix of A, B, C, etc. to indicate wiring closet A, B, or C. If wiring

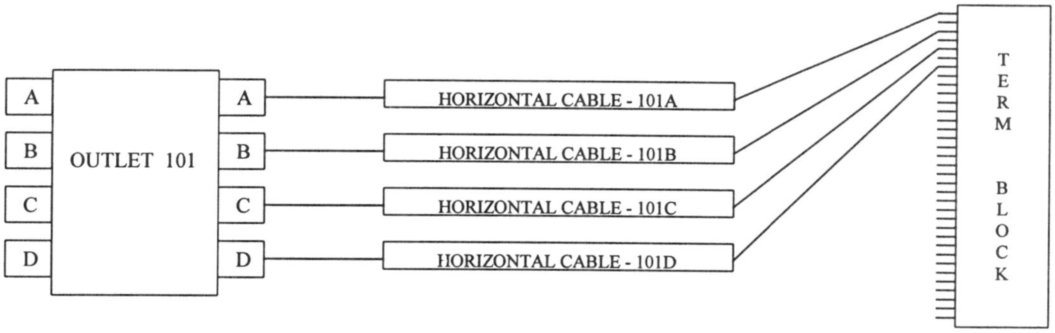

Figure 21.18: Office outlets set up with multiple connectors.

closets are located in specific geographical locations in a building, N, E, S, or W could be used to indicate the closet in the North, East, South or West section of the building. Some examples would be REN1A101 and REN1B101, or REN1E101, REN1W101, etc.

All of the these approaches consider the use of a sequential numbering plan so that as additional outlets are added they are given the next number in sequence. If outlets 101 through 178 were initially installed, the next new outlet would automatically become 179 and would be located next to 178 on the termination block, making it easier to locate for future changes.

These methods are preferred over using room numbers to identify outlets. If during the initial installation, four outlets are installed in Room 101 and are labeled 101A, 101B, 101C, 101D, the next outlets on the wiring closet termination block would be 102A, 102B, 102C, 102D. In order to add any additional outlets to room 101 (101E, 101F, 101G, or 101H), these new outlets might be terminated at the end of the wiring block where outlets 178A, 178B, 178C, 178D might be located. As more and more outlets are added, it may be difficult for the technician to locate the horizontal cable for an outlet, especially if outlets are added as they are randomly needed. The wiring block might have horizontal cables in any order, such as 101E,F,G,H, 114E,F,G,H, 127E,F,G,H, 103E,F,G,H, etc. Another problem with using room numbers is when a renovation makes three rooms where two rooms used to be. Do you now call the room in between these two rooms 101A? What do you call the outlet, 101AA? If you currently have outlets labeled by room number, you most likely have come across this type of problem.

Some systems create their own outlet names through the use of combining the Outlet Type with a sequential number. Be sure to discuss any current naming schemes for your outlets or jacks with the Vendor to make sure you do not have to use their outlet names, which may mean relabeling your existing outlets to correspond with the CMS-assigned names.

The final step in completing your catalog of outlets or jacks is to develop a method for labeling each outlet. Be sure each outlet is labeled on the cover. It is a good idea to label inside the outlet or label the horizontal cable itself in case building renovations require that faceplates or covers be removed. It is likely that the horizontal cable will have the same name as the outlet, which of course should be labeled on the wiring block or patch panel in the wiring closet. In offices where a building/floor prefix is used to make outlet numbers unique in the CMS database, it is not necessary to label each cover or faceplate with the prefix. Refer to Chapters 3 and 4 for more specific requirements for cable labeling.

Termination blocks are items such as 66 blocks, 110 blocks, Krone blocks, BIX blocks, etc. on which cables and ports may be terminated in wiring closets. These can be tracked as separate items or as a part of the cable or port record. Tracking of termination blocks as a independent component in the circuit can greatly impact the set-up, loading and ongoing maintenance of the CMS records. The following will help to provide some perspective on different methods for handling termination blocks.

Tracking of termination blocks as separate entities on a block, row and pin/pair position basis greatly increases the number of records and connections in the database. To illustrate the different steps involved, consider the two examples described below for creating a 300-pair backbone cable.

In the first example shown in Figure 21.19, the 66 blocks on each end of the cable are tracked as separate components and there are multiple steps for creating the cable in the database. First, a set of wiring blocks must be created at each end of the cable (the Main and Horizontal Cross-connects). Then you must create the cable itself. Finally, you must connect the 300 pairs at each end of the cable to the series of 66 blocks at each end.

In the second example shown in Figure 21.20, the 66 block terminations on each end of the cable are not tracked as individual records but as attributes of each cable pair. This reduces the number of steps for creating the cable in the database.

The first example (Figure 21.19) results in thirteen item records and 600 connections in the database before any circuits are created for this cable. With some CMS software, this makes it difficult to determine spare or available capacity for the cable since all pairs are terminated on both ends. It also requires assigning connections and circuits based on the block ID and pin/pair position rather than the cable name and pair or port name. This means the technicians must think and work in terms of block, row and pin/pair for all connections.

It becomes very critical that block, row and pin/pair are clearly labeled since the reference for cross-connects in the CMS records will be on that basis, and not based on the cable and pair or port name which is represented by the block positions. The blocks should also be labeled with pin counts which are easy to identify so technicians are not wasting time counting the pairs or pins.

The second example (Figure 21.20) results in one item record with 300 available connection points. Connections are created as circuits are entered in the database. This method provides an easy way to determine spare or available capacity on the cable since the number of actual connections identifies the number of cable pairs in use. It requires that all cable names and pair counts be clearly labeled for easy identification by the technician. Treating the termination hardware as an attribute of the 300-pair backbone cable, and labeling the blocks and rows to correspond with the cable name and pair counts, greatly reduces the number of records and connections in the database

Some CMS products have the capability

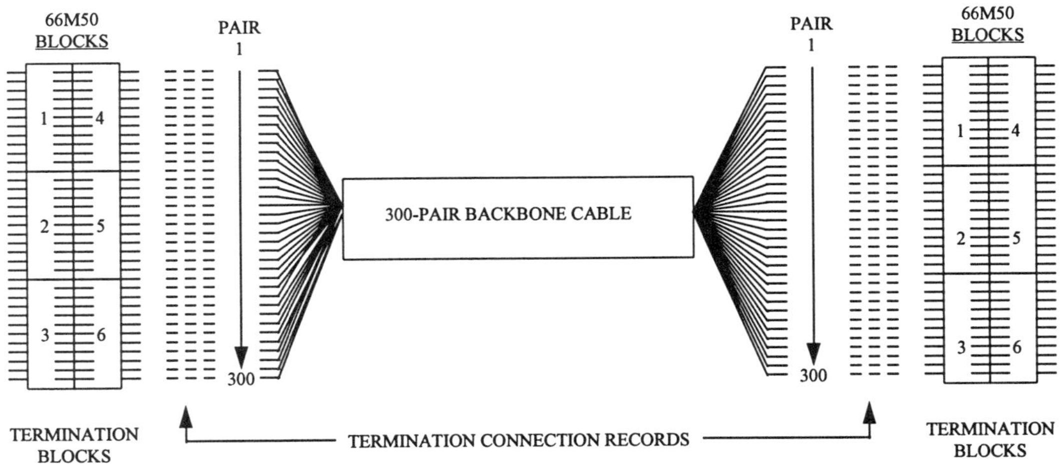

Figure 21.19: Termination blocks tracked as separate entities.

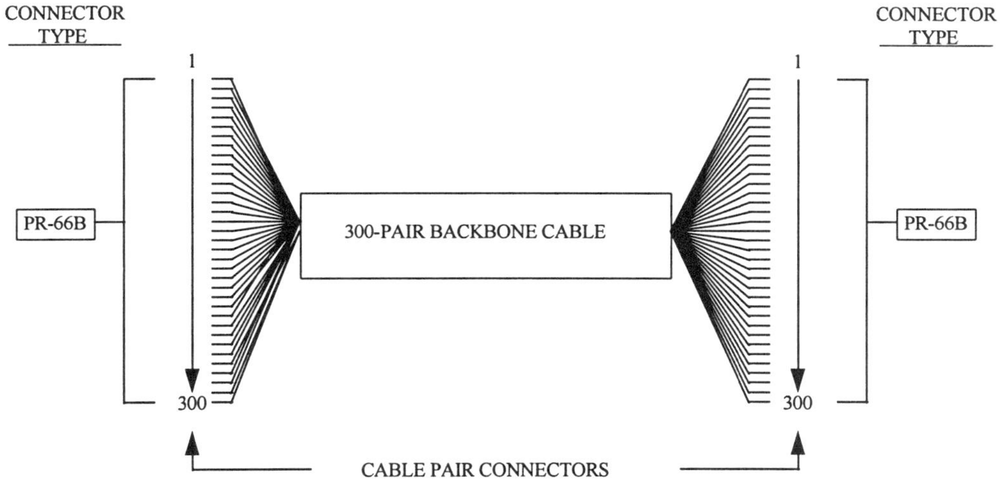

Figure 21.20: Termination blocks tracked as attributes of the cable.

to generate a circuit based on beginning and ending point, and often require the use of termination blocks to determine the path and connection points along the circuit. Determine if automatic circuit generation is a high priority in your list of requirements, and have the vendor outline the elements that are required to generate the circuit connections.

Once you have determined whether or not to track termination blocks as separate components, you will need to define their individual configurations. There are many different ways to represent them in the CMS. Blocks can be made up of connectors called pairs, pins, or ports, depending on the use of the particular block. Which method you choose will depend on the terms most commonly used and understood by technicians and other personnel in your organization. Some systems will have their own configurations for termination blocks but will still require information as to how many connectors and of what type are to be created for the termination block.

Some examples of different types of blocks which you may have in your wiring closets are listed in Table 21.22. Each set of blocks has a corresponding number of pairs, pins or ports to which connections will be made. Defining the types and numbers of connectors for each type of block will be required as each block is created in the CMS database. The database will need to know how many connectors will be available for connections, and which types of connections will be made to the back and front or left and right of the block.

It is possible to represent multiple blocks as a single block. A column of four 66M50 blocks could be represented as a single block with Pairs 1 to 200, or Pins 1 to 400. This approach reduces the number of blocks which must be individually identified, created, and tracked, and makes it easier to search for available pairs on a cable which is terminated on multiple blocks. A single query on the 200-pair (400-pin) block would identify all spare capacity for the cable.

Be sure to discuss the handling of termination blocks with the CMS vendor before

- 66 BLOCKS
- KRONE BLOCKS
- BIX BLOCKS
- 110 BLOCKS
- MULTIFLEX
- MODULAR PANELS

Table 21.22: Examples of termination blocks.

selecting a system. Some systems cannot look for available cable pairs based on a cable name if pairs are terminated on multiple blocks at each end. It may require looking at each individual block to identify the spare pairs.

After determining the configuration for each type of termination hardware, a method of naming for the individual devices must be developed. The simplest approach is to use existing naming conventions wherever possible, adding a building/floor or building/wing prefix designation to make each termination block unique. Often, the termination blocks are labeled based on their position as you enter the wiring closet, such as Block 1, 2, 3, 4, etc. as you go from top to bottom and left to right. These block numbers can be used, but may need a building/floor or building/wing prefix designation in order to differentiate the blocks in one wiring closet from the blocks in another wiring closet.

If you are working with a new installation, where no naming/numbering convention exists, some numbering schemes are provided here for your consideration.

- The simplest approach is to begin with first set of blocks to your left as you enter the wiring closet, and start with BL01, BL02, BL03, etc. as you move down the column. Go to the next column to the right and number these from top to bottom. If you are working with free-standing racks, start with the racks which are closest to the door, and go from top to bottom and left to right. The same numbering scheme can be used for each wiring closet, provided a building/floor prefix can be combined with the block number to create unique names on each floor, such as REN1EBL01, ADM01BL01, etc.
- You can use some type of indicator in the block name as to the type of block, such as 66MB (66 Block), 110B (110 Block), BIXB (BIX Block) or KRBL (Krone Block), followed by a sequential numbering scheme. A building/floor prefix can be combined with the block number to create unique names on each floor, such as REN1E66MB01, ADM0166MB01, etc.
- Another approach is to include some type of indicator in the block name as to the type of cable or ports contained in the blocks, such as SB01 (Station Block 1), RB01 (Riser Block 1), VB01 (Voice Block) DB01 (Data Block 1), MB01 (Mainframe Block 1), TB01 (Token Ring Block 1), EB01 (Ethernet Block 1), etc. Some examples would be REN1ESB01, ADM01SB01, etc. Otherwise, all termination blocks could begin with the letter "B", which may be beneficial when creating reports or performing queries on termination blocks from a general table of devices.

These approaches consider the use of a sequential numbering plan so that as additional blocks are added, they continue following the previous number. If Station Blocks SB01 through SB24 were initially installed, the next new block would automatically become SB25 and would be located next to termination block SB24 in the wiring closet. Again, the same numbering scheme can be used for each wiring closet, provided a building/floor prefix can be combined with the block number to create unique names on each floor.

Some systems create their own device names through the use of combining a device type with a sequential number. Be sure to discuss any current naming schemes for your termination blocks with the vendor to make sure you do not have to use their termination block names, which may mean relabeling the termination blocks to match the naming in the CMS. If this is the case, all termination blocks must be created in the CMS and labeled to match the CMS name before any inventory of connections is started. You will want to be sure the inventory records reflect the proper naming conventions.

You must develop a method for labeling each piece of termination hardware. Be sure the termination block is clearly labeled with the corresponding CMS block name. Sometimes this requires labeling each termination block individually, or labeling the backboard where the blocks are located. It is not necessary to label each termination position if their identifiers could be determined by a standard convention.

In wiring closets where a building/floor prefix is used to make the termination hardware names unique, it is not necessary to label each block with the prefix, however it should be indicated somewhere near the top of all the termination hardware. Refer to Chapters 3 and 4 for more specific requirements on termination hardware labeling.

The decision to track patch panels should be based on what the technicians normally refer to when handling requests for Moves, Adds and Changes. Tracking patch panels as separate devices adds another component to the circuit, but provides a means of changing the port connector to reflect the correct connection type. For example, certain ports could be accurately represented as an RJ45 jack rather than as part of an Amphenol connector. This will make for easier pre-assignment where patch cords are used.

Patch panels may be created as the only representation of the control device containing the actual ports. For example, rather than tracking an MAU attached to the patch panel, only the panel itself is defined with the MAU port names used as the names for the connectors on the patch panel. All circuits are therefore considered to start or terminate at the port on the patch panel. The port names assigned to the patch panel should then identify the type and ID of each port on the controlling device to which the patch panel is connected. Remember, each separate component that you track in the CMS records must have a unique label which should be meaningful to the person using the records.

Some CMS products treat patch panels the same way as termination blocks. The decision to track patch panels increases the number of records and connections in the database. As with termination blocks, you should consider several points to help determine whether patch panels should be tracked as separate entities or as components of the cable or hub.

Determine whether patch panels are currently labeled and referred to in terms of panel name and port position, or whether network port names or addresses are used. If the technicians and network support groups commonly refer to connections based on port name rather than the panel and position, it may be best to treat the patch panels as attributes or components of the

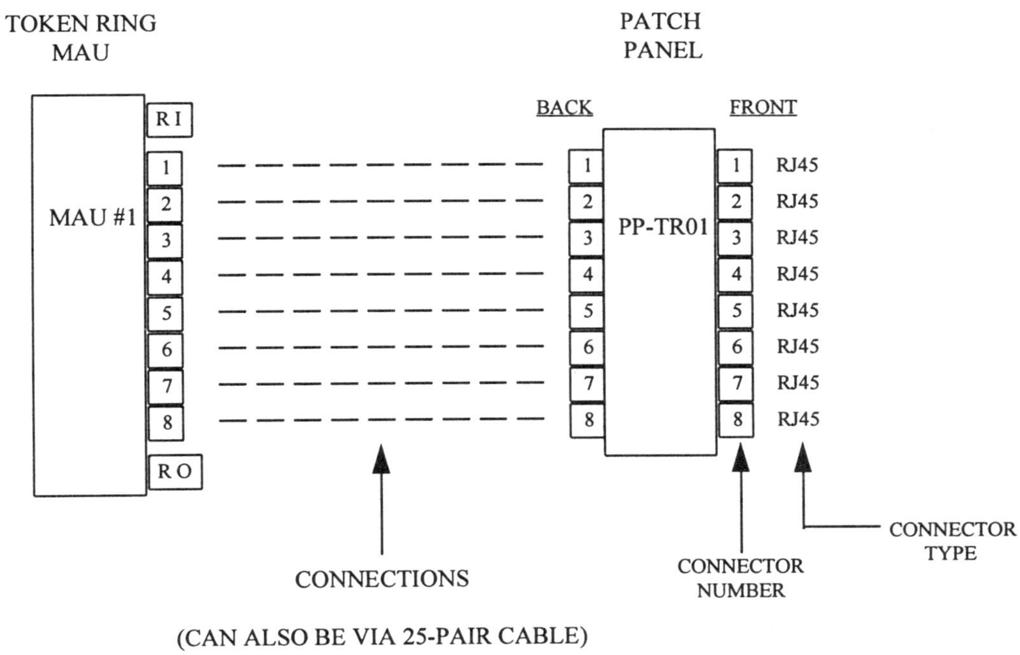

Figure 21.21: Patch Panel tracked as a separate entity.

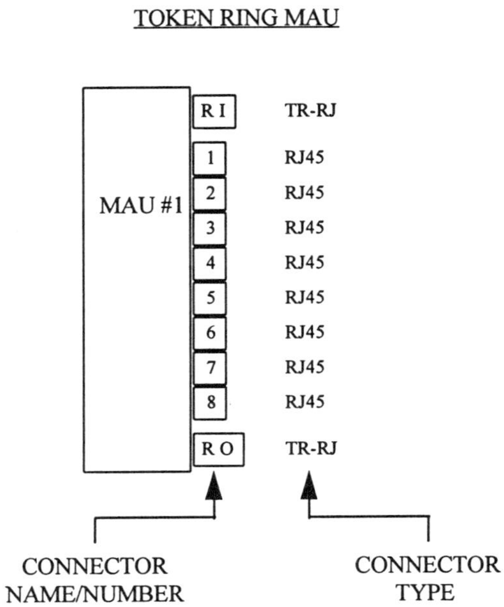

Figure 21.22: Patch Panel tracked as an attribute of the port.

controlling equipment or ports rather than as separate entities. An example might be a token ring MAU which is connected via baluns and multipair cable to a modular patch panel. The patch panel is labeled with the token ring port names represented by the MAU. One approach would be to treat the MAU and modular panel separately, while showing each Data Connector on the MAU attached to an RJ45 jack on the panel. As an option, you can treat the patch panel as an extension of the MAU and not track it separately, while simply showing the ports on the MAU as RJ45 connectors. In other words, the physical panel connectors are used to define the MAU connection points. Figures 21.21 and 21.22 demonstrate the different approaches to tracking patch panels.

Once you have determined whether or not to track patch panels as separate components, you will need to define their individual configurations. Connectors on a patch panel are usually referred to as ports. The type and number of connectors (ports) which will be defined on each patch panel will vary depending on the types of patch panels used. Each patch panel type will have a corresponding number of ports to which connections will be made. You will need to define the types, quantity, and numbering of connectors for each type of patch panel to be tracked. The database will need to know how many connectors will be available for connections, and which types of connections will be made to the back and front or left and right sides of the patch panel.

Some CMS products will have their own configurations for patch panels but will still require information as to how many connectors (ports) are to be created with the patch panel. Some systems will require you to specify the port type (such as RJ45, 110, BNC, etc.) and the port name (such as 01, 02, 03 or A1, A2, A3, etc.). You may be required to specify the number of ports, but may not be able to specify the type of port or assign individual port names to the ports on the patch panel.

Figure 21.23 shows an example of a patch panel which is used with a 32-port 3270 Controller. The CMS software may create 32 ports on the patch panel numbered 1 to 32, but may not allow you to specify each of those ports as representing the 8-digit identifiers LG1D1100 through LG1D1131. This makes it difficult to relate the controller and panel positions, such as when PP005-26 (Patch Panel 5, Port 26) is used on the panel to represent the controller port LG1D11, Port 25. On the other hand, naming the panel positions according to a specific set of ports (network addresses) will require major changes if the panel is later used for a different purpose. The CMS vendors should specifically identify any automatic naming or numbering conventions which are created by the software.

For the most part, organizations do not normally assign a name or identification label to the patch panel itself. The Port names or numbers might be labeled or indicated, but the actual panel rarely gets its own name or label. Where controlling device names (network addresses) will not be used for naming patch panels, a method of naming each panel will be needed. Note that even if a separate name is assigned to the panel itself, the panel positions can be given a sequence number or be named after the network address of the attached port.

The simplest approach is assign patch panel identifiers in a similar fashion as for termination blocks. Assign numbers based on the panel location as you enter the wiring closet, such as Panel 1, 2, 3, 4, etc. going from top to bottom on a rack and then moving left to right from rack to rack. If this type of numbering can be used, it may require a building/floor or building/wing prefix designation in order to differentiate the patch panels in one wiring closet from the patch panels in another wiring closet. Some naming conventions for patch panels are offered below for your consideration. You may want to start with a similar concept and adapt it to fit your individual environment. Try to adopt one approach and use it consistently throughout your facilities.

o You can begin with the first set of patch panels to your left as you enter the wiring closet, and start with PP01, PP02, PP03, etc. as you move to the right. The same numbering scheme can be used for each wiring closet, provided a building/floor prefix can be combined with the patch panel number to create unique IDs on each floor. Some examples would include REN1EPP01, ADM01PP01, FAC01PP01,

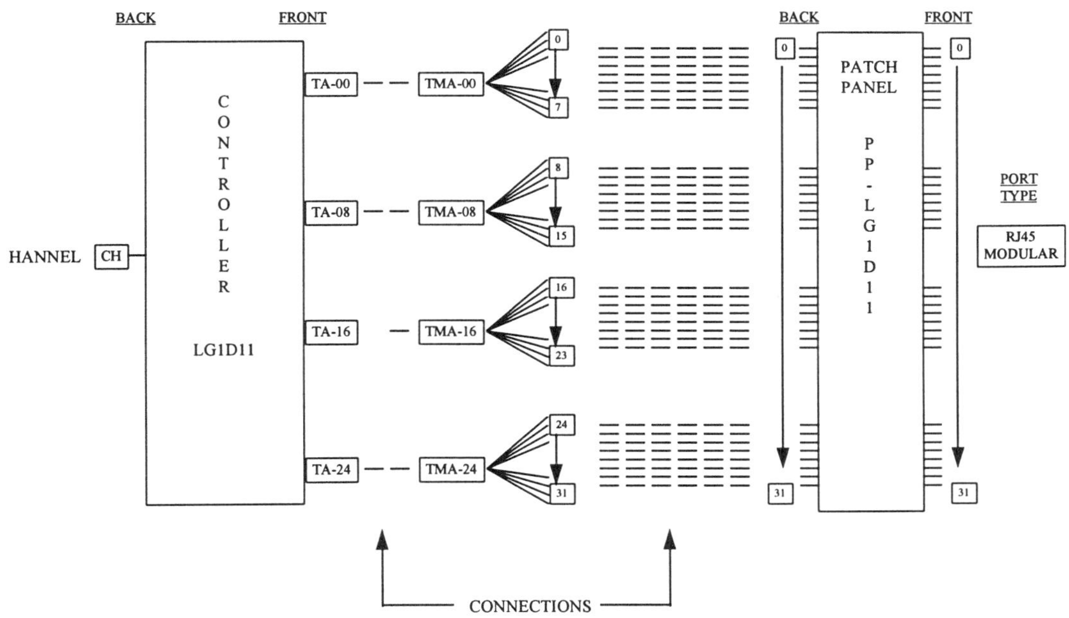

Figure 21.23: Patch Panel with 32 Ports for Controller LG1D11.

etc. This approach can be used for almost any type of patch panels.

- o Another approach is to include some type of indicator in the patch panel name as to the type of ports contained in the patch panel, such as SP01 (SNA Panel 1), EP01 (Ethernet Panel 1), TP01 (Token Ring Panel 1), DP01 (DECNet Panel 1), etc. Some examples would include REN1ESP01, ADM01SP01, FAC01SP01, etc.
- o You can include some type of indicator in the patch panel name as to the type of patch panels, such as ORT24-1 (Ortronics 24-Port Panel 1), ORT36-1 (Ortronics 36-Port Panel 1), IBM64-1 (IBM 64-Port Panel 1), etc.
- o Yet another option is to give the patch panel a name similar to the controlling device to which it is attached, adding a suffix or prefix to represent the patch panel, such as PP-LG1D11 to represent the Patch Panel for a 3270 controller whose name is LG1D11.

All of the these approaches consider the use of a sequential numbering plan so that as additional patch panels are added they continue to following the previous number. If patch panels PP01 thru PP10 were initially installed, the next new patch panel would automatically become PP11 and be located next to patch panel PP10 in the wiring closet. The same numbering scheme can be used for each wiring closet, provided a building/floor prefix can be combined with the patch panel number to create unique names on each floor.

Some systems create their own device or equipment names through the use of combining a device type (such as PP for Patch Panel) with a sequential number, or strictly a straight sequential number scheme regardless of the device type. Be sure to discuss any current naming schemes for patch panels with the Vendor to make sure you do not have to use their patch panel names, which may mean relabeling your patch panels to match the CMS names after the patch panel has been created in the CMS database. If this is the case, all patch panels must be created in the CMS database prior to any labeling work, and definitely before any inventory of connections is started. You will want the inventory records to reflect the proper naming conventions for the patch panels and ports.

The final step in completing the catalog of patch panels is to come up with a method for labeling each patch panel and its ports. Be sure the patch panel and its ports are clearly labeled with the corresponding CMS names. Sometimes this requires two sets of labels on a patch panel, one to represent the CMS device name and additional labels to indicate the actual ports (logical network port names or the controlling device to which the patch panel is connected).

In wiring closets where a building/floor prefix is used to make patch panel names unique, it is not necessary to label each panel with the prefix, however the prefix identifier should appear in a visible location in the wiring closet. Refer to Chapters 3 and 4 for more specific requirements on labeling.

Pathways

The final area to cover in the catalog is pathways, the conduit and cableways that contain backbone or horizontal cable. Each of the CMS products has a different method for dealing with conduits and pathways. The decision to track this information should depend largely on what information is available regarding conduit types and the actual routing of pathways, since the gathering of pathway information after-the-fact can be a significant chore. If the information regarding the size, material, contents, and routing of each pathway is already documented and available, it is a good idea to include it in the CMS so that the records are complete and in one location.

Tracking conduit and pathway information may be of value at a later date when new cables need to be run, or when pathway information is required to plan a building renovation or expansion. This is not the type of information which would be required and used on a daily basis, since routine Moves, Adds and Changes do not routinely involve working with cable pathways.

Some systems will require the conduits and pathways to be created in the database prior to adding cables, so that the routing of cables can be specified as the cables are created. Other CMS products do not require documenting the conduits and pathways, so they can be added after the cable information has been loaded. Be sure to ask the Vendor to specify any required order that the items must be loaded into the database, especially in the case of conduits and cableways. While the information required for pathways is relatively basic, the best approach is to wait until you have selected the final CMS system, and then determine what information is necessary for tracking pathways based on the architecture of the software.

If you determine you want to add conduit and cableway information to the CMS records, start by making a list of the information you currently have available. If this information is already in some other document or database, obtain a copy of the report for later reference. Some basic information to begin with is shown in Table 21.23.

Conduit and pathway items will need to have names which uniquely identify them. If there are existing names for these items, it is best to utilize those names, since construction drawings may reference these names, as well as other building files which are maintained by the building facilities and maintenance groups.

If there are no existing names for your conduits and cableways, you may want to consider one of the following options, or a variation which best meets your requirements.

- The simplest approach is to begin with an identifier for pathways or cableways, such as CW or CD, and then add a numbering scheme beginning at the Main Equipment Room with the numbers increasing as you move out to the other equipment rooms and Telecommunications Closets. This would provide additional information in that conduits and pathways with higher numbers would be located further away from the Main Wiring Closet or Equipment Room. Some examples would be CW01, CW02 or CD01, CD02, etc. As an alternative, you can base the naming on certain key spaces where pathways end. For example, all pathways running from the Main Equipment Room can carry the designation CWMER01, CWMER02, and so forth. Other selected riser rooms or equipment rooms could also be referenced, so that every pathway name indicates its originating space.
- You can include some type of indicator in the conduit or pathway name as to the type of conduit or type of cables contained in the pathway. Some examples would be CW-R01 (Riser Cableway), CW-F01 (Fiber Cableway), CW-V01 (Voice Cableway), etc. A similar approach could be taken based on ownership of the pathways, as in the case of Telco or private-property ownership of outside conduits.
- Another approach is to include some type of indicator in the pathway as to the type or size of conduit or cableway, such as Metal, PVC, 2" or 4", etc. Some examples would be C-4M01, (4" Metal Conduit) or C-2P01 (2" PVC Conduit).

All of the these approaches consider the use of a sequential numbering plan so that as additional conduits of a specific type in a given location are added, they continue on following the

PATHWAY NAME	PATHWAY SIZE	MATERIAL	CABLES
CWADM-1	4"	PVC SCHEDULE 80	1214
CWADM-2	4"	PVC SCHEDULE 80	NONE
CWADM-3	4"	PVC-SCHEDULE 80	FIBER
CWADM-4	4"	PVC SCHEDULE 80	C.O. TRUNK
CWADM-R1	4"	STEEL	RISER 1
CWADM-R2	4"	STEEL	RISER-2

Table 21.23: Basic information regarding pathways.

last number. If conduits CW01 thru CW10 were initially installed, the next new conduit would automatically become CW11. Any combination of the above naming conventions can be used as long as the names assigned have a specific numbering scheme with indicators to provide some information about the type or location of the pathway. The main thing is to keep it as simple as possible, but give the names some meaning.

Some systems assign their own naming conventions for cableways and conduits based on the type or location. Be sure to ask the CMS Vendor to indicate if their software utilizes any sequential numbering or naming assignment. If you select a CMS which utilizes a sequential numbering system to assign conduit and cableway names, it would be beneficial to label the pathways wherever accessible so there is a link to the information in the CMS records. If you must use pathway names assigned by the CMS, you will need to label or relabel the pathways after they have been created in the CMS. The relationship of cables to pathways must then be based on the assigned cable and pathway names.

The final step in completing your catalog of pathways is to develop a method for labeling each conduit or cableway. If you are tracking pathways in your database, it is very valuable to make sure the conduit or cableway names in the database match the labels in the wiring closet, computer room, data center, or wherever accessible. If access to some pathways requires a ladder, or the pathways are otherwise hard to reach, it might be best to prepare the labels in advance and then attach them when the pathways are being accessed for routine maintenance or some other purpose, rather than requiring a special effort.

PREPARING A REQUEST FOR PROPOSAL (RFP)

22

Having developed a good sense of your requirements and documented them in a comprehensive way, you could proceed directly with an informal evaluation of CMS candidates. However, there is significant value in performing a more rigorous evaluation by preparing and issuing a Request For Proposal (RFP). We will first consider the advantages of this approach listed in Table 22.1 and then cover how to prepare your RFP, including the specific areas which should be included. A sample RFP is provided at the end of the chapter for you to use as a reference and template for your own RFP.

The Value of an RFP

Using an RFP can be very beneficial in helping you to compare and evaluate candidate CMS systems as well as to be sure you know what you're getting. The RFP provides a written document which contains a clear description of your requirements from a functional as well as maintenance standpoint. In other words, it forces you to be disciplined and thoughtful in finalizing these requirements. It assures you have covered all areas impacted by the recordkeeping system, and provides a method to review the needs with other individuals or organizations to get their agreement and to be sure you have properly represented their needs and requirements.

The RFP provides a common ground on which to compare the candidate CMS products. It lets you compare "apples to apples" as the saying goes. It requires the Vendors to address each area and functionality requirement in a standard format, and often uncovers more obscure features that may help or hinder your connectivity management process. It is hard to evaluate and compare systems based on product literature and demonstrations alone. Vendors are more than willing to show their strong features but avoid discussing the weak points of their product. Too many organizations have seen demonstrations and thought the product was exactly what they wanted, only to find out after installing the product that it did not meet their expectations.

By issuing an RFP, you now have a document from the Vendor which indicates areas where they indicated compliance with your requirements. If in working with the system you find the product does not perform as indicated, you have the documentation to use in working with the Vendor to resolve the discrepancy. This puts you in a much better position than having the sales person replying that you misunderstood what they meant.

- CLEARLY DEFINED REQUIREMENTS
- DIRECT SYSTEM COMPARISONS
- VENDOR ACCOUNTABILITY
- DIRECT PRICE COMPARISONS
- INCREASED UNDERSTANDING

Table 22.1: Benefits of using a Request For Proposal (RFP).

• COVER LETTER
• INSTRUCTIONS TO BIDDERS
CONDITIONS
INSTRUCTIONS
PROPOSAL FORMAT
AWARD PROCESS
PRICING
ASSUMPTIONS
• TECHNICAL SPECIFICATIONS
INTRODUCTION
VENDOR QUALIFICATION AND SUPPORT
TECHNICAL REQUIREMENTS

Table 22.2: General sections of the RFP.

Another advantage of the RFP is the ability to obtain a firm quote from the Vendor based on the application platform and number of users you require. This provides a method to weigh the differences in functionality from one system to another versus the price of their software. Often, you will also get a better price from the Vendor if they know they are bidding against other Vendors for your business.

Preparing the RFP and evaluating the responses will also be a tremendous educational opportunity for you. By considering the strengths and weaknesses of each system, and seeing the trade-offs that must be made to accomplish certain functional objectives, you will be much better prepared to optimize the system you select. You will be prepared to handle weak areas or deal with undesired features. Your approach to cataloging, for example, will be much more effective if you understand the functions of the selected CMS. Finally, the RFP lets the Vendor know what they can expect from you, and that you are serious and committed to the success of the project.

Preparing The RFP

The first step in preparing the RFP is to separate your requirements into the different categories listed in Table 22.2, which will help to capture all your requirements and organize them in the most logical order. Start with the lists of items that you created while developing a records structure, such as for cables and circuits. Include all of the items, even some of the ones which may have been determined to be optional. Sometimes you'll discover it's a bonus which helps in selecting the final system. Be specific in your description of each requirement to be sure the Vendor has no question as to what the requirement means. Some of the categories used in preparing an RFP are listed below, along with a description of the types of requirements which might be included under each section.

A cover letter should accompany the Request For Proposal, which is usually issued by the Purchasing Department. This cover letter should include any company policies on responding to an RFP, as well as the name and address where responses are to be sent, and any special instructions regarding information to provide on the outside envelope containing the response. Other items which should be included in the Cover Letter are listed below:

- o Due Date & Time for responses
- o Name, Telephone Number & Fax for Vendor questions
- o Number of copies of responses to be submitted
- o Estimated Date of Award

- o Anticipated date for execution of the purchase order
- o Anticipated CMS installation date
- o Additional Bidder Instructions (Bidder's Meeting, etc.)
- o Terms & Conditions of Sale

The first part of the RFP itself is an Instructions to Bidders document, covering the general conditions and instructions for the Vendor while explaining something about the RFP format and the award process. Refer to the example for details. It is best to provide a pricing sheet with the RFP which breaks down the pricing by number of users, estimated hardware costs, software costs by module, any additional costs for installation, training and documentation, as well as fees for customization or database set-up assistance. By providing the pricing breakdown the way you prefer, it is easier to compare the total costs for each system. Each Vendor has a different way of quoting their prices, and it is not always clear what is included with the total cost and what services are additional.

Another item you may want to consider is providing a sheet where the Vendor can indicate any assumptions they are making while responding to your RFP. This helps to identify at an early point any areas which need clarification during the evaluation process.

In the Technical Specifications document the first section should be an Introduction, which is a good place to include a brief description of the type of system you are looking for. This section will include a summary of the scope of the facilities to be tracked in the CMS, and how you plan to use and maintain the records such as through the use of a Service Order or Trouble Ticket, etc. You may want to include descriptions of the different types of communications networks and wiring configurations, and indicate how many users will need to access the system. Much of the information developed in Chapters 13 and 14 will be useful.

The second section, Vendor Qualification and Support, should cover your requirements with respect to installation, training, documentation, troubleshooting guides/references, maintenance and support, warranty, licensing agreements, customization policies, and Vendor references. It is beneficial to have the Vendor supply information about their Customer Support Services and charges, as well as their policies for warranty and maintenance. This type of information helps to ensure the vendor you select will be around to support the product once you have purchased it, and that their level of support will be satisfactory.

The Technical Requirements section is where you should indicate all the requirements of what the CMS product must be capable of tracking, how it should display the desired information, as well as how Service Orders, Trouble Tickets and reports are handled. Other items to include in this section are printing requirements, validation tables, data loading, etc. In this section, it is a good idea to provide an area for the Vendors to indicate their level of compliance (full, partial or non-compliance) with each requirement. Listed below are some basic categories you may want to consider:

- o General Information - Single/Multi-User; CAD/Graphics; Network Platform; Database Platform; Password Protection; Authorization Levels; Dial-In Capability; Response Times; Help Feature/Tables, Hardware Requirements for CMS Software (Equipment Type, Hard Drive Space, Memory, Speed, Ram, etc.).

- o Inventory/Asset Tracking - Equipment Types including all field names to be tracked about each equipment type; catalog of equipment types, status (reserved, spare, in-use).

- o Service Order - Information (fields) to be included in the Service Order; circuit information to be included; access to inventory information; circuit creation within Service Order; reporting requirements (open, pending, closed, by Technician, etc.); printed copy requirements.

- o Trouble Tickets - Information (fields) to be included in the Trouble Ticket; circuit information to be included; update of circuit connections; trouble history tracking; reference to open service orders; reporting requirements (open, pending, closed, by Technician, by trouble type, etc.); printed copy requirements.

- o Connectivity Components - Basic circuit elements; cable types and configurations; circuit, cable, and equipment naming requirements; cable routing (cableways, conduits, pathways, etc.); bridging of circuits; cable pair counts and status; controlling equipment and ports; dataset structures (lists of fields/information to be tracked about each component) including desired field lengths; reporting requirements.

- o Circuit Functionality - Creation of circuits; required components of a circuit; circuit connections; port connections; cable-to-cable connections; cable-to-port connections; cable-to-jack or outlet connections; desktop equipment to jack/outlet connections; cable path designation; warnings for duplicate connections; circuit display/trace format; bulk connections and disconnects; warning for low availability of cable pairs/ports.

- o CAD/Graphics - CAD-based system or database with CAD-link; fully integrated modules; compatible CAD-type files (DXF, etc.); CAD-functionality; Icons and Symbols; drawing names/IDs; printing/plotting requirements.

- o Reports - Standard Reports; User-Defined Reports; Report-Writer; Printing Requirements for Reports; Downloading of Reports; ASCII report format.

- o Validation/Look-Up Tables - Key fields or components to be validated or have look-up tables available; wild-card feature for queries.

- o Data Entry Requirements - Bulk creation of cables, cable pairs, outlets/jacks and ports; ability to download information from existing systems; order required for data loading.

- o System Back-Up/Archiving - Procedures for System Back-up; archiving of old service orders/trouble tickets; process for restoring data.

- o Network/System Interface - Interfaces to other systems (Network Management; Voice Switch; Billing System; E-Mail) for downloading or uploading information.

Finally, if abbreviations or initials are used throughout the RFP to indicate types of communications systems, company names, equipment types, etc. it is a good idea to provide a glossary at the end of the RFP which defines each of the abbreviations to help the Vendor in responding to your RFP.

Sample RFP

In the sample RFP which follows, we have included as many requirements as possible in order to give you an idea of the types of requirements you might want to include in your RFP. You may want to start with this sample, and select those areas which are applicable to your environment and CMS requirements. Do not use the sample exactly as written, and consult with your Purchasing and Legal Departments for their specific requirements.

You will note, there is a space provided next to each of the requirements for the Vendor to indicate Full Compliance, Partial Compliance, and Non-Compliance. This makes it easy for the Vendor to respond to your RFP, and assures there is no confusion as to which requirement they are addressing.

You will also note there is a line at the bottom of each page for the Company Name or initials of the individual completing the RFP and the date. This is helpful in case the pages get mixed up with other Vendor responses, or if there is a question on functionality at a later date. It provides a method to clearly identify that the specific Vendor completed that page.

ABC Company

Health Foods and Beverages

January 4, 1999

Mr. Sales Agent
CMS Vendor
123 Cable Management Drive
Cabletown, USA 00000-0000

Dear Mr. Sales Agent:

INVITATION TO BID
INQUIRY #IT99-15 - CABLE & CONNECTIVITY MANAGEMENT SYSTEM

You are invited to submit a competitive bid and technical proposal for a Cable and Connectivity Management System (CMS). Enclosed with this Invitation to Bid you will find the following items:

Instructions to Bidder
Technical Specifications Document
Terms & Conditions of Sale Document

A Bidder's Meeting WILL NOT be held for this RFP. Vendor presentations as requested by ABC Company will be scheduled for the Week beginning February 15, 1999.

Your proposal must be in strict accordance with the Instructions to Bidders. The Proposal form has been structured to facilitate your responding to the individual specifications. Blank lines and tables are provided which must be filled in, and certain attachments are required to be added by you.

Your proposal is to be submitted in accordance with the attached specifications and is to be submitted in **triplicate**. All three (3) copies shall be signed by an authorized representative of your firm, sealed and submitted to the address below. All bids must be received by **12:00 Noon on Monday, February 4, 1999**.

Mr. Samuel McAdams
ABC Company
Purchasing Department MC-213
123 Painters Lane
Landisville, ME 00000-0000

123 Painters Lane, Landisville, Maine 00000-0000 *(207)123-1000*

Bid Invitation Letter #IT99-15 January 4, 1999
Page 2

The Bid Reference Number #IT99-15 must appear in the lower Left Hand Corner of the envelope containing your bid response.

Technical questions relative to this RFP may be submitted to Mr. Michael McMichael at (207) 123-1234; address as follows:

Mr. Michael McMichael
ABC Company
Information Technologies, MC-235
123 Painters Lane
Landisville, ME 00000-0000
Phone: (207) 123-1234
Fax: (207) 123-1235

Any questions regarding the preparation and submittal of your proposal shall be directed to the author of this letter, indicated below.

Very truly yours,

ABC Company

Mr. Samuel McAdams
Purchasing Agent - Information Technologies
(207) 123-1122

Enclosures

123 Painters Lane, Landisville, Maine 00000-0000 *(207)123-1000*

INSTRUCTIONS TO BIDDERS

REQUEST FOR PROPOSAL

BID NO. IT99-15

CABLE AND CONNECTIVITY MANAGEMENT SYSTEM (CMS)

January 4, 1999

ABC Company
Landisville, ME

Proprietary and Confidential

(EXAMPLE)

Table of Contents

			Page No.
1	STATEMENT OF PURPOSE		
	1.1	Purpose/General Overview	2
2	PROPOSAL CONDITIONS		
	2.1	Acceptance of Proposals	2
	2.2	Cover Letter	2
	2.3	Executive Summary/General Product Description	2
	2.4	Term of Support	2
	2.5	Terms & Conditions	3
	2.6	Confidentiality	3
3	PROPOSAL PROCESS INSTRUCTIONS		
	3.1	Proposal Time Table	3
	3.2	Due Date	4
	3.3	Delivery of Proposals	4
	3.4	Proposer Questions	4
	3.5	Bidder's Conference	4
	3.6	Vendor Presentations	5
	3.7	Proposal Costs	5
4	PROPOSAL FORMAT AND CONTENTS		
	4.1	Compliance/Response Format	5
	4.2	Pricing	6
	4.3	Assumptions	6
5	EVALUATION & AWARD PROCESS		
	5.1	General	6
	5.2	Evaluation Criteria	6
	5.3	Award Notification	6
	ATTACHMENT - Project Pricing		7
	ATTACHMENT - Assumptions		9

1.0 STATEMENT OF PURPOSE

This Request For Proposal (RFP) will result in the selection of an automated Cable and Connectivity Management System (CMS) to monitor and track installed communications cable, controlling equipment, and associated connections for ABC Company in Landisville, Maine. ABC desires a CMS which will support its variety of networks, including voice, SNA, Ethernet, Token Ring, AppleTalk, and special in-house data circuits. The system will be used for tracking all major copper and fiber cables and connection points, tracking service order activity, assigning routes for new circuits and cables, identifying remaining cable capacity and spare network ports, tracing problem circuits, and providing wire path information for routine Moves, Adds & Changes, and for use by the Help Desk in managing trouble calls.

2.0 PROPOSAL CONDITIONS

2.1 Acceptance of Proposals

Only proposals that follow the format and content instructions of the RFP will be considered.

2.2 Vendor Cover Letter

Vendor shall submit with their Proposal a cover letter on Vendor's letterhead certifying the accuracy of all information in their proposal and certifying that the proposal will remain valid for 6 Months from the date the proposal is submitted. This letter must be signed by the responsible official in the Vendor's organization.

2.3 Executive Summary/General Product Description

Vendor shall submit with their Proposal an Executive Summary or General Product Description including a detailed description of the individual modules/segments, characteristics, capabilities and technical specifications for the proposed system. This summary shall include a table of Fields by Data Set, with the field lengths.

2.3 Term of Support

The term of the Vendor support and maintenance for the CMS Product will initially be for one year from date of purchase.

2.4 Terms & Conditions

The attached Terms & Conditions (_______) will apply to this RFP.

Instructions to Bidders - Cable & Connectivity Management System January 1, 1999
ABC Company (SAMPLE) Page 3

2.5 Confidentiality

All information provided to vendors is considered private and is not to be disclosed to any person or parties not directly involved with this RFP. Vendor is not to disclose the existence of this specification and any confidential information herein to any third party or to any Vendor employees other than those who have a need to know such information, without the written consent of ABC Company.

3.0 PROPOSAL PROCESS INSTRUCTIONS

3.1 Proposal Time Table

RFP Issue Date	January 4, 1999
Proposal Responses Due	February 1, 1999
Vendor Presentations	Week of February 15, 1999
Notification of Bid Results	March 8, 1999
Initial System Installation(1)	Week of April 5, 1999
Final System Acceptance(2)	90 Days from completion of Initial System Installation

(1) Initial System Installation includes, but is not limited to the following elements. Completion of these 5 Steps will mark the point of Initial System Acceptance by ABC Company.

a) Installation & Set up of all Hardware supplied by Vendor in order to support the proposed system;
b) Loading of the Proposed CMS onto appropriate hardware;
c) Testing of the CMS software to ensure full operation and verify all functionality;
d) Testing of the system from the multiple points on the ABC Network to ensure all access points are trouble-free and multiple users can access the system simultaneously;
e) Training and documentation for at least 2 (two) ABC personnel shall be completed as part of this "Initial System Installation" milestone.

(2) Final System acceptance will be upon completion of all training, receipt of documentation, and 90 days of applied testing by ABC, following completion of the initial system installation.

3.2 Proposal Due Date

Proposals are due by **12:00 Noon** (Eastern Standard Time) on **(Monday, February 1, 1999)**. Proposals received after this date will be returned.

3.3 Delivery of Proposals

Vendor is to submit three (3) copies of their proposal to:

Mr. Purchasing Agent
ABC Company
Purchasing Department MC-213
123 Painters Lane
Landisville, ME 00000-0000

3.4 Questions

For questions about this RFP, or requests for additional information, contact:

Mr. Michael McMichael
ABC Company
Information Technologies
123 Painters Lane
Landisville, ME 00000-0000
Phone: (207) 123-1234
Fax: (207) 123-1235

Written, rather than verbal, questions and requests are preferred. Any responses to Vendor questions, or changes in scope which arise as a result of Vendor questions, will be communicated in writing to all vendors.

3.5 Bidder's Conference

A Bidder's meeting will be held for this RFP as indicated below. It is (is not) mandatory that all bidders attend this meeting.

Date:
Time:
Location:

(OR)

A Bidder's Conference will NOT be held for this RFP.

3.6 Vendor Presentations

Vendor may be required to meet with the evaluation team to provide a demonstration of their CMS product. The meeting, if required, will be scheduled by to take place during the week of February 15, 1999. Vendor is requested to be prepared to have the proper personnel available during this time in order to provide a live demonstration of their system.

ABC may request a visit or conference call to a client site where similar equipment and the CMS product being proposed are operating or have been provided.

3.7 Proposal Costs

All costs associated with preparing proposals in response to this RFP are the sole responsibility of the proposers.

4.0 PROPOSAL FORMAT AND CONTENTS

4.1 Compliance/Response Format

So that ABC Company can fairly evaluate all proposals, we require a uniform proposal format. Each section is listed, along with the contents required, in the "Technical Specifications Document" attachment.

Proposals must be provided in loose-leaf binders.

Proposals shall include a point-by-point response to each of the items in Section II of the Technical Specifications Document, and for each of the items in Section III indicate the level of compliance, as specified below. If Vendor can meet the requirement, using another segment or option with the database, they shall indicate how this would be accomplished, in detail sufficient for evaluation.

Full Compliance - Completely Meets or Exceeds Requirement

Partial Compliance - Partially Meets the Requirement; Include an Explanation of where the system fails to meet the requirement, and any alternative proposed in order to comply with the specification.

Non-Compliance - Cannot meet Requirement.

4.2 Pricing

All costs related to the CMS software requirements (including any operating system software, database engine software, communications software, CAD/Graphics Software), file server/PC hardware requirements (including CPU, monitors, terminals, plotters, printers), software license fees, training, installation, documentation, etc., are to be listed on the Attachment entitled "Project Pricing." Include a breakdown of all segment/modules of the software, and hardware items. Costs for hardware and installation are for ABC informational purposes, and does not imply ABC will purchase required hardware from the software vendor, or have Vendor provide installation.

4.3 Assumptions

Vendor shall identify any assumptions made in preparing this proposal, and shall list these assumptions on the Attachment entitled "Assumptions."

5.0 EVALUATION & AWARD PROCESS

5.1 General

Proposals will be objectively evaluated by a qualified team based primarily on how well the product functionality meets the technical specifications documented in this RFP. Interviews of bidder personnel may be required to ensure our understanding of their proposal and their understanding of our requirements. All bidders will be notified when the final selection has been made.

5.2 Evaluation Criteria

Following are the criteria that will be used in evaluating the proposals:

Technical Performance	____%
Vendor Responsiveness	____%
Price	____%
Vendor Reputation/Support	____%

Technical Performance will be evaluated based on our assessment of the vendor's understanding of our requirements, and the vendor's ability to meet the quality assurance and product functionality.

5.3 Award Notification

Following are evaluation process and final system selection, all bidders will be notified as to whether or not they were the successful bidder.

ATTACHMENT - PROJECT PRICING

File Server/PC Hardware	
CPU	$________________
Monitor	$________________
Terminal(s)	$________________
Modem	$________________
Special Power Supply	$________________
Printers/Plotter(s)	$________________
Back-Up Equipment	$________________
Hardware Installation	$________________
Other________________	$________________
________________	$________________
________________	$________________
Software	
Operating System	$________________
Application Software (If Required)	$________________
CMS Software	$________________
CAD/Graphics Software (If Required)	$________________
Communications Package	$________________
Back-Up Software	$________________
Software Installation	$________________
Other________________	$________________
________________	$________________
License Agreements	
CMS Vendor	$________________
Operating System/Application	$________________
Other ________________	$________________
________________	$________________
Other Costs:	
Upgrades/New Releases	$________________
Customization	$____________/man day
Enhancements	$________________
Training	$________________
Documentation	$________________
Maintenance Fees	$________________
File Conversion/Download Programs	$________________
________________	$________________
________________	$________________

PRICING GUARANTEED THROUGH: _____/_____/_____

Pg. 2 of 2

ATTACHMENT - PROJECT PRICING

Software - Breakdown By Module

Modules:

______________________	$______________________
______________________	$______________________
______________________	$______________________
______________________	$______________________
______________________	$______________________
______________________	$______________________
______________________	$______________________
______________________	$______________________
______________________	$______________________
______________________	$______________________
Total CMS System Cost	$______________________

Comments:____

____

____

____

PRICING GUARANTEED THROUGH: ______/______/______

ATTACHMENT - ASSUMPTIONS

Vendor shall indicate below any and all assumptions on which their response to this Request For Proposal is based.

TECHNICAL SPECIFICATIONS

REQUEST FOR PROPOSAL

BID NO. IT99-15

CABLE AND CONNECTIVITY MANAGEMENT SYSTEM (CMS)

January 4, 1999

ABC Company
Landisville, ME

<u>Proprietary and Confidential</u>

(EXAMPLE)

Table of Contents

			Page No.
I	INTRODUCTION		
	I.1	Project Scope	2
	I.2	System Scope	2
	I.3	Workflow Process	3
II	VENDOR QUALIFICATION & SUPPORT		
	II.1	Installation Procedures & Schedule	4
	II.2	Equipment Specifications	4
	II.3	Training/Documentation	5
	II.4	Maintenance/Support	5
	II.5	Customization/Enhancement Policy	6
	II.6	Warranty	6
	II.7	Licensing Agreements	6
	II.8	Company Information	7
	II.9	References	7
	II.10	Summary of Vendor Response Requirements	8
III	TECHNICAL REQUIREMENTS		
	III.1	General	9
	III.2	Asset Tracking/Inventory Control	11
	III.3	Service Order Module	13
	III.4	Trouble Ticket Module	15
	III.5	Connectivity Components	16
	III.6	Circuit Functionality	23
	III.7	Graphics Capability	26
	III.8	Reports	28
	III.9	Validation & Look-Up Tables	30
	III.10	Data Entry	31
	III.11	System Back-Up and Archiving	32
	III.12	Network System Interfaces	33
	III.13	Billing Functionality/Interfaces	34
IV	GLOSSARY		35

I INTRODUCTION

I.1 Project Scope

The cable management system for ABC Company (ABC) will be implemented for all of ABC sites in the State of Maine. The sites include both industrial and commercial facilities and are supported by over 3,000 employees. There are approximately 5,000 voice lines including local business lines, and several PBX's and an estimated 6,000 data connections in various communications networks, and many special in-house data circuits.

All copper and fiber outside plant cable and information pertaining to utilization of these cables shall be tracked. All inside plant cable, including distribution systems within buildings (wiring closets, computer rooms, main distribution rooms) shall be tracked. The scope of the project consists of four major commercial sites and 5 smaller single-building sites. There are approximately 45 buildings, served by an estimated 25,000 cable pairs associated with the outside plant and local inside cabling. There are an estimated 6,000 workstation jacks/outlets, consisting of 4 outlets each.

I.2 System Scope

The fundamental purpose of the Cable Management System (CMS) CMS is to track the individual circuits and capacity of the cable system and network controlling equipment at all ABC locations. ABC recognizes that tracking cable capacity requires individual circuits to be tracked against the backbone cable on a pair-by-pair basis, and intends to do this in as simple a fashion as may be allowed by the selected system. The circuit components are described in full detail in Sections III.5 and III.6, are based on associating a circuit name, an office jack and/or outlet wiring, a sequence of cable and cable pairs, and a port location on a controlling device.

Each of these elements is treated as a specific data set, or database table. This model design forms the basis for this Request For Proposal (RFP) and will be the reference point against which candidate systems will be evaluated.

It is intended that desktop equipment will for the most part be considered the beginning point of a circuit, whether it be a phone, computer, printer or terminal. The Equipment ID/Name of the desktop equipment (either a Phone Extension or ABC Company Tag Number) would be considered as the Circuit ID/Name. Asset

Tracking will be included for desktop equipment and controlling equipment. All control equipment such as 3270-type controllers, Ethernet Hubs, Token Ring Hubs, Terminal/File Servers, Multiplexers, etc., shall be described similarly as Controlling Devices and identified by unique ID number, device type and location. Each controlling device must have a set of user-defined ports which can be assigned to circuits, and each port shall be individually named by the user, rather than sequentially by the system.

Finally, some circuits will be located in temporary buildings or trailers, which may change locations, or be removed completely. The system will need to accommodate bulk deletion or changes for these circuits.

I.3 Workflow Process

The Service Order module will be used to track service requests, individual tasks, and for creating new circuits and assignments of jacks, ports and cable pairs for the Moves, Adds & Changes (MAC) process. Service Orders will be opened/created by a CMS Administrator, with the Administrator changing/creating new circuit connections in the database and providing old and new circuit information to the Technician. All field connectivity changes made by the Technicians who will supply any changes/updates to the circuit information back to the CMS Administrator, who will then be responsible for updating the circuit information in the CMS database, and closing the service order. This provides for tracking of scheduled physical changes and auditing to ensure proper updates are made to the database.

II VENDOR QUALIFICATION & SUPPORT

II.1 Installation Procedures & Schedule

II.1.1 Vendor shall include as a part of their Proposal, a copy of their normal schedule for installation and support of the proposed CMS. This shall be specified in terms of weeks (or days) after receipt of Purchase Order.

II.1.2 Vendor shall provide as a part of their Proposal, a copy of their installation and testing procedures used in the installation process and initial set-up for the proposed PC/server hardware and CMS software. This information is for informational purposes and does not imply ABC will request the Vendor to provide hardware and hardware set-up and installation.

II.1.3 ABC requires an installation date of the selected CMS Software within four (4) weeks of the Issue of a Purchase Order.

II.2 Equipment Specifications

II.2.1 Vendor shall specify if Vendor is offering to supply as part of the proposed system, the required hardware/ equipment required to support their CMS product, and shall indicate all costs for this equipment on the Attachment entitled "Project Pricing." Vendor shall indicate whether they are an authorized reseller of the equipment Vendor proposes to provide. ABC reserves the right to make the final decision on whether to purchase the required equipment separately or from the Vendor.

II.2.2 Vendor shall provide specifications for all equipment required to operate the proposed CMS, and any optional equipment which would enhance operation. This equipment for the main system platform/server is to include, but is not limited to CPU's, Monitors, Terminals, Servers, Modems, Special Power Supplies, Plotters and Printers. For all CPU equipment, Vendor shall specify minimum and optimum requirements for RAM, hard drive capacity, co-processor, and other internals.

II.2.3 Vendor shall indicate the type of equipment and any special memory, configurations or hard-drive size required to support system at the local workstation location, if operating in a LAN environment requires special equipment set-up or configurations at the local user level.

Technical Specifications - Cable and Connectivity Management System January 1, 1999
ABC Company (SAMPLE) Page 5

II.3 Training/Documentation

II.3.1 Vendor shall submit with their Proposal, a list of all types of documentation manuals which will be provided with the proposed system, also indicating the frequency with which manuals are updated.

II.3.2 A minimum of 6 copies of all User Manuals, Trouble-shooting Guides, Training Manuals and associated System Documentation shall be provided with the purchase of the proposed CMS software.

II.3.3 Vendor shall indicate how updates and revisions to their user manuals/documentation is handled and how new documentation is distributed to users (complete new manuals, revised sections only, paper copies, computerized disks, etc.).

II.3.4 Vendor shall provide full training for a minimum of 6 ABC personnel, on ABC premises. If training is to be at another specified location, Vendor shall so indicate in the Bid Response. Vendor shall also indicate if there is any additional cost for training, broken down by person, by day. Additional costs must be noted in the Attachment entitled "Project Pricing."

II.4 Maintenance/Support

II.4.1 Vendor shall include with their Proposal a copy of their Software Maintenance Policy.

II.4.2 Vendor shall provide User Support in the form of a Help Line, which must be available on a 24-hour, 7-day/week basis. A maximum 2-hour response time must be met in responding to requests for assistance and support of the proposed CMS.

II.4.3 If 24-Hour Help Line is not standard, Vendor shall indicate how after-hours maintenance and support is handled, and if there are any additional costs involved.

II.4.4 Vendor shall include with their Proposal a copy of their policy regarding upgrades/new releases, including any costs for the upgrades, installation or new documentation. These additional costs must also be noted in the Attachment entitled "Project Pricing."

II.4.5 Vendor shall indicate in their Proposal if Vendor requires Dial-In access to the proposed system for purposes of support and/or maintenance. Vendor shall indicate the type of communications software required to support this access, if required. Any costs associated with the communications software must be noted in the Attachment entitled "Project Pricing."

II.4 Maintenance/Support (Cont'd)

II.4.6 Vendor shall indicate in their Proposal if there is a User's Group for the proposed system, and the activities associated with the User's Group, such as meetings, newsletters, etc.

II.5 Customization/Enhancement Policy

II.5.1 Vendor shall include with their Proposal a copy of its policy for Customization and Enhancement requirements.

II.5.2 Vendor shall include with their Proposal a list of enhancements currently under development, with an expected date for release. Any prices/costs for these enhancements shall be stated in the Attachment entitled "Project Pricing." Prices for customization shall be stated in terms of cost per man day.

II.6 Warranty

II.6.1 Vendor shall include with their Proposal a copy of their standard warranty for the proposed system. This warranty, outside of a separate maintenance agreement, shall be provided at no additional cost and shall begin on the day of system acceptance, and shall continue for one year or the manufacturer's standard warranty, whichever is longer.

II.6.2 Warranty shall cover all defects in workmanship and material. During the warranty period, the entire system must be kept in proper operating condition at no additional labor or material cost to ABC. Major system defects shall be fixed within 72 hours.

II.6.3 ABC may require, at its option, that a copy of the source code for the system software be placed in escrow, in case for any reason, the manufacturer should discontinue this product or cease to do business. This shall be at no cost to ABC to exercise this option. Vendor shall indicate agreement with this requirement.

II.7 Licensing Agreements

II.7.1 Vendor shall include with their Proposal a copy of Vendor's software license agreement which is required with the purchase of proposed CMS. Costs for the license agreement must be noted in the Attachment entitled "Project Pricing."

II.7 Licensing Agreements (Cont'd)

II.7.2 Vendor shall include in their Proposal any requirements for third party Operating System, Application Software or Database Engine Software Purchases or Licenses that are required to support the proposed system. Vendor shall indicate the Version or Release No. for the appropriate operating system/software required. Costs for the software purchase/license agreement must be noted in the Attachment entitled "Project Pricing." Vendor shall indicate whether they are an approved reseller of such application.

II.8 Company Information

II.8.1 Vendor shall provide with their Proposal the following information regarding company background and business summary.

Size of Company (Number of Employees)
Longevity (Date Company was Founded)
Length of Time Proposed CMS has been on the Market
Total Number of CMS Product Installations
Other Product Lines

II.9 References

II.9.1 Vendor shall provide with their Proposal a list of at least three client references where the proposed CMS has been installed and is being used to manage facilities of comparable size and scope to ABC and is using comparable functionality. For each reference, include the following information.

Company Name and Address
Contact Name, Title and Phone Number
Number of Circuits/Connections
Number of CMS System Users
Date of Installation

II.10 Summary of Vendor Response Requirements

Following is a Summary of the Items which must be included with Vendor's Proposal as a part of Vendor's Response to Section II.

II.1.1	Copy of Installation Schedule
II.1.2	Copy of Installation & Testing Procedures
II.1.3	Confirmation of Installation Date Within 4 Weeks of Bid Award
II.2.1	Indicate if Vendor is Offering to Supply Equipment, and if Vendor is an Authorized Reseller
II.2.2	List of Equipment Specifications to support CMS
II.2.3	Equipment/Specifications for Local Workstation
II.3.1	Provide List of Documentation Manuals
II.3.2	Confirmation to Supply 6 Sets of Manuals
II.3.3	Process for Issuing Documentation Revisions/Updates to Users
II.3.4	Confirmation of Training for 6 Employees
II.4.1	Copy of Maintenance Policy
II.4.2	Indicate User Support/Help-Line Hours
II.4.3	Indicate Process for After-Hours Support
II.4.4	Upgrade/New Release Policy
II.4.5	Indicate if Dial-In Access Required & Software
II.4.6	Describe User Group Activities
II.5.1	Copy of Customization/Enhancement Policy
II.5.2	List of Customization/Enhancements Underway & Costs
II.6.1	Copy of Standard Warranty & 1-Year Commitment
II.6.2	Confirmation to Provide 24-Hour Correction of Major Defects
II.6.3	Agreement to Place a Copy of Source Code for CMS Software in Escrow, if awarded Purchase Order
II.7.1	Include Copy of Software License Agreement
II.7.2	Indicate Third Party Operating Software or Licenses Required
II.8.1	Provide Requested Company Information
II.9.1	Provide 3 (Three) Client References, as described

Technical Specifications - Cable and Connectivity Management System January 1, 1999
ABC Company (SAMPLE) Page 9

III TECHNICAL REQUIREMENTS

		Level of Compliance		
III.1 General		Full	Partial	Non
III.1.1	It is desired for the System to be accessible from any IBM-compatible desktop equipment.	____	____	____
III.1.2	Vendor shall list all operating systems on which their product runs. (Unix, DOS, etc.)	______________		
III.1.3	Vendor shall list all relational databases the proposed CMS system will operate under. (Oracle, SyBase, Informix, Ingres, Etc.)	______________		
III.1.4	Vendor shall list the protocol(s)their proposed system utilizes.	______________		
III.1.5	It is desired that the System operate on a file server in a PC environment running on a Token Ring/Ethernet LAN.	____	____	____
III.1.6	System shall be capable of networking together master and back-up file servers which may be located at different sites.	____	____	____
III.1.7	System shall operate under Windows, utilizing Windows 3.1 or higher.	____	____	____
III.1.8	System shall be a multi-user system, capable of supporting up to 20 simultaneous on-line users, from multiple sites, anywhere on the LAN.	____	____	____
III.1.9	System shall be capable of separating records for different sites into separate files within the same CMS database on the central file server.	____	____	____
III.1.10	System shall be user-friendly with menu options displayed and also provide the capability for experienced users to bypass menus and go directly to the desired module/screen.	____	____	____

Company Name:______________________________Date:______________

		Level of Compliance Full	Partial	Non
III.1	General (Cont'd)			
III.1.11	System shall have a Help Function, available as a Menu Option or Help Key, with help information specific to the screen or module in which the user is working at the time.	____	_____	____
III.1.12	ABC requires a maximum response time of 3-5 seconds between screens when adding records or performing system inquiries.	____	_____	____
III.1.13	System shall provide for different authorization levels, according to the individual authorization ID, including levels for Read/Display Access Only, Add New Records, Change Existing Records, Remove Records. Vendor shall indicate if authorization levels can be controlled at the module or dataset level. Module____DataSet____	____	_____	____
III.1.14	System shall have security protection capability, where users shall be assigned authorized IDs and specific passwords which must be verified by the system before accessing the CMS.	____	_____	____
III.1.15	System shall provide a field for indicating when records were entered into the database, or last updated.	____	_____	____
III.1.16	System shall provide a field for indicating which user entered a record into the database, or the last user who revised the record.	____	_____	____
III.1.17	System shall provide File Locking, so that when a record is being modified by one user, the same record cannot be accessed for change by another user.	____	_____	____

Company Name:_________________________________Date:________________

Technical Specifications - Cable and Connectivity Management System — January 1, 1999
ABC Company (SAMPLE) — Page 11

		Level of Compliance		
III.1	General (Cont'd)	Full	Partial	Non
III.1.18	System shall indicate when a cable pair or controlling device port has been reserved or has pending work associated with it. A status field for any pending changes shall be provided on all appropriate screens where updates can be made as described in the data set structures. The status for these components shall be assignable either manually or through the service order.	____	_____	____
III.1.19	System shall allow data entry and system operation from Remote Dial-In Access to the File Server or through a host PC on the network.	____	_____	____
III.1.20	System shall provide "Print Screen" capability for capturing the displayed screen to a local or remote printer.	____	_____	____
III.1.21	Vendor shall provide with their proposed system a Trouble Shooting Reference Guide which lists common system problems and error messages, and the action steps required to resolve the problem.	____	_____	____
III.1.22	System shall be compliant with the TIA/EIA 606 Standards for Cable Management Systems.	____	_____	____
III.2	Asset Tracking/Inventory Control			
III.2.1	The system shall be capable of maintaining a current inventory of desktop equipment (including phones, PCs, Terminals, Printers, Fax Machines, Modems, Answering Machines, etc.) and Controlling Equipment) including Controllers, Multiplexers, Bridges, Routers, Terminal Servers, LAN Hubs, etc.	____	_____	____

Company Name:_________________________________Date:_______________

		Level of Compliance		
III.2 Asset Tracking/Inventory Control (Cont'd)		Full	Partial	Non

III.2.2 Inventory Module shall be capable of tracking the following information about each inventory item, on the native equipment inventory screen:

Desired Field Name	Field Lengths	Full	Partial	Non
Equipment ID/Tag No.	15	____	_____	____
Type/Model	10	____	_____	____
Department/Division	10	____	_____	____
Network Address	17	____	_____	____
Drawing Reference No.	8	____	_____	____
User	20	____	_____	____
Phone No.	12	____	_____	____
Building	10	____	_____	____
Floor	3	____	_____	____
Room	7	____	_____	____
Software / Components	10	____	_____	____
Serial No.	12	____	_____	____
Vendor	20	____	_____	____
Part Number	20	____	_____	____
Purchase Cost	8	____	_____	____
Installed Date	Date (XX/XX/XX)	____	_____	____
Comments	40	____	_____	____
User Defined Field(s)	12	____	_____	____

III.2.3 The inventory module should be validated against a "master catalog" file with user-defined Equipment Classes and Types. This inventory module shall prohibit a user from entering inventory items which do not meet the Master Catalog definitions. ____ _____ ____

III.2.4 Vendor shall indicate the number of software items/ features (Adjunct Equipment, Components, etc.) which can be tracked against an inventory item. No.:_____________

III.2.5 The Inventory Module shall track the status of each equipment item as to In Use, Spare, Reserved. ____ _____ ____

III.2.6 The Inventory Module shall have a means of indicating if a particular inventory item has a pending Service Order open against it. ____ _____ ____

Company Name:___________________________________Date:_________________

Technical Specifications - Cable and Connectivity Management System — January 1, 1999
ABC Company (SAMPLE) — Page 13

		Level of Compliance Full	Partial	Non
III.2	Asset Tracking/Inventory Control (Cont'd)			
III.2.7	The Inventory Module shall have a means of indicating if a particular inventory item has an Open Trouble Ticket against it.	____	____	____
III.3	Service Order Module			
III.3.1	System shall have a Service Order function which provides for tracking of requests for moves, adds and changes.	____	____	____
III.3.2	The Service Order module shall allow for Service Orders to be created and accessed from any Workstation Access point or dial-in location.	____	____	____
III.3.3	The system shall be capable of generating a report of all Open/Pending Service Orders, with the following fields:			
	User Name	____	____	____
	Circuit ID (Extension)	____	____	____
	Equipment ID	____	____	____
	Due Date	____	____	____
	Technician	____	____	____
	Description of Work	____	____	____
III.3.4	The Service Order Module shall have the capability to provide a report of all open/ closed service orders by technician.	____	____	____
III.3.5	The system shall permit cancellation/removal of pending service orders.	____	____	____
III.3.6	The system shall allow for changes/updates to be made to the database without creating a service order. These types of updates would involve changes resulting from trouble or problem resolution not related to a specific service order.	____	____	____

Company Name:________________________________ Date:______________

		Level of Compliance		
III.3	Service Order Module (Cont'd)	Full	Partial	Non
III.3.7	The Service Order shall have the capability to create and/or delete the entire circuit, from desktop equipment or office jack, to controlling equipment port, including all cables and devices connections along the path.	____	____	____
III.3.8	The Service Order shall have the capability to add and/or delete equipment to the Database Inventory.	____	____	____
III.3.9	The Service Order Report shall have the capability to print out all connectivity instructions for the technician, including both old circuit (cross-connect) information and new circuit information for the technician.	____	____	____
III.3.10	The Service Order should provide the option of whether or not to incorporate circuit information into the Service Order, thereby permitting connectivity changes and updates to be made to the system, either at the opening of the Service Order, or after the technician has completed the work.	____	____	____
III.3.11	When an order is entered, a status of reserved or pending shall be assigned to all database components associated with the service order. When the service order is closed, status shall automatically change to Used or Spare.	____	____	____
III.3.12	The Service Order Module shall be capable of indicating levels of activity, such as the number new installs or disconnects generated in a month, or a high percentage of similar type service requests.	____	____	____
III.3.13	The Service Order shall contain, at a minimum, the fields listed in the Service Order Data Set Table at the end of Section III.5. Vendor shall indicate exact field names and lengths and which are required fields, and shall specify all other required fields in the Service Order Module of the proposed CMS.	____	____	____

Company Name:______________________________Date:______________

Technical Specifications - Cable and Connectivity Management System — January 1, 1999
ABC Company (SAMPLE) — Page 15

			Level of Compliance		
III.4	Trouble Ticket Module		Full	Partial	Non
	III.4.1	System shall have a Trouble Ticket module for handling and tracking the status of trouble calls.	____	____	____
	III.4.2	The Trouble Ticket module shall allow for Trouble Tickets to be created and accessed from any access point or dial-in location.	____	____	____
	III.4.3	The system shall be capable of generating a report of all Open/Pending Trouble Tickets, with the following fields:			
		User Name	____	____	____
		Circuit ID (Extension)	____	____	____
		Equipment ID	____	____	____
		Location	____	____	____
		Received Date	____	____	____
		Technician	____	____	____
		Trouble Code	____	____	____
		Description of Trouble	____	____	____
	III.4.4	The Trouble Ticket Module shall be capable of providing a report of all open/closed trouble tickets by technician.	____	____	____
	III.4.5	The Trouble Ticket Module shall be capable of providing trouble history reporting, based on the following fields/options:			
		Circuit ID	____	____	____
		Equipment ID	____	____	____
		Problem Code	____	____	____
		Problem Resolution	____	____	____
	III.4.6	The Trouble Ticket shall have the capability to provide complete circuit information on the problem circuit or equipment as a part of the printed trouble ticket. This will include originating equipment and/or office jack, to controlling equipment port, including all cables and devices connections along the path.	____	____	____

Company Name:______________________________ Date:_______________

III.4 Trouble Ticket Module (Continued)		Level of Compliance Full	 Partial	 Non
III.4.7	The Trouble Ticket Module shall have the capability to update the database with changes to the circuit, resulting from the resolution of the trouble ticket.	____	_____	____
III.4.8	The Trouble Ticket Module shall be capable of indicating levels of activity, such as a high number of troubles involving defective equipment or wiring or a high number of trouble tickets in a specific area.	____	_____	____
III.4.9	The Trouble Tickets shall contain, at a minimum, the fields listed in the Trouble Ticket Data Set Table at the end of Section III.5. Vendor shall indicate exact field names, field lengths, which of these are required fields, and shall specify all other required fields in the Trouble Ticket Module of the proposed CMS.	____	_____	____
III.5 Connectivity Components				
III.5.1	Names for Circuit ID, Desktop Equipment ID, Jack/Outlet, Terminating Block, Cable and Controlling Devices shall be user-defined.	____	_____	____
	Vendor shall indicate if proposed system uses any automatic naming convention, and whether such naming can be overridden.	Automatic________ Over Ride ________		
III.5.2	System shall allow for Circuit IDs or Equipment Names (of up to 15 characters) to be created by ABC which uniquely identify each circuit and can be assigned to either a jack/outlet or directly to a cable pair.	____	_____	____
III.5.3	System shall be capable of tracking all inside cables, including local cables between wiring closets, and location of bridges and splices, conduits, cableways, and wiring poles.	____	_____	____
III.5.4	Jacks and station wiring shall be tracked on a pair-by-pair basis, capable of identifying which cable pair of the station wiring is used for a circuit.	____	_____	____

Company Name:_________________________________Date:________________

		Level of Compliance		
III.5	Connectivity Components (Cont'd)	Full	Partial	Non
III.5.5	System shall be capable of tracking performance characteristics of cables, and date when performance testing was conducted.	____	_____	____
III.5.6	Cable Name, Cable Type, From and To Locations, Pair Count, and List of Pairs for each cable must be tracked. System shall track the pair-by-pair status of each cable pair (Used, Spare, Pending, Defective), thereby providing a means of tracking the remaining capacity/availability (%) of any given cable.	____	_____	____
III.5.7	System shall be able to provide a listing of all jacks/outlets in a given room and indicate their status as to Used, Available or Reserved.	____	_____	____
III.5.8	System shall be able to provide a listing of all cables (including Horizontal/Station Cables, House Cables, Fiber Cables) originating or terminating in a specified Building, Floor & Room, whether it be an office location or a wiring closet.	____	_____	____
III.5.9	System shall permit for specific pair counts to be assigned to a given cable, from one location to another, and not be required to begin with Pair 1. For example, Cable HC-SE12, (400-Pair Cable) with Pairs 401 to 800.	____	_____	____
III.5.10	System shall be capable of indicating all conduits and cableways between two points, and the fill capacity of each.	____	_____	____
III.5.11	System shall be capable of tracking Ground Bus Systems.	____	_____	____
III.5.12	System shall allow for multiple associations to a single jack or outlet, so that a splitter can be used with a jack/outlet to handle two connections against the same jack/outlet.	____	_____	____

Company Name:_________________________________ Date:________________

		Level of Compliance		
III.5	**Connectivity Components (Cont'd)**	Full	Partial	Non
III.5.13	The system shall define Controlling Device in the following way:			
	Device Name	____	____	____
	Device Class/Type	____	____	____
	Device Model	____	____	____
	Port List/Names	____	____	____
	Department/Division	____	____	____
	Network Address	____	____	____
	Building	____	____	____
	Floor	____	____	____
	Room	____	____	____
III.5.14	Each Type of Controlling Device shall have a user-defined number of ports with port-types specified by the User, which can be assigned to circuits.	____	____	____
III.5.15	Port Names for each of the Controlling Devices shall be named by the user, rather than a sequential number (1, 2, 3, etc.) by the system. This structure will be used to track 3270-type controllers, Ethernet/Token Ring Servers, Token Ring Hubs/ Cards, and AppleNet controllers, etc., and to assign ports to circuits, as well as provide for tracking of available ports when creating new circuits.	____	____	____
III.5.16	System shall provide a listing of all spare ports of a given port type, based on a specified Equipment Name or location, such as a building, or wiring closet room number.	____	____	____
III.5.17	The system shall allow for the association of a Circuit Name, rather than a physical device, to a Horizontal/Station Cable or Outlet for situations where an in-house data circuit is involved.	____	____	____

Company Name:______________________________ Date:________________

Technical Specifications - Cable and Connectivity Management System January 1, 1999
ABC Company (SAMPLE) Page 19

		Level of Compliance		
III.5	Connectivity Components (Continued)	Full	Partial	Non
III.5.18	System shall provide for a Comments or Description field of 40 characters in each of the data set structures, for additional information, which is not specifically identified by a given field. This comments field shall be included for:			
	Circuit IDs	____	_____	____
	Equipment ID	____	_____	____
	Jack	____	_____	____
	Cable	____	_____	____
	Controller/Port Device	____	_____	____
	Service Order	____	_____	____
	Trouble Ticket	____	_____	____
III.5.19	Vendor shall indicate if "Comments" field has the capability to be queried on, based on certain "key" words contained in the field.	____	_____	____

Company Name:_______________________________________Date:___________________

III.5 Connectivity Components (Cont'd)

III.5.20 System shall provide for basic circuit elements including Circuit ID and/or Desktop Equipment ID, Jack/Outlet, Terminating Block, Cable/Pair and Controller/Port. These circuit elements shall be considered standard database fields, according to the following Data Set Structure Tables. Vendor is to indicate conformance with each field listed in these tables and provide alternatives based on the proposed system's established structure. For each Data Set, Vendor shall indicate the exact field name, field length, field format, if it is a required field, and any additional required fields for each Data Set. Specified field lengths shall be the minimum length which is displayed on any screen, circuit trace or printed report.

Field Name [1,2]	Desired Field Length [1]	Level of Compliance Full	Partial	Non
Service Order Number	10 (A/N)	____	____	____
Date Received/Entered	(XX/XX/XX)	____	____	____
Requester Name	20 (A/N)	____	____	____
Department/Division	10 (A/N)	____	____	____
User Extension (Circuit ID)	15 (A/N)	____	____	____
Building *	10 (A/N)	____	____	____
Floor *	3 (A/N)	____	____	____
Room *	7 (A/N)	____	____	____
Requested Completion Date	(XX/XX/XX)	____	____	____
Description of Work	80 (A/N)	____	____	____
Technician Assigned *	3 (A/N)	____	____	____
Completed Date	(XX/XX/XX)	____	____	____
Comments/Notes *	40 (A/N)	____	____	____
Trouble Ticket Number	10 (A/N)	____	____	____
Date Received/Entered	(XX/XX/XX)	____	____	____
Requester Name	20 (A/N)	____	____	____
Department/Division	10 (A/N)	____	____	____
User Extension (Circuit ID)	15 (A/N)	____	____	____
Building *	10 (A/N)	____	____	____
Floor *	3 (A/N)	____	____	____
Room *	7 (A/N)	____	____	____
Description of Problem	80 (A/N)	____	____	____
Problem Code	4 (A/N)	____	____	____
Technician Assigned *	3 (A/N)	____	____	____
Completed Date	(XX/XX/XX)	____	____	____
Resolution Code	4 (A/N)	____	____	____
Comments *	40 (A/N)	____	____	____
Additional Note Screen*	Variable	____	____	____

Company Name:______________________________ Date:______________

Technical Specifications - Cable and Connectivity Management System January 1, 1999
ABC Company (SAMPLE) Page 21

III.5 Connectivity Components (Cont'd)

Field Name [1,2]	Desired Field Length [1]	Level of Compliance Full	Partial	Non
Circuit ID/Name	15 (A/N)	____	____	____
User	15 (A/N)	____	____	____
Circuit Type	20 (A/N)	____	____	____
Equipment Type *	10 (A/N)	____	____	____
Reference No.	15 (A/N)	____	____	____
Network Address *	17 (A/N)	____	____	____
Drawing Reference No.	8 (A/N)	____	____	____
Department/Division *	10 (A/N)	____	____	____
Status *	3 (A)	____	____	____
Building	10 (A/N)	____	____	____
Floor	3 (A/N)	____	____	____
Room	7 (A/N)	____	____	____
Last Updated By	3 (A/N)	____	____	____
Date Last Updated	(XX/XX/XX)	____	____	____
Comments *	40 (A/N)	____	____	____
User Defined Fields/Components *				
Software	8 (A/N)	____	____	____
Monitor	8 (A/N)	____	____	____
Features	8 (A/N)	____	____	____
Adjunct Equip.	8 (A/N)	____	____	____
Jack/Outlet ID	15 (A/N)	____	____	____
Site	4 (A/N)	____	____	____
Floor	3 (A/N)	____	____	____
Room	7 (A/N)	____	____	____
Type *	4 (A/N)	____	____	____
Pair Count/List *	4 (N)	____	____	____
Status *	1 (A)	____	____	____
Vendor *	20 (A/N)	____	____	____
Part Number *	20 (A/N)	____	____	____
Date Installed	(XX/XX/XX)	____	____	____
Comments *	40 (A/N)	____	____	____

Company Name:______________________________________Date:__________________

III.5 Connectivity Components (Cont'd)

Field Name [1,2]	Desired Field Length [1]	Level of Compliance Full	Partial	Non
Cable	12 (A/N)	____	____	____
Type	10 (A/N)	____	____	____
From Building	10 (A/N)	____	____	____
From Floor	3 (A/N)	____	____	____
From Room/Closet	7 (A/N)	____	____	____
To Building	10 (A/N)	____	____	____
To Floor	3 (A/N)	____	____	____
To Room/Closet	7 (A/N)	____	____	____
Pair Count/List	4 (N)	____	____	____
Pair Status	1 (A)	____	____	____
Vendor *	20 (A/N)	____	____	____
Part Number *	20 (A/N)	____	____	____
Date Installed *	(XX/XX/XX)	____	____	____
Comments *	40 (A/N)	____	____	____
Interconnect Block ID	15 (A/N)	____	____	____
Device Type *	10 (A/N)	____	____	____
Block Pin Count/List *	4 (N)	____	____	____
Building *	10 (A/N)	____	____	____
Floor *	3 (A/N)	____	____	____
Room *	7 (A/N)	____	____	____
Comments *	40 (A/N)	____	____	____
Equipment ID	15 (A/N)	____	____	____
Equipment Class	15 (A/N)	____	____	____
Equipment Type/Model	10 (A/N)	____	____	____
Network Address *	15 (A/N)	____	____	____
Department/Division *	10 (A/N)	____	____	____
Reference No. *	8 (A/N)	____	____	____
Vendor *	20 (A/N)	____	____	____
Part Number *	20 (A/N)	____	____	____
Port Names/List	15 (A/N)	____	____	____
Port Status	1 (A)	____	____	____
Building	10 (A/N)	____	____	____
Floor	3 (N)	____	____	____
Date Installed	(XX/XX/XX)	____	____	____
Room *	7 (A/N)	____	____	____
Comments *	40 (A/N)	____	____	____

Company Name:______________________________________Date:__________________

Technical Specifications - Cable and Connectivity Management System — January 1, 1999
ABC Company (SAMPLE) — Page 23

III.5 Connectivity Components (Continued)

Notes to Section III.5.20:

(1) Indicate Field Name and Length, If Different
(2) Indicate Required Fields with "R" to the Right of Field Name

(*) Indicates use of field must be optional.

A - Alphabetic
N - Numeric
A/N - Alpha/Numeric

III.6 Circuit Functionality	Level of Compliance Full	Partial	Non
III.6.1 System shall allow cable pairs on any given cable to be used in a circuit flowing in either direction from the cable endpoints. Once a cable is created in the database with a Beginning Point and Ending Point, system shall allow circuits to be created which originate at either the Ending Point or the Beginning Point of the cable.	____	_____	____
III.6.2 Use of ports as a termination point in any given circuit shall be optional. Circuits must be able to be terminated at any given wiring closet, or MDF, without having to connect to a PBX port or data port.	____	_____	____
III.6.3 System shall provide for the creation of circuits and termination of circuits at any given point along the connectivity path, including beginning in an office or specific wiring closet, and terminating at another wiring closet, or controlling device, without specifically requiring a jack/outlet or termination device or port.	____	_____	____
III.6.4 System shall allow up to 8 (eight) sequential cables to be contained in a single circuit, with up to 4 (four) parallel cable pairs per cable/circuit.	____	_____	____

Company Name:_________________________________ Date:________________

		Level of Compliance		
III.6	**Circuit Functionality (Cont'd)**	Full	Partial	Non
III.6.5	System shall be able to design and suggest the best cable path, including cross-connects, between any two given points, and also seek additional alternate routes, if specified. Vendor shall specify how they will address this requirement.	____	____	____
III.6.6	System shall provide all Wiring Closet Connections for a given station/horizontal cable against any cable and pair(s), or controlling equipment port.	____	____	____
III.6.7	System shall allow multiple associations to a desktop device such as a PC where multiple data connections may exist.	____	____	____
III.6.8	System shall allow multiple associations to a given cable pair, such as with bridged circuits.	____	____	____
III.6.9	System shall provide for tracking of circuits which terminate at multiple points, such as Bridged Numbers.	____	____	____
III.6.10	System shall provide a warning, with over-ride capability, when a user is attempting to assign a Cable Pair, Circuit ID, Controller Port, etc. which is already assigned to another circuit.	____	____	____
III.6.11	System shall provide for bridging of cable pairs to multiple locations.	____	____	____
III.6.12	System shall be able to trace the entire wire path for any given circuit, based on Circuit ID, Jack/ Outlet, Station/Horizontal Cable, Terminating Block/Pin, cable and pair, controller/port, wiring closet.	____	____	____

Company Name:________________________________ Date:______________

Technical Specifications - Cable and Connectivity Management System | January 1, 1999
ABC Company (SAMPLE) | Page 25

III.6 Circuit Functionality (Cont'd)		Level of Compliance Full	Partial	Non
III.6.13	When creating or tracing circuits, system shall provide for the following associations:			
	Circuit ID/Equipment ID to Jack	____	_____	____
	Circuit ID to Cable/Pair	____	_____	____
	Jack Wiring Pair to Cable/Pair	____	_____	____
	Cable/Pair to Cable/Pair	____	_____	____
	Jack Wiring Pair to Controller/Port	____	_____	____
	Cable/Pair to Controller/Port	____	_____	____
	Jack Wiring to Block/Pair Position	____	_____	____
	Cable/Pair to Block Pair/Position	____	_____	____
	Block/Pair Position to Block/Pair Position	____	_____	____
III.6.14	System shall trace circuits with a schematic display of the circuit (in a block diagram), in either direction, from originating or terminating point.	____	_____	____
III.6.15	System shall be able to print a copy of the circuit both schematically and in text format, showing where the circuit originates, the path it takes, and where it terminates.	____	_____	____
III.6.16	System shall be capable of displaying the entire circuit path on a single screen, with "Print Screen" capability.	____	_____	____
III.6.17	System shall provide for the direct association of cable pairs to cable pairs at appropriate concurrent locations in the database. Vendor shall specify how they will address this requirement, if it is non-standard for the proposed system.	____	_____	____
III.6.18	System shall be capable of handling bulk connects and disconnects of cable pairs or Controller Ports when removing cables or moving Controlling Equipment.	____	_____	____

Company Name:_________________________________Date:________________

		Level of Compliance
III.6	**Circuit Functionality (Cont'd)**	Full Partial Non
III.6.19	System shall provide a warning or flag when a low availability of spare cable pairs or controlling equipment ports has been reached. Vendor shall indicate if the warning level is system-defined, or user-defined.	____ ____ ____ System-Defined:___ User-Defined:_____
III.6.20	System shall at a minimum trace from desktop equipment to controlling device port. Optional tracing from desktop through intermediate controlling equipment to backbone equipment is desired. An example would be a Mainframe/SNA Circuit which may be traced from the terminal through the controller to the FEP.	____ ____ ____
III.7	**Graphics/CAD Capability**	
III.7.1	The system shall be a fully integrated Graphics-Database Cable Management System.	____ ____ ____
III.7.2	System shall be capable of utilizing existing AutoCAD drawings (.DWG) or DXF files.	____ ____ ____
III.7.3	The link between the database and the graphics shall provide for bi-directional flow of information between modules. Any updates/changes made through either module must automatically update the other module.	____ ____ ____
III.7.4	System shall have the ability to update CAD drawings (lines, walls, icons, text) through the CMS, without having to maintain/revise the drawings in a separate CAD application. Vendor shall specify how this requirement will be met, if non-standard for the proposed system.	____ ____ ____
III.7.5	Vendor shall provide a list of all CAD functions available in the proposed CMS System. (Attach List)	________________ ________________ ________________ ________________

Company Name:________________________________Date:________________

Technical Specifications - Cable and Connectivity Management System — January 1, 1999
ABC Company (SAMPLE) — Page 27

			Level of Compliance		
III.7	Graphics/CAD Capability (Cont'd)		Full	Partial	Non
	III.7.6	Vendor shall provide a list of standard icons which are supplied with the proposed CMS system. (Attach List)	______		
	III.7.7	System shall provide the option to create user-defined icons and associate them to database components.	___	___	___
	III.7.8	System shall be capable of selecting a drawing by entering a building name or description of the drawing, without having to rely on an index number of the drawing.	___	___	___
	III.7.9	System shall be capable of displaying a site or building drawing, and then be able to select on a building or floorplan, and display that drawing.	___	___	___
	III.7.10	System shall be able to generate printed copies of selected drawings with cable routes and component icons displayed, zoomed views with cable routes and component icons displayed, or selected drawings with only cable routes of a specified circuit shown.	___	___	___
	III.7.11	System shall be capable of creating DXF files with specific cable routes or component icons displayed as desired.	___	___	___
	III.7.12	System shall be capable of printing drawings to both printers and plotters, depending on size desired.	___	___	___
	III.7.13	System shall be capable of identifying the drawing name/number associated with a Circuit ID on the Circuit ID record.	___	___	___
	III.7.14	System shall be capable of printing drawings as indicated below:			
		To a local printer/plotter,	___	___	___
		To a remote LAN printer/plotter	___	___	___
		To a file for printing later	___	___	___

Company Name:______________________________Date:______________

		Level of Compliance Full Partial Non
III.7 Graphics/CAD Capability (Cont'd)		
III.7.15	Vendor shall specify all printing equipment required to meet Item III.7.14.	______
III.8 Reports		
III.8.1	Vendor shall provide a list of all standard reports generated by the proposed CMS. (Attach List)	___ ___ ___
III.8.2	System shall provide a report of all equipment, by type, location, or department/division.	___ ___ ___
III.8.3	System shall provide for User-Defined Custom Reports, or a report-writer function for creating custom reports. Vendor shall state alternatives to meeting this requirement.	___ ___ ___
III.8.4	System shall provide a Report of all Open/Pending Service Orders with the following information:	
	User Name	___ ___ ___
	Circuit ID (Extension)	___ ___ ___
	Equipment ID	___ ___ ___
	Due Date	___ ___ ___
	Technician	___ ___ ___
	Description of Work	___ ___ ___
III.8.5	System shall provide a Report of all Open/Pending Trouble Tickets with the following information:	
	User/Requester Name	___ ___ ___
	Circuit ID (Extension)	___ ___ ___
	Equipment ID	___ ___ ___
	Date Received	___ ___ ___
	Technician	___ ___ ___
	Trouble Code	___ ___ ___
	Description of Trouble	___ ___ ___

Company Name:______________________________Date:______________

Technical Specifications - Cable and Connectivity Management System January 1, 1999
ABC Company (SAMPLE) Page 29

		Level of Compliance		
III.8 Reports (Cont'd)		Full	Partial	Non
III.8.6	System shall provide a report which lists all cable pairs for a specified cable, with the status of each pair, (Used, Spare, Reserved/Pending or Defective), and the Circuit ID associated with each pair in use. This report shall also provide the remaining available capacity (%) for the specified cable.	____	____	____
III.8.7	System shall provide a list of Ports by Controlling Device/By Status or By Type/By Location (Used/ Reserved/Spare/Defective).	____	____	____
III.8.8	Vendor shall indicate printing capabilities for the proposed CMS, for purposes of printing reports, Service Orders and Trouble Ticket information. These shall include:			
	Printing at the local terminal	____	____	____
	Printing to a local shared printer	____	____	____
	Printing to a remote LAN printer in another building/another site	____	____	____
III.8.9	Vendor shall indicate where selection of print destination is made, i.e., CMS, LAN Software Profile, PC Configuration File, Windows, Etc.	____________		
III.8.10	System shall be capable of creating reports in ASCII format for importing into other database files.	____	____	____

Company Name:______________________________ Date:______________

		Level of Compliance		
III.9	**Validation Functionality & Look-Up Tables**	Full	Partial	Non
III.9.1	System shall have Validation Tables for certain key fields, indicated below:			
	Buildings	____	_____	____
	Cable Types	____	_____	____
	Service Types	____	_____	____
	Equipment Classes	____	_____	____
	Equipment Types/Models	____	_____	____
	Vendor shall indicate all other Validation Tables for the proposed CMS.	________________		
		________________		
		________________		
		________________		
III.9.2	System shall have Validation Tables which prevent the creation of two cables or devices with the exact same name or number.	____	_____	____
III.9.3	System shall have a Look-Up feature associated with the validation tables and key fields of component data sets listed below. This Look-Up feature shall be available on all data entry screens and query screens.			
	Jack/Outlet Number	____	_____	____
	Cable Name	____	_____	____
	Circuit ID	____	_____	____
	Extension Number	____	_____	____
	Service Order Number	____	_____	____
	Trouble Ticket	____	_____	____
III.9.4	User shall be able to narrow down the list of options provided by the Look-Up tables through entering the first letter or number in the field.	____	_____	____

Company Name:________________________________ Date:________________

Technical Specifications - Cable and Connectivity Management System — January 1, 1999
ABC Company (SAMPLE) — Page 31

		Level of Compliance		
III.9	Validation Functionality & Look-Up Tables (Cont'd)	Full	Partial	Non
III.9.5	System shall provide for a Wildcarding feature for query purposes. This feature shall allow for a portion of a field being completed, such as the beginning 2-3 characters, or the ending 2-3 characters, with the remainder of the field completed with a wild-card key, or substitution character. The CMS shall then supply a list of database entries matching the information entered in the field. This query capability shall be available for the following fields:			
	Circuit ID (Extension)	____	____	____
	Jack/Outlet Number	____	____	____
	Cable Name	____	____	____
	Service Order Number	____	____	____
	Controlling Device/Port	____	____	____
	Building	____	____	____
	Equipment Type	____	____	____
III.10	Data Entry			
III.10.1	System shall provide for bulk creation and loading of Cable Pairs when entering cables into the database. When entering a 300-Pair cable with pairs 301 to 600, the system shall generate the available cable pairs in a single step, for selection when creating circuits.	____	____	____
III.10.2	System shall be able to generate bulk cross connects of a range of cable pairs to a termination block and row, to a controlling device or to another range of cable pairs.	____	____	____
III.10.3	System shall be capable of utilizing flat files from existing databases, (especially those noted below) for loading information into the CMS product.			
	Excel	____	____	____
	Lotus	____	____	____

Company Name:____________________________________Date:__________________

		Level of Compliance		
III.10	Data Entry (Cont'd)	Full	Partial	Non
III.10.4	Vendor shall describe the process & format required for loading the CMS using ASCII Files from existing database files. Vendor shall also provide a list of the database formats which can be downloaded and an estimate of time associated with loading of a sample (typical) ASCII file. (Use space below or attach procedures.)	____	_____	____

____
____
____
____
____
____
____
____
____
____
____

III.10.5	If a program/file must be developed by the Vendor for downloading or converting existing files for use in loading the proposed CMS, Vendor shall indicate the cost for developing such a program in the Attachment entitled "Project Pricing."	____	_____	____
III.11	System Back-Up and Archiving			
III.11.1	Vendor shall provide as a part of the RFP response, a copy of the Procedures and Hardware Requirements for automatically backing up to a mass storage device, the system database and drawing files, application software, graphics software, and operating system software. Any costs associated with the Hardware required shall be noted in the Attachment entitled "Project Pricing."	____	_____	____
III.11.2	Vendor shall provide procedures for restoring data from the back-up to the System File Server.	____	_____	____

Company Name:___________________________________Date:_________________

Technical Specifications - Cable and Connectivity Management System — January 1, 1999
ABC Company (SAMPLE) — Page 33

		Level of Compliance		
III.11	**System Back-Up and Archiving (Cont'd)**	Full	Partial	Non
III.11.3	System shall be capable of having multiple servers at various sites, which can be networked together to provide alternate/back-up access if the link between sites is down.	____	_____	____
III.11.4	System must be capable of archiving service order and trouble ticket files after a period of 6 - 12 months to a tape back-up.	____	_____	____
III.12	**Network System Interfaces**			
III.12.1	ABC desires for the selected CMS to have interfaces to as many of its networks and management systems as possible. Vendor is to indicate if their system currently has links or interfaces to the following systems, or if a link is under development:			
	IBM Netview	____	_____	____
	Bay Networks Net Manager	____	_____	____
	Cabletron Spectrum	____	_____	____
	AT&T Voice Switch	____	_____	____
	Meridian Voice Switch	____	_____	____
	Rolm Voice Switch	____	_____	____
III.12.2	Vendor shall indicate if the above links/interfaces can handle downloads from these systems into the CMS system.			
	IBM Netview	____	_____	____
	Bay Networks Net Manager	____	_____	____
	Cabletron Spectrum	____	_____	____
	AT&T Voice Switch	____	_____	____
	Meridian Voice Switch	____	_____	____
	Rolm Voice Switch	____	_____	____

Company Name:______________________________________Date:__________________

Technical Specifications - Cable and Connectivity Management System January 1, 1999
ABC Company (SAMPLE) Page 34

		Level of Compliance		
III.13	Billing System Interfaces	Full	Partial	Non
III.13.1	Vendor shall indicate if proposed CMS product has a billing module for use in billing individual departments or users for services or connections.	____	_____	____
III.13.2	System shall be capable of downloading information to a billing system for chargebacks of services and connections.	____	_____	____
III.13.3	System shall be capable of generating a report format for use in billing individual departments/cost centers for service order activity on a monthly basis.	____	_____	____
III.13.4	System shall be capable of generating a report format for use in billing individual departments/cost centers for based on inventory items, or equipment which is in use.	____	_____	____
III.13.5	System shall be capable of providing on-demand and scheduled reports by Department, Equipment Type and Building location for billing purposes.	____	_____	____
III.13.6	System shall be capable of generating reports on individual charges by cost center, department or user which can be printed or saved to disk.	____	_____	____

Company Name:________________________________Date:______________

IV GLOSSARY

Term/Abbreviation	Definition
ABC	ABC Company
CER	Central Equipment Room
CMS	Cable Management System
CPU	Central Processing Unit
FEP	Front End Processor
IBM	IBM Compatible Personal Computer
MDF	Main Distribution Frame
MER	Main Equipment Room
PC	Personal Computer
RAM	Random Access Memory
RFP	Request For Proposal

EVALUATING RFP RESPONSES

23

Upon receiving back the Vendor responses to your RFP, you will need to evaluate them so that you get the best possible comparison of system, and identify the system which best meets the needs of your specific environment. This chapter will provide instructions on preparing an evaluation matrix, which is the key tool for comparing proposals on the different software products in a meaningful way. You will score the performance and pricing of each candidate CMS and identify the right system(s) for your organization. In addition, you might consider formal evaluations by team members during live product demonstrations, and prepare general lists of the advantages and disadvantages for each candidate CMS. Together, these documents provide a thorough and consistent basis for making a system selection. Further, this process provides tremendous educational benefits and will better prepare you to implement the selected system

Preparing an Evaluation Matrix

The evaluation matrix lists each of the specifications contained in your RFP in order to compare the CMS products on a common basis. The matrix should be created on an electronic spreadsheet such as Lotus, Excel, or QuatroPro so that points for each of the requirements can be totaled automatically for each of the systems and you can easily get an overall total score. If you do not have easy access to an electronic spreadsheet, the matrix can be prepared on paper but the scoring calculations will have to be done manually. A sample matrix is attached at the end of this chapter for your reference. You may want to prepare a separate sheet for pricing, which lists only the options indicated on the "Pricing Sheet." This will be helpful if the purchasing department requires a breakdown on the pricing comparisons. A sample pricing matrix is also included at the end of this chapter. Remember that these documents are only examples, and your versions will reflect the particular needs and requirements that you establish for your organization.

The first column of the matrix should be used for the Item/Specification Number, such as II.1.1, II.1.2, or however you numbered each of the specifications in your RFP. This will provide a reference point back to the RFP when comparing the different products against each specification. The second column can be used for a short abbreviation or reference as to what the specification involves. This avoids the need to refer to the RFP every time you want to discuss the specification item. The third column can be a narrow column, which can be used to flag or mark the more critical specifications which you have identified as being more important and possibly carrying more weight in the scoring. The fourth, fifth, sixth, etc., columns should be used to list each of the Vendors who responded to your RFP. You can list either the CMS Product or the Vendor Name, or both.

If you have not already done so, you must identify for yourself which of the technical specifications contained in the RFP are the most critical. This provides a means for directly comparing the different CMS products in the most important areas. When scoring the performance characteristics of each CMS, a higher rating is given when the candidate systems meet the more critical requirements - in other words, you can weigh the critical items more heavily.

Determine how you want to score each of the products for either meeting or partially meeting the individual requirements. For example, if a CMS has the "Full Compliance" column marked for a specific requirement, you might give them a score of "1"; if another CMS has the "Partial Compliance" column marked, you might give them a score of "0.5". A CMS which has the "Non Compliance" column marked would get a score of "0". These are only examples. You might want to use a scoring of "2" for Full Compliance, "1" for Partial Compliance and "0" for Non Compliance.

Once you have established the scoring for each of the specifications contained in your RFP, you will need to determine if you want to give a higher rating to those items which you have identified as the critical or most important items. You might want to use a score of "3" for Full Compliance, and "1.5" for Partial Compliance for those more critical items. If you decide to give a higher rating to the critical items, you will want to put some indicator such as an Asterisk (*) or Plus (+) mark in the column next to the Specification Item Number. This will act as a flag to remind you that those items get a higher rating, and also provides a means of sorting the items to provide a total scoring for each CMS product on the critical items only. Make a note at the bottom of the matrix as to what scoring table you are using. It will make for an easy reference when discussing the results with other team members and management when documenting the products which best meet your specifications.

Go through each of the RFP responses one by one, and fill in a score for each item contained in the RFP. Whether you give the Vendors a scoring for each area where they were to provide additional information, such as product literature, company information, references, etc. is optional. You may or may not want to include those items in the matrix. Sometimes, it helps to include them so you have a check list to make sure the Vendor has complied with all your requests. This helps when you are evaluating the Vendors on their responsiveness and cooperation with the bid instructions. How they respond and follow instructions tells you something about the Vendor's organization.

If there are matters for interpretation, be sure to discuss things fully with the team. Most Vendors are happy to provide clarification or detail upon written request. On the other hand, if a Vendor requests more information from you, it is usually required that the same information be provided to all Vendors in the bidding.

After you have gone through all the RFP responses, total the scores for each Vendor. The totals will help to rank the CMS products on a technical functionality basis. Double check each one to make sure you have given them the proper score for each specification. Use the column which has the Critical Item indicator to sort out the most important requirements, and total the scores for those items only for each Vendor. This will help to identify which products best meet your requirements for the critical items. This part of the evaluation deserves as much consideration as the overall score. A high-scoring CMS that misses key areas will not be a satisfactory system.

Vendor Presentations and Demonstrations

Once you have totaled the scores for technical performance for each of the CMS products, the next step is to see live presentations or demonstrations of the candidate systems. This is one of the best ways to do a final evaluation of the CMS software and to evaluate the Vendor's responsiveness, experience and knowledge. Some organizations may want to schedule these presentations during the time they are waiting for the Vendors to respond to the RFP. While this is an alternative, it often results in different understandings of your needs by the different Vendors. They will use their time with you to get a better understanding of how you will evaluate the systems and what they can do to make their product look more desirable. Due to the different questions which might be asked by different Vendors in these presentations, it is best to avoid any meetings with the Vendors until they have all responded to the RFP. This will also avoid scheduling demonstrations with Vendors that do not meet the minimum requirements of the RFP.

The next step is to determine if you want all the Vendors who submitted RFP responses to come in and provide a demonstration of their product. Based on the technical scoring and completeness of the RFP responses, you may determine you only need to see the top two or three systems.

If you decide to eliminate any of the CMS products based on the RFP response, document your reasons for eliminating that particular product. This will be beneficial when the Purchasing Department wants a substantive reason if the lowest bidder is eliminated. Price, while a factor in selecting a CMS, should be the one of the last reasons for elimination. If you want a successful recordkeeping system which will meet your needs, you do not want to select a product based solely on price. How well it meets your functionality requirements should be the key deciding factor.

Schedule the individual presentations for each of the final CMS candidates. It is usually sufficient to allow them approximately 2 hours for the actual presentation, with an hour or two for questions and further demonstration based on questions asked by the audience. Be sure to have the Vendors indicate if any special equipment or accommodations are needed for the presentation.

You may want to supply the Vendors with a list of the types of queries and functions you would like to see demonstrated during their presentations. You may also want to include specific items such as creating Horizontal Cables, Token Ring Connections, Riser Cables, Fiber Cables; Tracing a Circuit; Creating a Circuit; Creating a Service Order for a Phone Move; Creating a Report of all cable connections for a specific cable, etc. This will help to make sure that each Vendor demonstrates similar functionality so that you can see which one handles the different tasks better. Keep your requests fairly concise and specific, so they can be performed in the two-hour time frame.

Invite as many individuals from the team as possible to the presentations. Be sure to have a representative from each of the different groups or organizations who will be working with the CMS product such as technicians, help desk personnel, customer service personnel, management, facilities management, etc. The more individuals who see and evaluate the functionality on a first hand basis, the better your evaluation will be.

Prepare an Evaluation and Comments form to be completed by each individual attending the Vendor presentations. This will help to capture the comments of each attendee while it is fresh in their mind, and it also helps to trigger thoughts on what they specifically liked or disliked about each product or Vendor. Include areas for screen display, moving around the database, graphics capabilities, circuit creation and display, etc. You may want to score some items on a 1 to 10 scale to make it easier for individuals to complete the form and for summarizing all evaluation forms for comparison. A sample form is provided at the end of this chapter.

Upon completion of the Vendor presentations, you should be at a point to identify which of the CMS products will be your top candidates. To help in documenting the technical and visual performance of each CMS, you will need to summarize the results of the RFP Responses and Vendor Presentations.

Prepare an Advantages/Disadvantages sheet for each of the CMS Products based on CMS Team discussions. This will help the team focus on critical areas, and will help build consensus as to the final required functionality. Use the information submitted on the Evaluation and Comments forms and the results of the scoring in the RFP Response. A sample sheet is included at the end of this chapter for your reference. During discussions with the team, fill in any missing comments and add any questions which need to clarified with the Vendor.

If you decide to eliminate any of the CMS products based on the these Advantages/ Disadvantages, this will help to document your reasons for eliminating that particular product. This will be beneficial when the Purchasing Department wants a substantive reason if the lowest bidder is eliminated. Again, how well the CMS meets your functionality requirements should be the key deciding factor.

CMS/Vendor References

After you have scored each of the systems on technical merit and have seen actual product demonstrations, you have probably narrowed down your list of potential candidates. The next area which also helps to identify the best CMS is to check the client references provided by the Vendors. This is a very important part of the evaluation process. You will wish to know that the CMS has been properly represented, that it has been operating satisfactorily and with few bugs, and that the Vendor has been responsive in handling problems or questions for assistance. Be aware that some organizations will struggle with a CMS because of problems of their own making, and that this does not necessarily indicate problems with the CMS.

Prepare a brief list of questions you would like to ask the client references. You want to be sure to cover areas on product performance, vendor responsiveness, enhancements and upgrades, quality of documentation, user satisfaction, ease of use, response time, length in service, etc. The individual questions will vary depending on your environment and the type of system you are evaluating.

Select a CMS Team member to be the spokesperson for each of the references which are to be contacted. It is always good to contact at least two references for each candidate CMS. For best results, conduct the call to the client reference as a Conference Call. This way, all team members hear the reference's comments, and often additional questions are triggered which may be asked toward the end of the call.

Inform the client reference that you are currently evaluating CMS products, and that their name was provided as a reference for Product X, from Vendor XYZ, and that your CMS evaluation team would like to ask them a few questions, if they have a few minutes. Often, the reference will be glad to answer your questions or will transfer you to an active user who is more familiar with the product. Be sure to thank the Client Reference for their time and assistance.

Determining the Final Candidates

You have now completed all of the steps in evaluating the CMS products and Vendors. By this time, you should have a good idea of which specific systems are your first and second choices. Complete the following steps to finish your documentation on the performance of each CMS.

Update the Advantages/Disadvantages sheet for each CMS to include the comments made by the different Client References. Use these sheets as a basis for discussions with CMS Team members to eliminate any additional CMS candidates. Prepare a brief summary sheet indicating the reasons for eliminating each of the CMS products which have been deleted from the list of remaining potential systems. Again, this will be beneficial in documenting the reasons for your final choice.

By this time, there should be a clear indication as to the number one candidate. If there are two systems still equally qualified, then the decision can be made on price or a particularly strong aspect of one CMS or Vendor. If the best candidate still falls short in one or two key areas, the only remaining step which can help to narrow down the choice would be to consider whether customization of the software is appropriate to best meet the remaining requirements.

Identify the areas for each of the remaining CMS products where they do not meet or only partially meet critical requirements. Any areas which are lacking and which the CMS Team determines are necessary should be documented as items for customization. The Vendors should be contacted and asked to provide a quote on cost and timing to customize their product to meet these requirements. You may have to work with the Vendor a little bit to develop a complete and understandable description of the desired customization.

Issues of software customization are covered in Chapter 24. After customization is considered, the final purchase decision must be made. An optional final system review and selection approach is presented in Chapter 25.

ABC COMPANY - CMS EVALUATION MATRIX - PAGE 1

Spec.#	Brief Description	CMS #1		CMS #2	CMS #3	
Section II - Installation & Support						
II.1.1	Installation Schedule	3-4 Wks		4 Wks	4 Wks	
II.1.2	Installation & Testing Procedures	0	NI	1	0.5	
II.1.3	Installation Within 4 Weeks	0	NI	1	0.5	
II.2.1	Equipment to be Supplied By Vendor	0	NI	1	0.5	
II.2.2	Equipment Specifications - Server	0	NI	1	0.5	
II.2.3	Equipment Specifications - Local User	0	NI	1	0.5	
II.3.1	List of Documentation Manuals	0	NI	1	1	
II.3.2	6 Copies of Documentation	0	NI	1	1	
II.3.3	Documentation Revision Process	0	NI	1	1	
II.3.4	Full Training for 6 Employees	0	NI	1	1	
II.4.1	Maintenance Policy	0	NI	1	1	
II.4.2	24-Hour/7-Day Week Help Line	0	NI	1	0.5	
II.4.3	Non-Standard 24-Hour Support	0	NI	Standard	Non-Std	
II.4.4	Upgrade/New Release Policy	0	NI	1	1	
II.4.5	Dial-In Access Required For Support	0	NI	YES	YES	
II.4.6	User's Group Activities	0	NI	0	0	
II.5.1	Customization/Enhancements Policy	0	NI	1	1	
II.5.2	List of Enhancements Currently Underway					
II.6.1	Standard Warranty (One-Year)	0	NI	1	1	
II.6.2	24-Hour Response Major Defects	0	NI	1	1	
II.6.3	Copy of Source Code in Escrow	0	NI	1	1	
II.7.1	Copy of CMS License Agreement	0	NI	1	1	
II.7.2	Third Party Operating/Application	0	NI	1	1	NA
II.8.1	Company Information	1		1	1	
II.9.1	References	1		1	0	NI
	Section Summary/Total	2		18	14	

ABC COMPANY - CMS EVALUATION MATRIX - PAGE 2

Spec.#	Brief Description	CMS #1		CMS #2	CMS #3
Section III - Functional Requirements					
III.1 - General Specifications					
III.1.1	* IBM Compatible Equipment	2		2	2
III.1.2	Operating Environment	1		0.5	0.5
III.1.3	DOS				
	Unix	1		1	1
	NT	1		1	1
	OS/2			1	
	Other			1	
III.1.4	Protocol(s):				
	TCP/IP	1		1	1
	SPX/IPX			1	1
	Other				
III.1.5	* PC Environment on LAN	2		2	2
III.1.6	* Networking of Master & Back-Up Servers	0		2	2
III.1.7	* Windows 3.1	0		2	1
III.1.8	* Up to 20 Simultaneous Users	2		2	2
III.1.9	* Separate Database for Sites	1	C	2	2
III.1.10	Menu Options Displayed/By-Pass	0.5	C	1	1
III.1.11	* Help Function/Key - Screen Specific	2		2	2
III.1.12	* 3 - 5 Second Response Time	1		2	2
III.1.13	Levels of Authorization	Screen		Module	Dataset
	Control - at Module or Dataset	1		1	1
III.1.14	* Security - IDs & Passwords	0	NI	2	2
III.1.15	Date Entered/Last Update	1		1	1
III.1.16	Entered By User/Last User to Update	0		0.5	0
III.1.17	* File Locking on "In-Use" Records	2		2	2
III.1.18	Reserved Status (Manual & S.O.)	0.5		0.5	1
III.1.19	Dial-In / Data Entry & Operation	1		1	1
III.1.20	Print Screen Capability	1		1	1
III.1.21	Troubleshooting Reference Guide	1		1	1
III.1.22	EIA/TIA Compliant	0.5		1	1
	Section Summary/Totals	20.5		30.5	29.5
III.2 - Asset Tracking/Inventory Control					
III.2.1	Current Inventory - Desktop & Control	1		1	1
III.2.2	Inventory Fields & Length (Total 18)	0.5		0.5	1
III.2.3	Master Catalog File with Verification	0.5		1	0.5
III.2.4	No. of Software Items/Features (Info only.)	15		NI	Unlimited
III.2.5	Status - Used, Spare, Reserved	0.5		1	1
III.2.6	Pending Service Order Indicator	0.5		0	0
III.2.7	Open Trouble Ticket Indicator	0		0	0
	Section Summary/Totals	3		3.5	3.5

ABC COMPANY - CMS EVALUATION MATRIX - PAGE 3

Spec.#	Brief Description	CMS #1		CMS #2		CMS #3
III.3 - Service Order/Move, Add, Change						
III.3.1	* Service Order for Tracking Moves/Changes	2		2		2
III.3.2	Dial-In Access/Creation of Service Orders	1		1		1
III.3.3	Open/Pending Order Report (Total 6)	1		1		1
III.3.4	Report of Orders by Technician	1		1		1
III.3.5	Permit Removal/Cancellation of S.O.	0.5		1		1
III.3.6	Updates/Changes Outside S.O.	0		1		0
III.3.7	* Create or Delete Entire Circuit in S.O.	2		0.5	UD	0
III.3.8	Add/Delete Inventory Components	1		1		1
III.3.9	Printed Copy Include Tech'n Instructions	0.5		1		1
III.3.10	Optional Inclusion of Circuit Connections	0.5	UD	1		1
III.3.11	Management of Status Indicators	1		1		1
III.3.12	Activity Indicators by S.O. Type	0	C	0		0.5
III.3.13	Data Set Fields/Tables	0.5		1		1
	Section Summary/Totals	11		12.5		11.5
III.4 - Trouble Ticket Module						
III.4.1	Trouble Ticket for Managing Trouble Calls	0.5	NDI	1		1
III.4.2	Dial-In Access & Any LAN Connection	0.5	NDI	1		1
III.4.3	Open/Pending Ticket Report (Total 8)	1		1		1
III.4.4	Report of Trouble Tickets by Technician	1		1		1
III.4.5	* Trouble History Reporting By Field (Total 4)	2		1	OM	2
III.4.6	Provide Complete Circuit Info. with Ticket	1		1		1
III.4.7	* Capability to Update Circuit Changes	2		2		1
III.4.8	Indicate Levels of Activity Types	0.5		1		1
III.4.9	Contains Fields Indicated in III.5.20					
	Section Summary/Totals	8.5		9		9
III.5 - Connectivity Components						
III.5.1	Component Names/IDs - User-Defined	1		1		1
III.5.2	Circuit IDs up to 15 Characters	1		1		1
III.5.3	Cables/Bridges/Splices/Conduits/Pathways	1		1		1
III.5.4	Station Wiring - Pair-By-Pair Basis	1		1		1
III.5.5	Cable Performance & Test Date	1		1		1
III.5.6	Cables By Name, Type, From/To & Capacity	1		1		1
III.5.7	List All Jacks in a Room with Status (U/A/R)	1		1		1
III.5.8	Provide Listing of All Cables in a Room	1		1		1
III.5.9	Allow User-Specified Cable Counts	1	UF	0	C	1
III.5.10	List Conduits/Cableways & Fill Capacity	1		1		1
III.5.11	Tracks Ground Bus Systems	0.5		1		1
III.5.12	Use of Splitters/Multiple Connections	1		1		1
III.5.13	Controlling Equipment Tracking (Total 7)	1		1		1
III.5.14	User-Defined No. of Ports/Controlling Equip.	0	NI	1		1

ABC COMPANY - CMS EVALUATION MATRIX - PAGE 4

Spec.#	Brief Description	CMS #1		CMS #2		CMS #3
III.5.15	User-Defined Port Names/Controlling Equip.	0	NI	1		1
III.5.16	Listing of Spare Ports - By Type/Location	0		1		1
III.5.17	Use of Circuit ID vs Equipment	1		1		1
III.5.18	Comments/Description Field (Total 7)	0.5		1		1
III.5.19	Query Capability on Comments Field	0		1		1
III.5.20	DataSets - Service Order (Total 14)	0.5	FL	0.5	C	1
III.5.20	DataSets - Trouble Ticket (Total 15)	0.5	FL	1		1
III.5.20	* DataSets - Circuit I.D. (Total 21)	1	FL	1		2
III.5.20	DataSets - Jack/Station Cable (Total 15)	0.5	FL	1		1
III.5.20	DataSets - Cable (Total 13)	0	NI	1		1
III.5.20	DataSets - Interconnect Block/Panel (Total 7)	0.5	FL	1		1
III.5.20	DataSets - Controlling Equipment (Total 14)	0.5	FL	1		1
	Section Summary/Totals	17.5		24.5		27
III.6 - Connectivity Components						
III.6.1	Cable Pairs/Circuits Flow in Either Direction	1		1		1
III.6.2	User or Ports as Termination Points - Optional	1		1		1
III.6.3	Creation/Termination of Circuits at Any Point	0		1		1
III.6.4	Up to 8 Cables (X-Conn's) & 4 Pairs/Circuit	1		1		1
III.6.5	* Design/Suggest Best/Alt. Path Incl. X-Conn's	1		1		2
III.6.6	Provide All Wiring Closet Connections/Jack	1		1		1
III.6.7	* Allow Multiple Connections to Desktop Equip.	2		2		2
III.6.8	Allow Bridged Connections to Cable Pair	0.5		1		1
III.6.9	Track Circuits with Multiple Termination Pts.	0		1		0
III.6.10	Warning on Attempt to Use Assigned Pair/Port	1		1		1
III.6.11	* Bridging of Cable Pairs to Multiple Locations	2		2		2
III.6.12	Trace Entire Circuit from any Point on Circuit	0.5		0.5		1
III.6.13	Associations - Per Listing (Total 9)	1		1		1
III.6.14	* Provide Schematic Display (CLR) of Circuit	2		2		2
III.6.15	Print Circuits - Both in Text & Schematic	1		1		0
III.6.16	Display Entire Circuit w/Single Print Screen	0		0	C	0
III.6.17	Allow Connection of Cable/Pair to Cable/Pair	0		1		1
III.6.18	Bulk Connect/Disconnects for Cable/Eqpt.	0		1		1
III.6.19	Warning - Low Availability of Pairs/Ports	1		1		1
III.6.20	Trace Stopped or Continued at Control Equip.	1		1		1
	Section Summary/Totals	17		21.5		21

ABC COMPANY - CMS EVALUATION MATRIX - PAGE 5

Spec.#	Brief Description	CMS #1		CMS #2		CMS #3	
III.7 - Graphics/CAD Capability							
III.7.1	* Fully Integrated Graphics-Database	1		2		2	
III.7.2	Utilize Existing AutoCad Drawings	1		1		1	
III.7.3	* Bi-Directional Link Between Graphics/Database	1		2		2	
III.7.4	Update Capability - Walls/Lines/Icons/Text	0		0		1	
III.7.5	Provide List of CAD Functions	1		1		1	
III.7.6	Provide List of Standard Icons	1		1		1	
III.7.7	Ability to Create User-Defined Icons	1		1		1	
III.7.8	* Select Drawing by Name/Description	1		2		2	
III.7.9	Display Site Drawing & Select Drawing of Bldg.	0		1		1	
III.7.10	Print Drawings w/Cable Routes/Circuits	0		1		1	
III.7.11	Create DXF Files with Cable Routes/Circuits	0.5		1		1	
III.7.12	Print Drawings Both to Printer & Plotter	1		0		1	
III.7.13	Identify Drawing Associated with Circuit/Eqpt.	0		0.5		1	
III.7.14	* Printing: Local / Remote LAN / File	1		1		2	
III.7.15	Specify All Printing Equipment Required	1		1		0	NI
	Section Summary/Totals	10.5		15.5		18	
III.8 - Reports							
III.8.1	Provide List of all Standard Reports	0	NI	1		0	NI
III.8.2	Equipment - By Type/Location/Department	1		1		1	
III.8.3	* User-Defined Custom Report/Report Writer	1		2		2	
III.8.4	Report of Open Service Order (Total 6)	1		1		1	
III.8.5	Report of Open Trouble Tickets (Total 7)	0.5		1		1	
III.8.6	List of Pairs w/Status, Circuit ID, % Usage	0		1		1	
III.8.7	Ports - By Controller/Status/Type/Location	1		1		1	
III.8.8	Printing Reports - Local /Remote/Remote LAN	1		1		1	
III.8.9	Indicate Print Destination	CMS		CMS/LAN		LAN	
III.8.10	* Create Reports in ASCII Format	2		2		2	
	Section Summary/Totals	7.5		11		10	
III.9 - Validation & Look-Up Tables							
III.9.1	Validation Tables (Total 5)	1		0.5	C	1	
III.9.2	Validation Prevents Duplicate Names/ID's	0.5		1		1	
III.9.3	Look-Up Vs. Validation Tables (Total 6)	1		1		1	
III.9.4	Ability to Narrow Down Look-Up Options	0		1		0.5	
III.9.5	Wild-Carding Feature (Total 8)	0.5		1		1	
	Section Summary/Totals	3		4.5		4.5	

ABC COMPANY - CMS EVALUATION MATRIX - PAGE 6

Spec.#	Brief Description	CMS #1	CMS #2	CMS #3
III.10 - Data Entry				
III.10.1	Bulk Creation of Cable Pairs - Single Step	1	1	1
III.10.2	Bulk Cross Connects to Pairs/Blocks/Equip.	0	1	1
III.10.3	* Utilize Flat Files to Load Database	0	2	0
	Excel		X	
	Lotus		X	
III.10.4	Describe Process for Loading Flat Files	1	1	0
III.10.5	Vendor Developed Conversion Required	YES Fee	Available	Available
	Section Summary/Totals	2	5	2
III.11 - System Back-Up & Archiving				
III.11.1	Procedures/Hardware for System Back-up	1	1	NI
III.11.2	Procedure for Restoring from Back-up	1	1	NI
III.11.3	* Back-up Server at Remote Site	2	2	0 NI
III.11.4	Able to Archiving 6-12 Month S.O./Troubles	1	1	1
	Section Summary/Totals	5	5	1
III.12 - Network/System Interface Links				
III.12.1	Interface Links to Following Systems:	0		0
	IBM Netview			
	Bay Networks Net Manager			
	Cabletron Spectrum			
	AT&T Voice Switch		0.5 PL	
	Meridian Manager			
	Rolm Voice Switch			
III.12.2	Download Capability From Following Systems:	0		0
	IBM Netview			
	Bay Networks Net Manager			
	Cabletron Spectrum			
	AT&T Voice Switch		0.5	
	Meridian Manager			
	Rolm Voice Switch			
	Section Summary/Totals	0	1	0
III.13 - Billing System Interfaces				
III.13.1	Billing Module for Chargeback of Services	0	0	0.5 UD
III.13.2	Download of Information to Billing System	0	1	1
III.13.3	Report for Service Orders By Cost Center	0.5	1	1
III.13.4	Report for Equip/Services By Cost Center	1	1	1
III.13.5	On-Demand/Scheduled Reports for Billing	0.5	1	1
III.13.6	Charges By Cost Center to Disk	0.5	1	1
	Section Summary/Totals	2.5	5	5.5

ABC COMPANY - CMS EVALUATION MATRIX - PAGE 7

Spec.#	Brief Description	CMS #1	CMS #2	CMS #3
Summary Sheet				
Section II - Installation & Support		2	18	14
Section III - Functional Requirements				
III.1	General System Operation	20.5	30.5	29.5
III.2	Inventory Module/Asset Tracking	3	3.5	3.5
III.3	Service Order	11	12.5	11.5
III.4	Trouble Ticket	8.5	9	9
III.5	Connectivity Components	17.5	24.5	27
III.6	Circuit Functionality	17	21.5	21
III.7	Graphics Capability	10.5	15.5	18
III.8	Reports	7.5	11	10
III.9	Validation & Look-up Tables	3	4.5	4.5
III.10	Data Entry Procedures/Process	2	5	2
III.11	System Back-up & Archiving	5	5	1
III.12	Network/System Interface Links	0	1	0
III.13	Billing System Interfaces	2.5	5	5.5
	Section III Totals	108	148.5	142.5
	Total Scoring	110	166.5	156.5
	* Critical Items - Totals	37	48.5	46

SCORING

Full Compliance - Critical Item	2
Full Compliance - Standard Item	1
Partial Compliance - Critical Item	1
Partial Compliance - Standard Item	0.5
Functionality Under Development	0.5
Non-Compliance	0

Notes:

C = Available Through Customization
FL = Field Lengths Do Not Meet Minimum
NI = Not Included/Indicated
NA = Not Applicable
NDI = No Dial-In Capability
NLP = Normal LAN Procedures
NR = No Remote Printing Capability
OM = With the Use of Other Modules
PL = Planned
UD = Under Development
UF = Done Through the Use of User-defined Fields

ABC COMPANY - CMS EVALUATION PRICING MATRIX - PAGE 1

Item Description	CMS #1	CMS #2	CMS #3
File Server Hardware:			
CPU			
Monitor			
Terminal(s)			
Server			
Modem			
Special Power Supply			
Printers/Plotter(s)			
Back-up Equipment			
Hardware Installation			
Other			
Other			
Hardware Subtotal	$0	$0	$0
Software:			
Operating System			
Application Software			
CMS Software (See Breakdown)			
CAD/Graphics Software (If Required)			
Communications Package			
Back-Up Software			
Software Installation			
Other			
Other			
Software Subtotal	$0	$0	$0
License Agreements:			
Vendor			
Operating System/Application			
Other			
Other			
License Fees Subtotal	$0	$0	$0
Total - Software & License Fees	$0	$0	$0
Total - Hardware, Software, & Licensing	$0	$0	$0

ABC COMPANY - CMS EVALUATION PRICING MATRIX - PAGE 2

Item Description	CMS #1	CMS #2	CMS #3
Other Costs:			
Upgrades/New Releases			
Customization			
Enhancements			
Training			
Documentation			
Maintenance Fees			
File Conversion/Download Programs			
Other			
Other			
	$0	$0	$0
Total System Cost			
Software - Breakdown By Module			
CMS Software System Cost	$0	$0	$0

CMS DEMONSTRATION - EVALUATION SHEET 1
FOR ABC COMPANY USE ONLY

Name: ______________________________ Vendor/Product: ______________________________

Demonstration	LOW <<======>> HIGH (Circle One)
1) What is your Overall Impression of the CMS?	1 2 3 4 5 6 7 8 9 10
a) Provides the Information you Need?	1 2 3 4 5 6 7 8 9 10
b) Easy to Use/Operate ?	1 2 3 4 5 6 7 8 9 10
c) Menu Options Obvious/Displayed ?	1 2 3 4 5 6 7 8 9 10
d) Easy to Query for Information?	1 2 3 4 5 6 7 8 9 10
e) Help Feature	1 2 3 4 5 6 7 8 9 10
f) Response Time	1 2 3 4 5 6 7 8 9 10
g) Screen Layouts (Easy to Read/Follow)	1 2 3 4 5 6 7 8 9 10
h) Look-Up/Validation Tables	1 2 3 4 5 6 7 8 9 10
i) User-Defined Fields	1 2 3 4 5 6 7 8 9 10
j) Comments Fields	1 2 3 4 5 6 7 8 9 10

Add'l Comments: ______________________________

____

____

2) General Impression of the Service Order	1 2 3 4 5 6 7 8 9 10
a) Easy To Follow/Use	1 2 3 4 5 6 7 8 9 10
b) Integration with other Modules	1 2 3 4 5 6 7 8 9 10
c) All Required Information Supplied/Tracked	1 2 3 4 5 6 7 8 9 10
d) Handling of Circuit Information (Creating, Changing, Deleting)	1 2 3 4 5 6 7 8 9 10

Add'l Comments: ______________________________

____

____

3) General Impression of Circuit Functionality	1 2 3 4 5 6 7 8 9 10
a) Easy to Follow	1 2 3 4 5 6 7 8 9 10
b) Handling of Jacks/Outlets	1 2 3 4 5 6 7 8 9 10
c) Handling of Major House/Riser Cables	1 2 3 4 5 6 7 8 9 10
d) Tracking of Controlling Devices & Port Info.	1 2 3 4 5 6 7 8 9 10
e) Circuits - Easy to Create/Delete	1 2 3 4 5 6 7 8 9 10
f) Ports Easy to Identify	1 2 3 4 5 6 7 8 9 10
g) Easy to Locate Spare Ports/Cable	1 2 3 4 5 6 7 8 9 10
h) Circuit Trace - Easy to Read/All on One Screen	1 2 3 4 5 6 7 8 9 10

Add'l Comments: ______________________________

____

____

CMS DEMONSTRATION - EVALUATION SHEET 2
FOR ABC COMPANY USE ONLY

	LOW <<======>> HIGH (Circle One)
4) General Impression of Performing Inquiries	1 2 3 4 5 6 7 8 9 10
a) Wildcarding Feature	1 2 3 4 5 6 7 8 9 10
b) Easy to Identify/Locate Controlling Equipment	1 2 3 4 5 6 7 8 9 10
c) Easy to Identify Jacks in an Office	1 2 3 4 5 6 7 8 9 10
d) Easy to Query for the Active Port Connection	1 2 3 4 5 6 7 8 9 10

Add'l Comments: ______________________________

____

____

5) Overall Drawing / CAD	1 2 3 4 5 6 7 8 9 10
a) Drawings Easy to Access/Identify	1 2 3 4 5 6 7 8 9 10
b) Time Required to Display Drawing	1 2 3 4 5 6 7 8 9 10
c) Link to Display Information from the Database	1 2 3 4 5 6 7 8 9 10
d) Ability to Add New Records from the Drawing	1 2 3 4 5 6 7 8 9 10
e) Ability to Make Changes to Drawing	1 2 3 4 5 6 7 8 9 10
f) Ability to Plot/Print Drawings	1 2 3 4 5 6 7 8 9 10

Add'l Comments: ______________________________

____

____

6) General Impression of the Trouble Tickets	1 2 3 4 5 6 7 8 9 10
a) Easy To Follow/Use	1 2 3 4 5 6 7 8 9 10
b) Integration with other Modules	1 2 3 4 5 6 7 8 9 10
c) All Required Information Supplied/Tracked	1 2 3 4 5 6 7 8 9 10
d) Handling of Circuit Information (Changes from Problem Resolution)	1 2 3 4 5 6 7 8 9 10
e) Trouble History Tracking	1 2 3 4 5 6 7 8 9 10

Add'l Comments: ______________________________

____

____

7) Links/Interfaces to Other Networks	1 2 3 4 5 6 7 8 9 10
a) Voice Switches	1 2 3 4 5 6 7 8 9 10
b) Data Network Management Systems	1 2 3 4 5 6 7 8 9 10

Add'l Comments: ______________________________

____

____

CMS DEMONSTRATION - EVALUATION SHEET 3
FOR ABC COMPANY USE ONLY

	LOW <<======>> HIGH (Circle One)
8) **General Impression of Report Capability**	1 2 3 4 5 6 7 8 9 10
a) Reports Easy To Create & Read	1 2 3 4 5 6 7 8 9 10
b) User-Defined/Custom Reporting Capability	1 2 3 4 5 6 7 8 9 10

Add'l Comments: ______________________________

____

____

9) **Overall Impression of Vendor Capabilities**	1 2 3 4 5 6 7 8 9 10

Add'l Comments: ______________________________

____

____

10) **What did you Like Most about the CMS System?**

____

____

____

11) **What did you Dislike Most about the CMS System?**

____

____

____

Add any other comments or questions in the area below.

____

____

____

____

____

____

____

____

____

CMS ADVANTAGES AND DISADVANTAGES
ADDITIONAL COMMENTS FROM TEAM DISCUSSIONS

CMS SYSTEM/ VENDOR	ADVANTAGES	DISADVANTAGES
CMS#1 XYZ Company	o Utilizes TCP/IP Protocol o Runs in UNIX, DOS & NT o Circuit Functionality flexibility o Can produce schematic Circuit Trace o Can utilize flat files from Excel & Quatro Pro o Good client support o Will provide new version as free upgrade with standard Maintenance Agreement o Appears easy to use and learn o Different Levels of Access/ Authorization	o Drawings must be updated/printed through separate CAD System o Currently does not track Cable Performance Characteristics o Circuit creation only provides one path
CMS#2 123 Company	o Best meets tracking of Information/ Fields for Circuits o Can run on Unix Platform o Has full CAD graphics capability o Can change Field Names/Labels on screens to match current naming o Had the best CLR Schematic Trace o Menus were clear & easy to read	o No separate Trouble Ticket Module o No warning on Low Availability of Cable Pairs/Ports o Cannot utilize flat files in loading database w/o Development Cost o Single Point of Connection to Single Point of Connection "No Bridging" of Cable Pairs/Circuits o CAD package is proprietary o New version will be an additional not free with Maintenance Agreement o Automatic Circuit Creation Functionality requires additional module

CMS ADVANTAGES AND DISADVANTAGES
ADDITIONAL COMMENTS FROM TEAM DISCUSSIONS

CMS SYSTEM/ VENDOR	ADVANTAGES	DISADVANTAGES
CMS#3 CMS Company	o Has full CAD graphic capability o Some Voice Switch interfaces o Strong Circuit Functionality	o No separate field for Floor o No way to limit Wildcard Look-up (ALL items will be displayed) o Does not use TCP/IP protocol o Cannot update circuit changes through Trouble Ticket o Documentation is difficult to follow o Graphics is not fully integrated with the Database in all Modules.

SOFTWARE CUSTOMIZATION

24

During the course of your software evaluations, it is possible that you have seen or selected a CMS that seemed particularly well-suited to your needs, but that certain reasonable modifications would make it work even better. We can now discuss how to prepare a request to the Vendor for software customization which might be desired to better meet your needs. Customization might include enhancements to functionality, changes in screen display or field names, adding fields or special custom reports, or special batch transfer programs. Table 24.1 shows a general list of frequent modifications.

Making fields and reports more consistent with existing recordkeeping systems is a common request. Changing field characteristics is usually straightforward and not too expensive. However, making major changes to table structures or modifying the core functionality can be fairly radical and is likely not worth the cost or the effort. Things to keep in mind when considering software customization are discussed below.

General Considerations

Customization can make it easier for users to work with a new CMS product by adding functionality which can save time, or by modifying field names to reflect the current terminology used by your organization. While customization can provide benefits in ease of use, it is often a costly activity. It requires the Vendor to change the source code of the software, and based on the extent of the customization, requires a certain amount of testing and debugging by the Vendor before it is provided to you for your final testing and acceptance.

Customization may result in higher maintenance and support fees, since the Vendor must maintain and support your version of the CMS software as well as their standard version. Your specialized version of the software may require that additional modifications be made each time the Vendor releases a new version of the standard software. If the standard software is

- FIELD AND RECORD MODIFICATIONS
- MENU CHANGES
- NEW OR ENHANCED FUNCTIONALITY
- SCREEN MODIFICATIONS
- REPORT FORMATS
- BATCH TRANSFERS
- SOFTWARE INTERFACES

Table 24.1: Possible areas for CMS customization.

enhanced, it requires enhancing your version, or creating another version of the new release specifically tailored for your organization. This can sometimes result in different problems or unpredictable behavior of the software.

Always be sure to have the Vendor indicate when they will have the desired customization completed and specify a separate cost for each item of customization requested. This way you can decide what items are critical and should be done immediately, and which items can be postponed until a later date.

If you decide to defer some customization until a later date, be sure the deferred items will not impact the records created in the database in the meantime. An example would be to add a Validation/Look-up Table for the Building, Floor and Room. If this customization is done at a later date, the records which have been created prior to the customization may cause errors or problems once the validation table is added. It may require revising a portion of the existing records to correspond with the new table. This can be done, but you need to anticipate the updates to existing records as part of the customization.

The general rule for customization is to provide the Vendor with as much specific detail as possible. This will guide the software developer and give you the best results in the shortest period of time. Changes that are vaguely defined result in "blue sky" projects that can cost a fortune and not produce the desired results. If the customization is fairly extensive, schedule regular meetings or correspondence with the Vendor for updates and feedback.

Specific Requirements

The simplest changes are those to fields. You can often modify the field length or type, as well as any validations or restrictions on the field. If you increase a field length, make sure it will still fit wherever it appears - on every screen, every drawing, report, and any other display formats. If it does not fit everywhere, decide whether this is acceptable. If it requires minor screen changes, indicate to the Vendor where on the screen the field is to be relocated.

It is also not very difficult to simply add a field or a new piece of information to a record in a particular table. You must specify the length and characteristics of the field, and should indicate where on the screen(s) you want it displayed. However, if you are expecting the CMS to do something with that field, operate on it in some fashion, then that could be a bigger task and would be considered a functional enhancement.

Moving commands from menu to menu, changing the command name or description, or eliminating commands can be reasonably done in many cases. Adding new commands usually involves development of new functions, such as initiating batch transfers, performing backups, or doing table maintenance. New commands with new functions represent significant changes, and should be considered carefully.

Changes that enhance the functionality of the system include items such as a specialized Service Order Number generation, adding billing functions for particular services, adding a new table, creating a special command option to perform specialized functions, and performing back-ups or archiving of Service Order and Trouble Ticket information. Be very specific and describe exactly what the new requirement is and how it should function. An example might be the development of a new asset management feature to indicate the depreciation on certain types of equipment. You will need to define the new table, a new screen, what information is to be included in the reports, and how the printed reports should appear. You will also need to define the depreciation calculations, how information will be initially entered, what modifications will be allowed, and how to limit access for security. It must be determined whether the calculations should be run automatically at specific times or be manually requested as required by the user.

In general, include a rough sketch or example of how new screens or tables should look, including exact lengths for each field, and whether you desire any reporting functions associated with the new table or screen. If adding a new table, be sure to list all screens or modules where you may need to access records or fields from the new table. Determine whether you will need to display and pick from this list (Look-Up), and/or whether an entry in this field must match an entry in the new table (Validation).

Simple screen modifications are frequently desired after working with a CMS for a little while. Users may prefer that the fields on the screen be changed to have a different label which more closely matches their normal terminology, or might prefer to have certain fields near the top of the screen for ease of data entry or queries. When requesting changes in the name for a field on the screen, be sure to indicate the current field name and the desired field name, and indicate if the change is to be made on every screen in the CMS database or only on certain screens. A sample request is shown in Table 24.2.

If changing the location of fields on the screen, it is helpful to provide the Vendor with a marked-up sketch of how you want the revised screen to appear. Indicate if any field lengths are to be increased or decreased on the screen, whether the field is to be alpha or numeric (or both), and the exact number of characters required in the field length. Also indicate if you want to be able to query on this new field. This will help the Vendor determine how much activity is required to modify the screen to your specifications.

Every CMS comes with canned reports meant to anticipate the most common ways that information will be retrieved from the database. In designing the various CMS products, no Vendor can anticipate each and every report format which might be required. Many CMS products use a built-in ad hoc report writer, a third-party report generator, or the SQL software for the specific database to handle these requirements as discussed in Chapter 32. However, if you still need a special customized report which can be run routinely from the menu, it can be beneficial to have the Vendor add this report to the standard report formats.

When requesting a customized report, be sure indicate all the fields of information which are to be contained in the report, and the order in which they are to appear. If you have an existing report which resembles the report you desire from the CMS software, include it with your customization request. Be sure to indicate if the printing can be compressed, and if the report is to be printed in portrait or landscape format. If the report is to be used for billing purposes or to upload information into another system, be sure to indicate the format in which the report must be created, such as ASCII text or Spreadsheet format.

Batch transfers from existing files to the CMS database or transfers from the CMS database to other systems often requires the creation of executable batch files with specialized commands. In some cases there are two executable files, one to create the original batch file and a second to import it into the other application. Executable files should be written in a native programming language where the originating application creates the batch file and the receiving application interprets it. The batch file structure must be common to both applications.

The Vendor may be able to create the batch programs if they are familiar with the particular system or database format which will share the CMS database information. Often, the creation of such batch transfer files will require someone from your organization to work with the

REQUEST NAME	DESCRIPTION
EXPENSE CODE FIELD NAME CHANGE	The EXPENSE CODE field name shall be changed to read "COST CENTER" on the following screens: Inventory - User Termination Hardware Inventory - Cable Service Order Screen Inventory - Device (Sample Screens Attached) The "COST CENTER" field shall validate and look-up against the current validation table for EXPENSE CODE. No changes need to be made in the EXPENSE CODE table. The current field name of EXPENSE CODE on the EXPENSE CODE Validation table can be left unchanged.

Table 24.2: Sample customization request for change in field name.

FROM COMPANY DIRECTORY	TABLE NAME	FIELD LENGTH	TABLE NAME	TO CMS DATABASE
LAST NAME	EMP_SCR	15	USER	USER, LAST
FIRST NAME	EMP_SCR	10	USER	USER, FIRST
PHONE	EMP_SCR	12	DEVICE	EXTENSION
DIVISION	ADM_SCR	10	USER	DEPARTMENT
BLDG	LOC_SCR	8	DEVICE	BUILDING
FLR	LOC_SCR	3	DEVICE	FLOOR
RM	LOC_SCR	7	DEVICE	ROOM

Table 24.3: Sample list of corresponding field names for batch file transfer.

Vendor in the development and testing of the batch programs. Provide the Vendor with as much information about the other system as possible, and whether the batch program is to export information from the CMS database, or import information into the CMS database.

Indicate all the fields, tables, and relationships which are involved in creating the batch file. Make a chart similar to Table 24.3 which lists the names for the fields in both the source and receiving systems, indicating the one-to-one correspondence of information - a string in this field will end up in that field. Sometimes the relationship between fields in sending and receiving systems is one-to-many, many-to-one, or many-to-many. Certain receiving fields may be generated by modifying a source field during the transfer. These cases may require special software code in the batch transfer file. The example in Table 24.3 is for transfer between a CMS and a Company Directory, where information from the existing directory is to be transferred to the CMS database. Note how the table clearly indicates where each piece of information comes from and where it goes, and the respective field lengths for each field. Indicate whether the batch transfer is to be initiated from the CMS or from the other system, and if this will be a routine function performed on a monthly or weekly basis.

If batch programs are desired to archive records from the database, be specific as to exactly what information must be archived including any specific tables or records, and if the information is to be archived in a report (readable character) format. If it is to be in report format, include a list of the fields in the order in which they are to appear on the report. Also specify whether the batch program is to create an electronic record, a printed report, or both.

Other batch program files might include a discrepancy report to compare information in the CMS database with information in another system, which would identify records where the information contained in the two systems does not match. The information could then be manually adjusted to update the incorrect records in the appropriate database. Sometimes, this type of batch program is easier to develop than a program to actually transfer information between systems.

The final and most formidable form of customization is to request a live, real-time interface between applications. Customization of this type would likely be extensive (several months) and fairly expensive. A study must be done to compare each application, from operating system to programming language to platform, to determine feasibility. If there is a reasonable fit, then a detailed definition of the interface must be developed. It is recommended that you have an in-house software analyst available to assist with such a project. The actual methods and process will be detailed and specific and are outside the scope of this book.

FINAL SYSTEM SELECTION

25

Having considered the native functionality of a number of CMS products, and having considered some possible customization of the leading systems, you are in a position to make a final decision if you have not already done so. We will discuss how to make the final system selection based on all of the criteria so far presented. You will need to take into account the overall technical performance, the Vendor's responsiveness and ability to support the product, demonstration evaluations, the advantages and disadvantages listed for each system, pricing, and any customization requirements. Be sure to discuss each item fully with the Team to reach a workable consensus. Recognize that the final selection is a subjective decision, as much to do with how people feel about the system as how it actually works. Feeling confident in a product is a good omen for successful implementation.

Prepare a summary chart similar to Table 25.1 to document the final evaluation results. Start in the second column and enter the name of each CMS still under consideration in the top row. In the first column, list each of the areas of consideration such as shown in Table 25.1. Rate each CMS package in each of these categories and then enter the rating in the appropriate column on the chart.

EVALUATION ITEM	SYSTEM A	SYSTEM B	SYSTEM C
TECHNICAL PERFORMANCE	9	7	8
VENDOR RESPONSIVENESS AND SUPPORT	7	7	9
DEMONSTRATIONS	9	6	7
ADV./DISADV. EVALUATIONS	7	9	7
TOTAL SCORE:	32	29	31
TOTAL PRICE	$XXX,XXX	$YY,YYY	$ZZZ,ZZZ
CUSTOMIZATION REQUIRED			
ITEM 1 (WEEKS/COST)	2/$5K		4/$7K
ITEM 2 (WEEKS/COST)		8/$11K	

Table 25.1: Evaluation summary for final system selection.

Ideally, technical performance should have the strongest impact on the system which you select as your final choice. You want to select a system which matches your requirements as closely as possible. Find the technical performance score for each of the candidates on the RFP response evaluation, divide each number by the highest possible score and multiply by 10. This puts each score on a 1 to 10 scale, and allows you to consider technical performance on an equal basis with the remaining items. If you wish to put greater emphasis on pure technical merit, multiply by 20 instead of 10 in the last step. Enter this number for each CMS in the first row.

Take final consideration of how well each of the Vendors followed the instructions in responding to the RFP, how the Client References rated the Vendors on their customer service and product support, and how responsive, helpful, and knowledgeable the Vendors were during the demonstrations. Then rate each of the Vendors on a scale of 1 to 10 (with 10 being excellent) and enter the ratings in the appropriate columns.

Using the forms completed by the attendees at each of the Vendor presentations, rate each CMS on a scale of 1 to 10 (with 10 being excellent). These evaluations are based on which system most appealed to the CMS Team from the standpoint of being easy to understand, work with and navigate through. Enter the ratings for each system in the appropriate columns.

Review the sheets which were developed on the advantages and disadvantages during Team discussions on each CMS. As best you can, give each system a score of 1 to 10, with 10 being highest in the proportion of advantages to disadvantages. Different advantages and disadvantages to not have to be given equal weight, and lack of critical items should be considered a grave disadvantage. Enter the ratings for each CMS in the appropriate columns. Leave a row where you can total the ratings for each CMS for the previous four items.

Skip a row, then enter the total system cost for each CMS, including hardware, software, and supporting items. If you wish, you could break the pricing down into components and indicate the cost of the CMS and support software, hardware/equipment requirements, maintenance fees and any additional costs for training and documentation separately.

The final item to consider is how much customization would be required for each of the CMS products to meet your specific requirements, and the timing required to complete the customization. The additional time for customization may be more of a penalty than the incremental cost. Enter the time and cost for each customized item on the summary chart under the appropriate columns. Some customization items may be required for more than one CMS.

The summary chart documenting the scores for each of the different areas, combined with the pricing and customization information should help to identify the system which should be your final selection. Total the CMS scores for the first four categories. The highest score should be the system which is best suited to meet the needs of your environment and which will be the easiest for members of your organization to adapt to and learn to use.

If the scores are extremely close, the Pricing and Customization Requirements should be the final decision factors. Of course, lower price is preferred if all other things are still equal. On the other hand, customization is not desirable because of added cost, delay, and maintenance. It will require additional time for someone in your organization to test and certify that it meets your expectations.

Once you have identified the final system, it is beneficial to prepare a one-page summary with the reasons for selecting that particular CMS. This will be helpful when working with management to obtain the proper authorization and commitment and will be required by your purchasing section to award the contract, particularly if the winner is not the lowest cost bidder. Once the final system has been selected, all Vendors should be notified as to whether or not they were the successful bidder. Refer to the next chapter for details on preparing the Purchase Order and License Agreement for acquisition of the CMS software.

THE PURCHASE ORDER AND LICENSING AGREEMENT

26

To complete the CMS selection process, purchase orders must be prepared, and in some cases a licensing agreement must be signed between you and the Vendor. We will discuss some helpful hints here for preparing the purchase orders and for understanding the software license. By clarifying certain limitations or expectations in the purchase order or licensing agreement, it makes for better working relations with the Vendor ongoing and helps to avoid surprises in the future. This chapter is simply to help you, and does not substitute for a thorough review of all contracts and agreements by your Legal and Purchasing Departments.

Preparing Purchase Orders

If purchasing equipment or hardware to support the CMS directly from a dealer or manufacturer, it is best to place the order as soon as possible and request the dealer to indicate expected delivery date. The lead time on certain hardware can be 4 to 6 weeks, and will impact the scheduled installation date for the CMS software. Obtain exact equipment and hardware specifications from the CMS Vendor, and list each requirement on the Purchase Order to be sure the hardware will be capable of properly supporting the CMS operation. Determine any warranty requirements and place them on the Order. Specify whether the you, the dealer, or the CMS Vendor will be responsible for set-up and installation of the hardware.

If the CMS Vendor is supplying the equipment or hardware, specify all the equipment which is to be included with the delivery and installation of the system. Determine any warranty requirements and place them on the Order. Specify that the Vendor shall be responsible for set-up, installation and testing of the hardware prior to loading of the CMS software.

Include in the Purchase Order to the CMS Vendor a listing of all the modules or segments of the CMS software which are to be supplied by the Vendor. Use the Software Breakdown submitted with the Pricing Sheets by the Vendor to create this list. Modules might include items such as Base Software, CAD/Graphics Module, Service Order Module, Trouble Ticket, Custom Report Writer, etc. If possible, make the Vendor's proposal (RFP response) part of the Purchase Order as the formal technical description of the product. If you have discussed any other additions or modifications with the Vendor, document these and make them part of the Purchase Order as well.

Installation instructions should be clearly stated in the purchase order, including the anticipated installation date. The instructions will vary depending on whether the CMS Vendor is providing the equipment as well as the software. Some CMS Vendors will ship you the disks and provide installation instructions. It is always best to have the CMS Vendor do the installation of the CMS software at your facilities. Having the CMS Vendor perform the installation avoids a lot or problems, and protects the warranty for defective performance of the software. It is also beneficial to have the CMS Vendor on site to deal with any error messages which might appear during the initial testing of the system, or if problems or conflicts occur when working with a network application.

Sometimes, another software application is required in order to operate the CMS, such as a database or CAD system, or a specific operating system might be required for your new equipment. It is best to purchase that software through the CMS Vendor and have the CMS Vendor do the installation if they are an authorized reseller. This avoids any errors in the installation or set-up of the operating system or application software, and helps for a smoother start-up of the CMS software. If the Vendor is not an authorized reseller of support software, be sure to have the Vendor indicate the specific versions or release numbers which are required for compatibility with the CMS. Your LAN group will usually be able to handle the installation of specific network software if the CMS Vendor does not provide installation. Otherwise, you may have an existing relationship or licensing agreement with a database or network vendor who should provide this service.

Be sure to specify in the Purchase Order the number of days of training (and the number of individuals to be trained) included with the purchase of the CMS software, as well as the dates for the training and whether that training is to take place on your premises or at the Vendor's facility. If there is an additional cost for the training, be sure to clearly specify the cost for the training and what is to be provided.

Specify in the Purchase Order what documentation will be provided with the CMS software, including titles of manuals when known and the number of copies required. If there is an additional cost for any documentation, be sure to clearly specify the nature of the documentation that is to be supplied for the additional cost.

If costs for any customization are included in the Purchase Order, specify exactly what items of customization are included, with the dollar amount listed separately for each item. It is also a good policy to include a copy of the customization quote from the Vendor, which details the activities to be performed as part of the customization.

A few CMS Vendors provide the source code for the software to you so that you can make desired modifications to screen formats, field names, etc. Since most CMS Vendors do not supply the source code, it is a good policy to indicate in the Purchase Order that a copy of the source code for the CMS be placed in escrow in case the Vendor should discontinue support of the product or cease to do business. You do not want to end up with a software product that cannot be modified or enhanced in future years. This is particularly important for networked applications, where network software or operating system enhancements are made which affect the operation of the CMS. You may want to consider including in the body of the Purchase Order (or in the Licensing Agreement) a statement similar to the one shown in Table 26.1.

Another item you may want to specify in the Purchase Order is a statement on the length of the warranty period, and that under the warranty the Vendor shall be responsible for maintaining the CMS software in operating condition, with major defects addressed and resolved within 72 hours. While this subject is covered in the following section on the Licensing Agreement, it is a good idea include it in the Purchase Order. A

ESCROW COPY OF SOURCE CODE

"Upon acceptance of this Purchase Order, CMS Vendor (Company Name) agrees to place a copy of the source code for CMS Product (Software Name) into Escrow for ABC Company (Your Company Name), in the event CMS Vendor (Company Name) should discontinue support of the CMS Product (Software Name)."

Table 26.1: Sample statement on placing a copy of CMS software source code in escrow.

sample set of specifications which might be included in a Purchase Order are shown in Table 26.2.

The Licensing Agreement

The CMS Vendor should have provided you with copies of their standard Licensing Agreement covering their CMS software as part of their RFP response. Any necessary supporting software, such as a database, that requires Licensing should have been indicated as well. These agreements should be reviewed prior to issuing a Purchase Order to the selected Vendor.

Copies of all Licensing Agreements

ITEM NUMBER	SPECIFICATION DESCRIPTION
1	Purchase Order includes the following CMS Software modules: Inventory, Directory Circuit Generator, Report Generator Connection Manager, Service Order Drawing Manager, Trouble Ticket Basic Software & Setup Utilities
2	Installation of the CMS software is to be completed by CMS Vendor on ABC Company's equipment at ABC Company facilities. Installation is to be completed no later than 7/15/94.
3	Three (3) full days of training will be provided at ABC Company facilities for up to 12 users. There is no additional cost for this training.
4	Upon acceptance of this Purchase Order, CMS Vendor is to place copy of CMS Software source code in escrow for ABC Company.
5	Warranty - CMS Vendor shall provide one full year of warranty from date of purchase order at no additional cost to ABC Company. Warranty shall cover all functional operation with major defects addressed and resolved within 24 hours.
6	CMS Vendor shall provide 12 copies of the User's Manuals for CMS Software.
7	Annual maintenance fee will be based on 15% of the purchase price of the CMS software, and shall be due annually (upon invoice from CMS Vendor), beginning one year from date of purchase order.
8	Customization included in this Purchase Order is for modification of the Expense Code field name, for a cost of $2,500, per quotation dated 4/15/94, from Mr. William Cameron of CMS Vendor.
9	Vendor's proposal dated 3/17/94, in response to RFP BID No. IT99-15 is attached to and made a part of this Purchase Order.

Table 26.2: Sample set of specifications for inclusion in purchase order.

should be reviewed by your appropriate Legal Division or Department. Most companies have a set of their own Terms and Conditions of Sale which may differ with statements in the Vendor's Licensing Agreements. Any modifications which need to be made to either the Terms and Conditions of Sale or the Licensing Agreement should be discussed and resolved before issuing the Purchase Order. Rarely is there a problem or conflict that cannot be worked out.

The Licensing Agreement should indicate the number of users which are covered by the agreement, as well as the cost for adding any additional users at a later date. Some Licensing Agreements are either single-user or multi-user. If a "multi-user" license does not state a specific number of users, make sure it clearly indicates that the license covers an unlimited number of users.

Some Licensing Agreements are based on the number of devices or pieces of equipment on which the CMS software is installed. This might be stated in terms of sites, locations or specific installations. If the license is based on the number of installations, make sure the Licensing Agreement clearly lists each location or piece of equipment covered by the agreement. It is also beneficial to have the Vendor include the cost for installing the CMS software at a new location at a future date. Some Vendors will have a reduced cost for adding another site or location, without having to pay the full list price for the software.

The warranty should be provided at no cost and should cover full system operation for a minimum of 90 days from date of acceptance by your company. A period of up to one-year is sometimes available and is preferred. The warranty should cover all defects in workmanship and material. During the warranty period, the Vendor shall be responsible for maintaining the CMS software in proper operating condition, with major defects addressed and resolved within a 72-hour period.

Annual fees for maintenance and support for the CMS software should be clearly stated in the Licensing Agreement. The maintenance fees are usually a fixed percentage of the purchase price of the software, ranging from 10 to 20%. The Licensing Agreement should indicate when these fees are due, whether the Vendor will invoice your organization for the Maintenance Fees annually, or whether it requires a Purchase Order to be issued by your company each year. A statement should be added to the Licensing Agreement which states that the Vendor must provide One Year's advance notice before increasing the rate for their Maintenance Fees. The Maintenance Fees may be adjusted over time based on the total dollar value of the CMS software purchased to date. The specific services and levels of support should be clearly stated in the Maintenance Agreement.

PART 3:
IMPLEMENTING A CABLE MANAGEMENT SYSTEM

INSTALLATION AND TESTING

27

Each CMS installation will include various combinations of hardware and software which must be installed and tested. There are a number of steps to follow during the initial setup and installation to test the basic operation of the system. Servers and PCs, as well as the network, communications, database, CAD, and CMS software all must be properly loaded and functional. Proper testing of the software prior to entering live data will save time, frustration and help to keep a positive attitude for members of your organization. This first phase of the implementation is one of the most critical.

In most cases, installation of hardware such as servers, PCs, and printers will require support from your network administration group and should be fully tested before loading any application software. Configure the hardware and install the network and/or device operating systems first, then run all the normal check disk tests, memory tests, network tests, and print tests to make sure the hardware and operating system are fully operational with no defects. Perform any PC upgrades such as changing hard disk drives or adding random-access memory to user machines, and test them fully as well.

The Project Manager or an appointed individual from your organization should work with the CMS Vendor to oversee the installation and provide any network support/access which might be required. This not only makes for a smoother installation, but also provides someone in your organization with additional information and insight into how the CMS software is loaded and what commands or set-up strings affect operation with the network software as well as any application software.

Next, install any communications software or network protocol software (sometimes called modules or drivers) such as TCP/IP or IPX, for example. At the same time, make any necessary changes to PC configuration or batch files that are required, as specified by the CMS Vendor. Sometimes this must be done on each user PC, particularly with the client-server systems. At this point, you should be following the installation sequence as directed by the CMS Vendor.

Some CMS software will require the loading of supporting applications, such as a database, a CAD system, or a report generator. You or your internal support group may be able to handle this ahead of time, or you can allow the Vendor to do this. The Vendor will know the best way to configure the supporting software properly, and has experience with its installation. If you will be using existing, currently licensed software, you may wish to check with the Vendor to determine if any configuration changes must be made. In a client-server database, the Vendor will likely need to set up the database domains and table structure. Finally, the Vendor should perform the actual installation of the CMS software.

The testing of the CMS software itself involves a number of steps each serving a different purpose, as will be described in the next several chapters. These steps are listed in Table 27.1 A fundamental objective of the testing procedures is simply to familiarize yourself with the data structures, naming conventions, and flow of the data entry steps. The major CMS products have gotten very sophisticated and will require some

• HARDWARE SET UP AND CONFIGURATION
• NETWORK CONFIGURATION
• USER NAMES AND PASSWORDS
• VALIDATION TABLES
• EQUIPMENT AND CABLE TYPES AND CONFIGURATIONS
• CIRCUIT CONNECTIONS
• CIRCUIT TRACE AND DISPLAY
• PRINTING AND REPORTING
• GRAPHIC SYMBOLS/ICONS
• SERVICE ORDERS
• TROUBLE TICKETS

Table 27.1: Elements of initial installation and testing.

general experience before you begin entering a lot of data. In this chapter, you will test the basic functionality of the CMS and the general flow of building the catalog or library of cable and equipment types. For example, you must determine the specific sequence of steps for creating a cable type - whether connectors must first be defined, or some type of end configuration first created. Some steps may be required through the CAD system rather than the database. By knowing the flow of these activities, you will be able to prepare paper documentation of your cable and devices that will make them easier to enter and create.

In the next chapter (28), you will take what you have learned about data structure and data entry flow and proceed to complete your catalog. This will require a more in-depth look at data structure and some adaptation of your initial catalog to fit the data structure of the specific CMS. You must determine how to create each type of cable and device in the database, and enter this information in the proper sequence. At this point, you must set up the database and drawings with the types of things that will be included.

In Chapter 29, you will go one step further and determine how individual items of each type (cables and devices) are created. You will create a set of individual items of each type. Working with items, you can add them, delete them, connect them, and report on them in a way that tests the overall functionality of the CMS and the consistency of your catalog. Specifically, this will tell you that the cable and equipment types you have defined actually make sense when put together as a circuit, and that useful information is presented on screens and reports. In other words, cable pairs are connected to pin locations, cards are properly connected to hubs, a 4-pair jack is connected to a 4-pair plug, etc.

The final step in catalog testing will be for the Service Order and Trouble Ticket modules as described in Chapter 30. Testing of these documents should indicate that appropriate and understandable information is presented in these documents with respect to the items and connections affected. This presentation includes both screen and printed report formats. Your "fine tuning" of the catalog will be complete when you are satisfied with this result.

You may notice some duplication of steps for each of these developmental phases, but they are necessary. For example, you will need to create some simple cables and devices in the next section for purposes of testing the flow of data entry. Later, you will create cables and devices again, but for the purpose of completing and testing your catalog. We suggest that you be patient, allow the Vendor to give you guidance, and take it one step at a time. Your results will be more satisfying and useful in the long run.

While the CMS Vendor is still at your facility, the following activities should be conducted to test the software performance and resolve any immediate problems or difficulties. As you execute these tests, note the response time for record retrieval and screen refreshes. These tests can be performed by either the CMS Vendor or a member of your organization, with the assistance and guidance of the CMS Vendor.

While many of these functions should be covered again during the normal training sessions provided by the Vendor, it is recommended that basic testing of the CMS functionality be done at the time of installation. Some additional areas which will be covered in this initial testing are listed in Table 27.2. If you have selected a CAD/Graphics based system, you may have to begin your testing by importing a drawing file to use as a working environment.

Test the database to see if it is case sensitive. Using any table, create one record using all CAPS, and then attempt to create a second record using lower case letters. If you do not receive a warning that this is a duplicate record, you will know the database is case sensitive. That is, the different cases appear as totally different entries to the CMS. Also test the query function to determine if queries must match the exact entry with respect to upper and lower case letters. If the database is case sensitive, it is recommended you enter all information in ALL CAPS, and that you train all users to do the same. That simply means put your "CAPS LOCK" button on before you enter the database. The CMS Vendor should be able to inform you of any case sensitive areas or tables, but you should check them nonetheless.

Be prepared with some sample data which can be used to test the CMS support tables. You will want to add a few items to each Validation Table, such as Sites, Buildings, Departments, Equipment Classes and Types, Patch Panels, Cable Types, Cost Codes, etc. so these tables can be tested as you work throughout the database.

The first thing to test is setting up new users and passwords in the database, with different levels of access or authorization. The first user to be added should be your own ID and password. For testing purposes, your ID and password should have unlimited authority to add, change, and delete records You will also want to set up at least one user with read or view access only. This will provide an opportunity to make sure those users with read only access levels cannot change or delete records. Be sure to add at least one additional user with change capabilities for purposes of testing the record locking if two users attempt to change the same record at the same time.

Now you are ready to begin working with the individual modules, screens and functionality of the CMS software. You will need to begin with validation tables, then equipment and cable types, followed by individual inventory items which can then be connected together to form circuits.

Add the first three buildings you will inventory to the building or location validation table. Remember to be consistent with the number of characters used in the abbreviation tables. If the location tables require floor and room designations as well, create three offices, one in each of the buildings, and three Telecommunications Closets, one in each of the buildings.

Add at least three department codes or abbreviations to the department or division validation tables. These can include your own department or division and then two departments which occupy the first building which will be inventoried. Do the same thing for any cost codes or expense codes which will be utilized in the database.

If the software tracks equipment types under a higher level of classes or categories, create three categories under which specific equipment types will be included. These might include phones, PCs, controllers, or hubs.

Create at least one type of equipment for each of those classes or categories. Some examples might include a Model PS2/90 PC, a D-7407 Phone, a Bay Networks 3000 Hub, or a 3174-31C Controller. This is where the Vendor will be able

- TEST FOR CASE SENSITIVITY
- IDENTIFY REQUIRED FIELDS
- DETERMINE FLOW OF DATA ENTRY
- TEST LOOK-UP TABLES
- IDENTIFY WARNING AND ERROR MESSAGES
- CHECK FILE LOCKING FEATURE

Table 27.2: Elements of functionality testing.

to assist in the proper sequence of tasks required to define the number and types of connectors on each side of these items. A PC may have 5 slots or connectors on the back, numbered 1 to 5, while a phone has a single RJ45 or RJ11 on the back or bottom. A controller might have 32 BNC ports or connectors and a network or line driver connector, where as a hub might have 12 slots. If you will be tracking the cards as separate items connected to hubs, create at least one card type, such as a Bay Networks 3502 card which has 12 ports (Ethernet) on the front.

Add at least three cable types, such as a 100-pair copper backbone cable, a 4-pair UTP horizontal cable, and a 12-strand fiber backbone cable. Again, the Vendor will be able to assist you in setting up the appropriate number and types of connectors for each cable type.

If required, add three types of termination hardware or patch panels such as a 66-block (with pairs 1 to 100 or Pins 1 to 200), a 110-Block (with pairs 1 to 25 or Pins 1 to 50) and an office outlet (with pairs 1 to 4 or pins 1 to 8).

If you will be tracking pathways, add at least two types of pathways, such as a 4" PVC backbone conduit and an overhead ventilated cable tray.

Add at least three records to any other validation tables which might exist in the database, such as service order types, problem types, or resolution codes.

After entering the sample data in the validation tables for locations and component types, you should now create at least one item of each type of component. These items could include: users/employees, sample desktop equipment, controllers, hubs/cards, horizontal cables, backbone cables, fiber cables, patch panels or distribution blocks, data racks/distribution cabinets, conduit/pathways, etc. As you create each item, use the look-up feature on appropriate fields whenever available. You should attempt to enter items which are not in these tables to test for proper validation or error messages. Check to see if a new item can be added to the Validation Table from the current screen, without having to return to a different section of the database. You should also create an item record with a duplicate name or ID to test error or warning messages which inform the user of a duplicate record.

Create a 66 block, a 110 block or panel and a controller or hub and card in the telecommunications closet in the first building. Also create a another 66 block in the telecommunications closet in the second building. Use your own naming conventions or simple names for test purposes, such as BL01, BL02, CTL01, HUB01, etc.

Create a 100-pair backbone cable which runs between the telecommunications closet in the first building and the telecommunications closet in the second building. Then connect the 100-pair cable to the 66 blocks in each of these closets. With some systems, this step may be handled by specifying the terminating devices at the time the cable is created. Use the "group" or "bulk" connection function if available.

Create a phone, a PC, two outlets and two horizontal cables in the office in the first building. Connect each of the horizontal cables to the appropriate office outlets and the 110 block in the Telecommunications Closet, if this was not done when creating the horizontal cables.

After creating each of the sample records, test the query and retrieval functionality. Use wildcards where available, and test the query capability based on entering query criteria in different fields on the screen.

There are several ways you may want to test the circuit create functionality of the system. First, create a circuit using the PC inventory item which you created. Use the automated Circuit Creation function, if available, in creating the circuit. This is where you can work with the Vendor to set up the appropriate circuit templates for both voice and data circuits. If there is no automatic circuit generation function, make each connection of the circuit individually, such as connecting the PC to the outlet or horizontal cable in the office. Then, connect pair 1 of the horizontal cable to the first port on the controller or hub and card in the Telecommunications Closet. After completing the circuit, use the query function to check if the status of the components has been changed to In Use.

Create another circuit using a telephone. Connect the phone to pair 4 of the outlet or horizontal cable. Then connect pair 4 of the horizontal cable to pair 1 of the backbone cable.

You may also want to attempt to create two incorrect type of connections to test any restrictions or warning messages associated with the software. Attempt to connect pair 2 of the horizontal cable to pair 1 of the backbone cable. This will test the error or warning message indicating the cable pair is already in use. Attempt to over-ride the warning message to see if the system will permit dual connections, such as in the case of a bridged circuit.

Next, attempt to connect the second connector or port on the back of the PC to the second port on the controller or hub and card. This will identify whether there are any restrictions which prevent a user from connecting items which are not in the same location. Some CMS software products have this restriction. It can be both beneficial and limiting.

Use the query function to retrieve one of the circuits which has been created and trace the circuit to show all connections from the beginning to end. If the functionality allows tracing from any point along the circuit, test the trace from any intermediate point along the circuit. Test the report and print screen capabilities of the Trace to see what information is provided on the printed copy.

Delete one of the circuits created earlier. Check to make sure the status for all components has changed back to available or spare.

Create a Service Order for a new installation of a Phone, and Computer for a new user or employee. This will test all functions of the Service Order by adding a new user and all their circuits. Use the look-up feature on appropriate fields wherever possible to retrieve data from the validation tables. Attempt to install a new horizontal cable in a new office in the first building as a part of this Service Order as well. This will indicate whether new cabling and inventory items can be added through the Service Order Module. If you cannot add a new horizontal cable through the Service Order, use the second outlet in the original office for the new connections. Request a printed copy of the Service Order, but do not close it out yet. Review the printed copy to make sure all the required circuit information is provided.

Create a Trouble Ticket for the Phone which was just installed through the Service Order. Check to see if any reference is provided which indicates there is a pending Service Order against the Phone. Request a printed copy of the Trouble Ticket. Check to make sure the circuit information for the phone is provided with the Trouble Ticket, if appropriate.

Use the Inventory module or segment to retrieve the PC record which is used in the circuits which were created. Attempt to delete the PC from the database. Check to see if the CMS software provides any warning or error message which indicates there is a connection associated with this item. If there is no warning and the CMS software deletes the item, check to see if the circuit connected to the item was also deleted. If the circuit was also deleted, check to see if the status of the cable pair(s)/port has been changed to Available.

If the selected CMS product has a CAD or Graphics module, you will need to perform some specific tests on the CAD interface. Have a sample drawing file available for importing into the database. This will not only test the import capabilities, but it will also provide an estimate of the time required to import drawing files. If possible, create the drawing name or location to match the building/floor location you used in creating your sample items and circuits. This will avoid any conflict in attempting to place an item on a building and floor which does not match the locations of the inventory records.

Send a copy of the drawing file to the local printer. This will test the printing capability as well as display the options you have for selecting the drawing/paper size and printer to which your drawing will be sent.

Place or insert on the drawing an existing icon for each of the items created as inventory records, such as the phone, PC, controller or hub, termination block, patch panel, pathways, horizontal cable, backbone cable, etc. This will identify if any symbols must be created or imported into the database to handle your specific equipment and cable types.

Select or pick the phone or PC and request a circuit trace for the item to see how the circuit is displayed in the Graphics environment. Zoom in on the Trace, or the area where the inventory item is located. Request a printed copy of the displayed area.

Standard report formats should be tested as well. Create several reports using both the standard report configurations and your own user-defined reports if available. Create a report of all equipment items and another report of all cables. You may also want to create a report of all cable connections to determine whether the report shows all cable connectors, or only those which are connected. This will provide an idea of how well the reports are displayed, and the time required to generate the report. Send the reports to the printer, and see how long each takes to reach the printer. Printing time will depend on the length of the report. Finally, request the reports be stored as a file or sent to disk. Later you may also want to import the file into a spreadsheet or text document, if possible, to see what formatting changes need to be made to utilize the data.

The following three access tests can be performed on your own, if time does not allow you to do this with the Vendor present. If the CMS is a network application, you should test the ability to access the database from multiple points. If some files are required to be loaded on each of the devices which will access the system, this will provide a test to determine whether there are any problems with the configuration of different PCs or terminals which must access the CMS database.

Once you have set up several different PCs or terminals for access to the CMS, set up at least two additional users with CMS IDs and passwords, and have them log into the database at the same time. This will identify any problems encountered when multiple users access the system at the same time.

Attempt to have two users request to change the same inventory item, such as the phone, PC, or horizontal cable connections in the first office. If there is no "change" functionality for an inventory item, attempt to create two Service Orders to move the phone at the same time. The system should prevent one of the users from accessing the record, or provide a warning message that the record is in use or being modified by another user.

If the CMS software is a network application, test the capability to direct the printing of a report or drawing to a printer at another location, other than the local printer. This may have to be done at the network level, rather than from inside the CMS itself. This will determine where the print driver for the CMS product is located, and how users can designate the desired printer.

If the CMS software is set up to handle imports of files or records from other systems or databases, have a sample file which might contain a few of the types of records which will need to be transferred to the CMS database. These types of records might include department names, expense codes, user names and locations, etc. Transfer the files using the proper batch transfer file or command. The CMS Vendor should be able to help you set up the transfer. After the information has been imported into the CMS database, check to make sure the correct fields were populated, and that the information is displayed properly.

There may be additional tests or functions, specific to the selected CMS package, which you or the Vendor may want to conduct to insure the CMS software is operating properly.

SETTING UP THE CATALOG AND DATABASE

28

Having selected a CMS to meet certain objectives, you must now finalize the formats and descriptions of the cable and devices (the catalog) upon which the system will be based. A method for naming the individual items must also be finalized. Modifications may be required in the length of cable or equipment names, and this may require some adaptation of the naming conventions themselves. Adjustments might be required in configuring or naming of cables, controlling equipment or termination hardware, drawing files/names, etc. The terminology of the selected CMS must be used for specific elements, and you will later prepare documentation on this basis. Configurations, or the definition of connection points on cables and devices. were originally covered in Chapter 6.

The basic testing described in the previous chapter should have given you a sense of the data structures and data entry flow of the selected CMS, and should help guide you through the required adjustments to your original catalog developed in Chapter 21. After setting up the catalog and database you will create specific items of each type in the next chapter, testing the configurations and naming conventions while making any final adjustments. You need to be confident in your catalog and the CMS functionality before starting entry of real cables, devices, and connections.

As you proceed with your final catalog development, keep in mind the objectives and guidelines you originally used so that the same objectives can be met in the final product. It is recommended that you work this catalog out on paper before entering anything into the database or drawing files. Some of the steps involved in setting up the database are listed in Table 28.1

You will want to check whether the CMS comes with an existing library of cable and equipment types. If so, obtain the documentation or printouts of each defined type and compare them to your initial catalog requirements. If the match is very good, you can save significant time by using the existing catalog. However, if you wish to adapt an existing library of types, or use selected components of that library, be very careful that new and existing component definitions are consistent at the connection level. For example, if you define cable on a "pair" basis then the equipment or hardware connections must represent "pair" terminations rather than a "pin" for an individual conductor.

Validation Tables and Support Information

Begin by making a list of all the validation and support tables which are contained in the CMS. Review the list of information for each of these tables to be sure you have all the required information. These tables will include lists of sites, buildings, floors, expense codes, etc. Review the list for each table to see if any modifications in abbreviations or names are required to accommodate the field length in the CMS. Check to make sure each list is consistent with respect to the number of characters in each abbreviation or name and that any additional description fields are also consistent. If there are any new tables, prepare the appropriate list of entries using the proper field length and a consistent, structured format.

- CHECK FOR EXISTING LIBRARY OF CABLE AND EQUIPMENT TYPES
- LIST ALL VALIDATION AND SUPPORT TABLES
- REVIEW DATA ENTRY SCREENS
- CHECK FIELD LENGTHS FOR CABLE AND EQUIPMENT NAMES
- REVIEW AND OBTAIN REQUIRED FIELD INFORMATION
- MODIFY CABLE AND EQUIPMENT CONFIGURATIONS
- MODIFY TERMINATION HARDWARE CONFIGURATIONS
- REVIEW PROCESS FOR CREATING PATHWAYS
- CHECK DRAWING IMPORT REQUIREMENTS
- REVIEW LIBRARY OF ICONS AND SYMBOLS

Table 28.1: Elements of setting up the database.

Add any new Validation Tables or other tables which must be developed for use in creating cable and equipment records, such as connector types, port types, termination types, etc. You may wish to prepare these lists while you are finishing the cable and equipment descriptions in the next section. Check the data entry screen for each Validation Table to make sure you have all the required information (each field) for each record.

You should enter any site, building, and floor tables at this time. Prepare lists of users, department code, and expense or cost codes in paper or electronic format, but do not put them into the database at this time. You will put a limited amount of this information into the database while testing the catalog in the next chapter, and can finish entering this information after testing is completed.

Cable and Equipment Types

Next, review the cable and equipment types and configurations which you developed in your initial catalog to determine if they are still appropriate for the selected CMS. Starting with cables, check the data entry screen for the cable types to make sure you have all the required information for each type of cable. Information such as manufacturer, part or specification number, cable jacket, number of cable pairs, type of terminations, color codes, level/category, etc. may be required when setting up the cable type in the CMS database. If the pair type, connector type, or termination type is validated from another table or list, enter each type on your list for the appropriate support table. These tables must be completed before the actual cable types can be created.

In some CMS systems, the cable types might be very basic, such as horizontal cable, riser cable, campus cable, fiber cable, coax, etc. With these types of CMS products, the number of cable pairs and specific pair counts are specified at the time each individual cable is created and is essentially independent of cable type. Other CMS products may require that a separate cable type be set up for each combination of cable type, number of pairs and actual pair count. For example, a different cable type must be created for each pair count such as "RC1-100", "RC101-200", and "RC201-300" for riser cables with pair counts 1 to 100, 101 to 200, and 201 to 300 respectively.

Make any modifications in the cable type names/identifiers, connectors, pair counts or termination hardware connections as appropriate to fit the specific system architecture. Create a new list of cable types on paper using the CMS format. Keep in mind that adaptation at the "connection" level must still be consistent with connectors or terminations on hardware and devices. If you have selected a CMS which creates system-assigned cable names or IDs based on combining a cable type abbreviation with a sequential number, be sure the abbreviations properly identify the style or use of the cable and are meaningful to the technician.

Enter the list of connector types and/or connection configurations that will be used to define the specific cable types in the appropriate tables, if required. Next, create each of the cable types you have identified in the cable catalog. Use look-up tables to specify previously entered connector types or connector configurations where appropriate.

As with the validation tables, consistency is extremely important when creating your cable types. Be consistent with regards to upper and lower case letters, dashes, hyphens, slash marks, spaces, etc. Also check to see how the CMS database treats the use of the “space bar”. Some CMS products will treat a space as a character and not a blank, so it is important to use the delete key to remove any characters at the end of a field rather than using the space bar to blank them out.

After you have created the cable types, request a printed report which you can review for inaccurate, incorrect, or missing information. This allows for corrections to be made before records are created using the cable type configurations.

Just as you did with cables, review the equipment types and configurations which you developed initially to determine if they are still appropriate for the selected CMS. Check the data entry screen for adding equipment types to make sure you have all the required information which must be entered for each record. Information such as manufacturer, part or specification number, system use, number and type of ports/connectors, etc. may be required when setting up equipment types in the CMS database.

Some CMS products require the definition of Classes or Categories of equipment prior to defining the equipment types themselves. These may be broad classifications such as PCs, Controllers, and Hubs under which specific equipment types will be listed. Some CMS products will also require the definition of port or connector types which will be used for specific equipment configurations. These port types might include items such as RJ45, BNC, Slot, ST, etc. of which there will be a specified number associated with each equipment type. If the port or connector types are validated from another table or list, enter each type on your list for the appropriate table. These tables must be completed before the actual equipment types can be created.

Make any modifications to the equipment type names or identifiers, connectors, and port names/types as appropriate to allow proper connections. Create a new list of equipment types on paper using the CMS format. Remember that adaptation at the “connection” level must still be consistent with connectors or terminations on cables. If you have selected a CMS which creates system-assigned equipment names or IDs based on combining an equipment type abbreviation with a sequential number, be sure the abbreviations you select properly identify the style or model of the equipment and are meaningful to the technician.

Enter any required classes or categories of control or desktop equipment, and the list of connector types, port types, and/or connection configurations that will be used to define the specific equipment types, if required, in the appropriate tables. Next, create each of the equipment types you have defined for both desktop equipment and control equipment. Some CMS systems have separate modules for control equipment and desktop equipment since controlling equipment may have a different set of characteristics. Some CMS products do not require defining the number of connectors or ports as part of the equipment type, but require that the type, number, and naming of ports or connectors be defined at the time each item record is created. Other CMS products will require the creation of a separate equipment type for each type of equipment consisting of different port/connector configurations, such as 32PORT-CTL for a Controller with 32 ports or 12SLOT-HUB for a Network Hub with 12 slots.

Be consistent with regards to upper and lower case letters, dashes, hyphens, slash marks, spaces, etc. Also check to see how the CMS database treats the use of the “space bar”. Some CMS products will treat a space as a character and not a blank, so it is important to use the delete key to remove any characters at the end of a field rather than using the space bar to blank them out.

After you have created all the entries for your equipment types, request a printed report which you can review for inaccurate, incorrect, or missing information. This allows for corrections to be made before records are created using the equipment configurations. It will be helpful later to prepare graphic diagrams of each type of cable and device as part of your system documentation. This step is covered in Chapter 33.

Termination Hardware

Review the termination hardware and patch panel types and configurations which you developed in your termination hardware catalog. Determine if the termination hardware/patch panels are best handled as separate entities with their own unique record, if they are best treated as an attribute or extension of the cable pair or controlling device, or if they can be ignored completely. Your testing in the previous chapter will indicate which option(s) the CMS will support. Check the data entry screen for adding types of termination hardware/patch panels to make sure you have all the required information for each record. Information such as manufacturer, part or specification number, system use, number and type of ports/connectors, etc. may be required when setting up types of termination hardware/patch panels in the CMS database.

As with the equipment types, some CMS products also require the defining of port or connector types (such as RJ45, BNC, Slot, etc.) in a separate table, and then specifying the number and types of ports for each type of termination hardware or patch panel. If the connector type or termination type is validated from another table or list, enter each type of connector or termination in the appropriate table. These tables must be completed before the actual termination hardware types can be created.

Make any modifications in the termination hardware/patch panel type identifiers, connectors, port names/types as appropriate to allow proper connections in the CMS system. Create a new list of termination hardware types on paper using the CMS format. Remember that adaptation at the "connection" level must still be consistent with connectors or terminations on cables and equipment. If you have selected a CMS which creates system-assigned names or IDs based on combining a hardware/panel type abbreviation with a sequential number, be sure the abbreviations you select properly identify the style or model of the hardware or panel and are meaningful to the technician.

After setting up any required port or connector types, create each of the termination hardware/patch panel types. Some CMS products will treat termination hardware and patch panels differently than equipment, while others will handle these items exactly the same as desktop and controlling equipment. For some CMS products, termination hardware and patch panel types may be very basic such as 66-Block, 110-Hardware, IBM Data Connector, etc. where the individual number and type of connectors or ports are defined at the time each individual termination device or patch panel record is created. The types may even be defined based on what they are used for such as Riser, Horizontal, etc. Other CMS products will require the creation of a separate termination hardware or patch panel type for each block or panel consisting of different port/connector counts, such as 66M25 for a 25-Pair 66 Block, 66M50 for a 50-Pair 66 Block, or ORT-36 for a 36-Port Ortronics Patch Panel, and IBM-64 for a 64-Port IBM Patch Panel.

Consistency is extremely important when creating your termination hardware types, since these are used by some systems in creating the system-assigned individual termination names or IDs. Be consistent with regards to upper and lower case letters, dashes, hyphens, slash marks, spaces, etc. Use the delete key to remove any characters at the end of a field, rather than use the space bar to blank them out.

After you have created all the entries for your termination hardware and patch panel types, request a printed report which you can review for inaccurate, incorrect, or missing information. This allows for corrections to be made before records are created using the termination hardware and patch panel type configurations in the database.

Pathways

Review the procedures for creating conduits and pathways in the CMS database. Each system will differ on how conduits and cableways are entered. Some CMS products require that all conduit and pathway information be created in the database before the cables are created. If this is the case, you will want to make sure you have properly identified each conduit and pathway before you start adding any cables, and that you have also identified which cables are contained in each conduit and pathway. For cable that is not in a pathway, check to be sure that the pathway is optional or else make up dummy pathway names where required.

Check the data entry screen for adding conduits and pathways to make sure you have all the required information for each record. Information such as size, fill factor, use, or construction material (steel, PVC, etc.) may be required when creating conduit and pathway types in the CMS.

Some CMS systems may require that you first set up some preliminary codes or tables for pathways such as use (such as voice, data, riser, fiber, horizontal, etc.), size (such as 2", 4", 6", etc.) and construction material (such as PVC, metal, etc.) prior to defining actual pathway types.

Review your naming conventions to see if any modifications are necessary to meet the field length restrictions of the system. If you have selected a CMS which creates system-assigned pathway names based on combining a conduit or pathway type abbreviation with a sequential number, be sure the abbreviations you have selected properly identify the style, size, or use of the conduit or pathway and are meaningful to the technician. Some systems may already have pre-defined types which you can use, rather than specifying your own.

Once you have set up any preliminary characteristics for pathways, you can then create the individual types. Again, each of the CMS products will handle the creation of conduits and pathways differently. Some CMS products will require that a separate type be set up for each combination of use, size and construction material, while other CMS products will require that these attributes be indicated at the time each individual pathway record is created.

Be consistent with the use of dashes, hyphens, slashes, spaces, etc. Use the delete key to remove any characters at the end of a field, rather than use the space bar to blank them out.

After you have created all the entries for your conduit and pathway types, request a printed report which you can then review for inaccuracies, or incorrect or missing information. This allows for corrections to be made before records are created using the pathway type configurations.

Drawing Files

If you have selected a system which has a Graphics/CAD module, you will want to review the list of drawing files and determine whether to use the existing file names or create new names for use in the CMS. Often, existing file names may reflect a drawing reference number which corresponds to construction files. You may wish to use drawing names which provide some reference to the site, building or floor which is represented by the drawing. Review the list of drawing files and make any adjustments to the file names by renaming the file, or copying it to a new disk with a new name.

Many CMS systems automatically assign the drawing a separate name in the CMS database based on the name of the file which is being imported. This is why it is important to rename or copy the existing file to the desired file name before importing the drawing into the CMS software. For example, if the file is imported as BUILDING.DXF, the CMS will use the drawing name "Building." Be consistent when creating new file names so that all drawing names are the same number of characters and contain the same information, such as a combination of the building and floor, or site, building and floor. You may wish to keep a list of old file names, and reference the old file name on the drawing record when you import it into the CMS database. That way, you will have a reference to the original construction drawing file for future revisions and updates.

Check the data entry screen for importing drawing files into the CMS database to make sure you have all the required information for each drawing record. Some CMS products have fields for information such as site, building, floor, description, last revision date, reference file name/number, etc.

Review the instructions in the CMS Reference Manual for loading drawings into the database. This may require exporting the drawing from the CAD system in a certain format, such as DXF, and then importing the drawing into the CMS. Some CMS products require that a drawing have a specific file USC origin or that some set-up scaling be done before exporting the file into DXF format. Each drawing may require a different USC origin. If so, you will have to first modify each drawing to reflect the desired origin and then perform the export.

After completing all the steps for importing the drawing file, display each drawing to make sure the drawing was transferred properly, that it is the correct size or scale, and that the text and drawing components are properly aligned. If any problems occurred during the drawing import, make sure the instructions were followed properly, and then check with the Vendor to determine how to resolve the problem before importing the remaining drawings.

After importing the drawings, run a report of all the drawing files. This will identify any inconsistencies in file naming, and will act as a reference list when working with the drawing files later on.

Drawing Symbols/Icons

If you have selected a system which has a Graphics/CAD module, you will want to review the list of symbols or icons which come as standard items with the CMS, if any. Review this list against your list of equipment to identify the specific items for which symbols/icons will need to be created and added to the CMS. You will need to determine whether the drawing assigns the icon automatically based on equipment class or type, or whether you can select the icon representing each type or item. If the CMS allows icons for each type or item, some types may use the same or different icons, such as to distinguish a PC from a dumb terminal.

Once you have identified the icons or symbols which are needed, discuss these requirements with the CMS Vendor to determine the best way to create and import these icons into the system. If you have someone in your organization who can work with CAD drawings and files, you may be able to export one of the existing icons or symbols from the CMS and modify it to create new or different icons.

Check the data entry record screen for importing icons/symbol files into the CMS database to make sure you have all the required information which must be entered for each drawing record. Some CMS products have fields for information such as the class or sub-category which the icon/symbol will be used for (PCs, Desktop Equipment, Termination Hardware, etc.).

Review the instructions in the CMS Reference Manual for importing icons/symbols into the database. Some CMS products require that a specific file USC origin or set-up scaling be created before exporting the icon file into DXF format.

Carefully follow the instructions provided and create or add any new symbols/icons to the database. As you add each item, be consistent in assigning a name/ID to the symbol/icon which exactly matches the equipment type or class which you defined in the catalog of equipment types. This is extremely important in making sure the right symbol will be displayed for each of your components, and that the system recognizes the symbol/icon for the appropriate equipment type.

After importing the new icon file, display each symbol or icon to make sure it was imported properly. The icon should be of a comparable size to other symbols or icons which were part of the standard CMS software. If any problems occurred during the icon drawing import, make sure the instructions were followed properly and then check with the Vendor to determine how to resolve the problem before importing any additional icons.

After importing all new icon files, display a list or run a report of all symbols or icons. This will identify any inconsistencies in their names or descriptions and will act as a reference list to make sure your symbols and icons match the list of equipment types to be displayed on the drawings.

TESTING THE CATALOG AND DATABASE

29

After you have completed the catalog and performed the initial database setup with respect to cable and equipment types, it is appropriate to test the combined functionality of the database and catalog by creating individual items and relating them together as circuit components. You must make sure things are properly set up and perform as expected before you begin adding numerous records and are depending on those records for daily activities. Proper testing of the database and catalog will identify areas where the configurations for cable or equipment types need revision or modification. By testing each of the activities which will be routinely performed, you will be able to create specific documentation and training materials unique to your organization and database catalog.

If you have selected a CAD-based CMS, the activities outlined below may need to start with the display of the floor plan drawing and then selecting an icon to create each new item.

All of the steps outlined are based on a full scale CMS implementation set up to track all components, from individual users to each type of control device and the individual port assignment as listed in Table 29.1. Depending on the potential scope and number of records in the CMS, all of these activities may not be appropriate for your individual implementation. Select those areas which apply to your individual recordkeeping needs.

The basis for testing includes first creating a small set of records on paper, then testing Validation Tables and support information, creating individual items, connecting these items together, creating entire circuits, and tracing through individual circuits. This exercise should also test any rules and restrictions for connecting different components or tracing the circuit path.

To properly test the setup and functionality of the database with your specific

COMPONENT	EXAMPLES
LOCATIONS	OFFICES, WIRING CLOSETS, MANHOLES
CABLES	HORIZONTAL, BACKBONE
TERMINATION HARDWARE	BLOCKS, OUTLETS, PATCH PANELS
CONTROLLING EQUIPMENT	HUBS, CONTROLLERS, SERVERS
DESKTOP EQUIPMENT	PCs, PHONES, PRINTERS
EMPLOYEES / USERS	OFFICE OR WORKSTATION OWNERS
PATHWAYS	CONDUITS, CABLEWAYS
DRAWINGS	SITE PLAN, FLOOR PLANS, ICONS

Table 29.1: List of catalog components to test.

catalog, you will need to enter enough records to test each type of cable, equipment and connection. This process will be handled through a number of steps, as listed in Table 29.2

Use data from a small area such as five offices, one local wiring closet, and an Equipment Room all in the building where you plan to begin implementing the CMS. Enter all the equipment, cables and termination hardware you plan to track in each of these areas. As you create each inventory item, (cables, desktop equipment, termination hardware, controlling equipment, etc.), use the Look-Up and Validation tables as often as possible to make sure they are complete, easy to use, and help to provide consistency throughout the data entry process.

Creating the Components

To begin, create five basic offices where employees currently reside. Some systems require the creation of a room number in a location table prior to creating any inventory items which refer to that specific room. If the CMS does not validate the room number against a pre-defined table, remember to be consistent in entering room numbers with respect to hyphens, dashes, slash marks, spaces, etc.

Create the horizontal cables in each of the offices. Use the look-up feature for selecting the cable type and any other fields where look-up tables are available. Be consistent in specifying the cable names, the ends of the cables, (such as End 1 or 2, Originating or Terminating, Beginning or Ending), room numbers, etc. Once you establish which end is the beginning or originating end, always follow that rule for all cables. If the CMS system has a feature for duplicating information from one record to another in creating similar items, use that feature to carry information forward from one record to the next.

If outlets or jacks are to be tracked as individual components, create each of the outlets or jacks for the five offices. These may be created as inventory items or as termination hardware, depending on the CMS database structure.

If you are planning to track users and employee information in the CMS, create three employee records for three of the five offices. Leave two offices empty for testing purposes.

Create at least one of each type of backbone cable (Fiber, Voice Riser, Data Riser, Tie Cable, etc.) which is located in the wiring closet (the Horizontal Cross-connect) where the horizontal cables are terminated. Again, remember to follow the rule for whether the originating end (End 1) of each cable is closest to the User's office or to the source of the circuit. Create at least one of each type of additional backbone cable, and terminate these in the Main Equipment Room where the other end of the first backbone (riser) cable is located.

Create one of each type of desktop equipment (PC, Phone, Terminal, Printer, Answering Machine, Fax, etc.) for each of the three office areas you created earlier. Whenever possible, use the table look-up features and the copy functions which can duplicate records from one new item to the next. Follow the naming/numbering conventions you developed as part of your cataloging. Leave the last two offices vacant for purposes of testing the Service Order functions.

Create one of each type of control device or network equipment which will be tracked in the CMS (Controllers, Hubs/Cards, Servers, etc.). For purposes of testing the connection features of the database, most of the equipment should be located in the wiring closet where the horizontal cables are terminated or in the Main Equipment Room where the other end of the backbone (riser) cable is terminated.

If you plan to track termination hardware as separate and unique components in the CMS, create at least one of each type of termination hardware (66 Blocks, 110 Blocks, BIX Blocks,

- CREATING INDIVIDUAL COMPONENTS
- CONNECTING BACKBONE COMPONENTS
- CONNECTING INDIVIDUAL CIRCUIT ITEMS
- TRACING THE CIRCUITS
- TESTING RULES AND RESTRICTIONS

Table 29.2: Elements of testing the catalog.

etc.). For purposes of testing the connection features of the database, you will need to create the minimum amount of each type to handle the entered horizontal cables, the backbone cables, and any of the controlling equipment which will be connected to termination hardware. Make sure the naming/numbering conventions you developed are appropriate for each type of termination device. The termination hardware should be located in the wiring closet where the horizontal cables are terminated or in the Main Equipment Room where the other end of the backbone cable is terminated. Create at least one additional block in a different wiring closet for testing purposes.

Create at least one of each type of patch panel which will be tracked in the CMS for connections to data or controlling equipment. For purposes of testing the connection features of the database, you will need to create at least the minimum amount of each type of patch panel to handle the connections for any controlling equipment items which you have created. Make sure the naming/numbering conventions you developed are appropriate for each type of patch panel. The panels should normally be located in the same room as the controlling equipment.

If appropriate, create the conduits and pathways which contain the horizontal cables to each of the five offices. Use the look-up feature for selecting the conduit or pathway type, size, and material as required. Also create any conduits or pathways containing the backbone cables between the local wiring closet and the Main Equipment Room or Main Distribution Frame. Use the look-up feature for selecting the conduit or pathway type, size, and material, as required. You may be required to specify the fill factor for the pathway at the time you create the pathway record.

Each of the CMS products will handle the cable-to-pathway relationship in a different manner. Some will require the cables to be identified at the time the pathway record is created, other products will require that the pathway be specified at the time the cable record is created, and some may relate cable to pathway in a separate cable routing module or screen after both cables and conduits have been created. Review the instructions provided in the CMS Reference Manual to make sure you create these components in the proper order and that they reflect the correct information. If the cables contained in each of the pathways are not specified as a part of either the cable record or the pathway record, complete the steps required to route each cable through the appropriate pathway. Depending on the CMS selected, this may also be handled through the CAD/Graphics module.

If you are working with a CMS that is managed primarily through a CAD format, display the drawing file which represents the area covering the five offices and the local wiring closet you created at the beginning of this chapter. Request the system to display all equipment on the current drawing through the use of a query. This will identify whether a separate step is required in order to place a symbol/icon for each component onto the appropriate drawing.

Some CMS products will automatically insert an icon on the drawing based on the room location of the inventory record when each component was created. This applies to CMS products which use a grid locator to identify each office area on a floorplan drawing. Automatic insertion will apply to equipment only, since cables need to be routed and located on the drawing in order to properly document the actual path. Other CMS systems will require that you complete an additional step of inserting or placing an icon in the appropriate location on the floorplan drawing for each piece of equipment you want to track. They will need to be located on the floorplan drawing, just as the cables need to be located in their proper location.

If the symbols/icons for equipment need to be inserted or placed on each floorplan drawing, complete the necessary steps of specifying the item to be inserted (through the query process if available), and then using the mouse to point the indicator to the appropriate location. Complete the above process until a symbol/icon for each equipment item has been inserted or placed on the drawing. This will be valuable in estimating the time required to populate the drawings and will also provide a good indication of whether the size of each of the symbols/icons is appropriate with respect to the size of the drawing display.

Follow the same process for inserting/routing the pathways on the drawing. Again, be careful to follow the procedures outlined in the Vendor's Reference Manual as to whether the pathway needs to be defined and located on the drawing prior to creating the cables routes on the

CABLE NAME	LOCATION / END	PAIRS	TO	TERM. HDWE. NAME	SIDE	POSITION
101V	BEGINNING	1		JK101V	BACK	1
101V	BEGINNING	2		JK101V	BACK	2
101V	BEGINNING	3		JK101V	BACK	3
101V	BEGINNING	4		JK101V	BACK	4

Table 29.3: Sample of a basic connection record for connecting a horizontal cables to an information outlet.

drawings, since the cables will often be routed through a pathway. After locating the appropriate pathways on the floor plan, insert the cables onto the drawing. Carefully follow the instructions outlined in the Vendor's Reference Manual for proper routing of cables.

Connecting the Components

Now that you have a sample base of components which will make up the backbone, you need to link those components together to form a complete circuit. The circuit must adequately represent how the network and cable plant is structured and connected together. Your sample segment should contain a representation of all cables types, termination hardware, controlling equipment and patch panels which form the backbone structure for carrying the individual circuits. The following steps are provided as a guide for testing connections among your catalog items and to determine if the CMS records properly document your network backbone and cable facilities.

If you have set up the outlets or jacks as separate components independent of the horizontal cables, the first step is to connect each pair of the horizontal cable to the appropriate pin position(s) on the outlet. Required information will normally include the cable name, cable end/location, cable pair(s), outlet name/ID, and the device side and/or connector(s)/position(s) as shown in Table 29.3. Some CMS products handle this step at the time each cable is created, requiring the user to specify the termination equipment on each end.

Follow the instructions in the Vendor's Reference Manual for connecting a cable to a device or termination block. This may require a separate record entry for each pair, or may be performed as a series of cable pair to termination connections in a single step or operation.

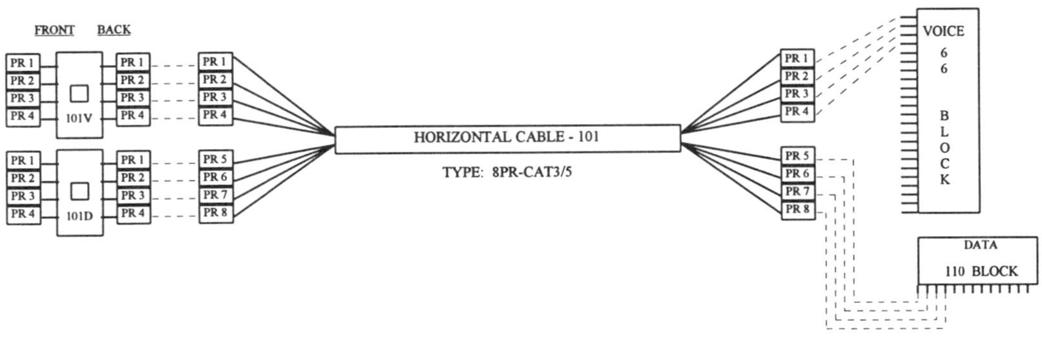

Figure 29.1a: Hybrid 8-pair horizontal cable consisting of four Category 3 pairs and four Category 5 pairs.

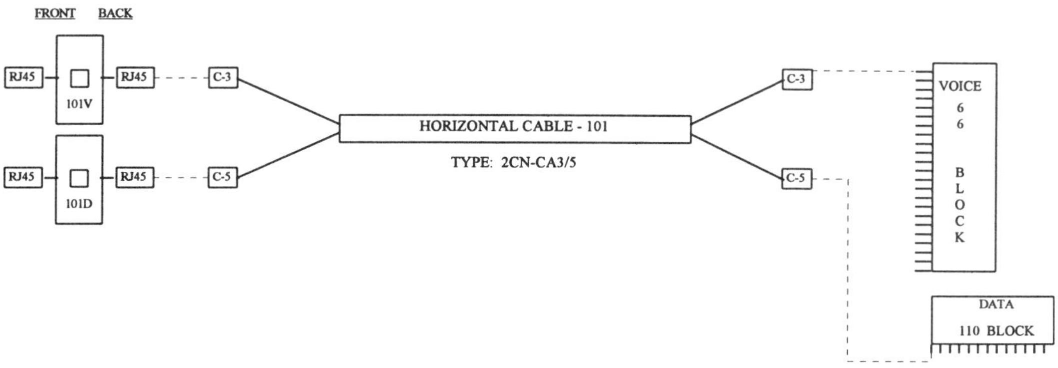

Figure 29.1b: Example of hybrid horizontal cable consisting of one Category 3 connector and one Category 5 connector.

Depending on whether you are tracking your horizontal cable as a single entity or as a combination of different cable types will impact how the connections are created in the database.

- o If the cable is a single entity, it may only require one connection from the horizontal cable to the connector (such as an RJ45) on the back of the outlet or jack.
- o If the cable is a composed of four cable pairs, it will require the connection of the four cable pairs to specific pin combinations or positions on the back of the RJ45 connector at the outlet or jack.
- o If the horizontal cable is a combination (hybrid) of different cable types such as a 4-pair Category 3 and a 4-pair Category 5, it may require specifying which pairs are to be connected to which connector(s), such as pairs 1-4 connected to the first outlet and pairs 5-8 connected to the second outlet as shown in Figure 29.1a. The connections shown in Figure 29.1b reflect a similar type of hybrid cable, where each of the included cables has a single connector with the Category 3 connector is connected to the first outlet and the Category 5 connector is connected to the second outlet.

If you are tracking the horizontal cable down to the pair level, be sure to connect the proper pairs to the corresponding pins on the outlet to match the wiring scheme in use, such as 568A or 568B as shown in Table 29.4. Specify the proper end of the cable to be connected to the outlet. Remember to follow the rule for whether the beginning/originating end, or End 1, is closest to the User's office or to the source of the circuit.

If you have set up the termination hardware as components separate from the cable itself, the next connection step is to connect each pair of the horizontal cables and backbone cables to the appropriate termination position(s) on the termination block. Some CMS products handle this step at the time each cable is created, requiring the user to specify the termination block and positions at each end. Follow the instructions in the Vendor's Reference Manual for connecting a cable to a device or termination equipment. Be sure to specify the proper end of the cable to be connected to the termination block. Remember to follow the rule for whether the beginning/originating end, or End 1, is closest to the User's office or to the source of the circuit.

CABLE PAIR	568A PINS	568B PINS
1	4-5	4-5
2	3-6	1-2
3	1-2	3-6
4	7-8	7-8

Table 29.4: Pin configurations for 568A and 568B outlets.

ITEM	CABLE			BLOCK	
NAME:	101V			SB01	
END/SIDE:	ENDING			LEFT	
STARTING PAIR/PIN:	1			1	
ENDING PAIR/PIN:	4			8	
RATIO:	1	(PAIRS)	TO	2	(POSITIONS)

Table 29.5a: Sample horizontal cable to termination hardware connection record.

ITEM	CABLE			BLOCK	
NAME:	RC-REN1E			RB01	
END/SIDE:	BEGINNING			LEFT	
STARTING PAIR/PIN:	1			1	
ENDING PAIR/PIN:	50			100	
RATIO:	1	(PAIRS)	TO	2	(POSITIONS)

Table 29.5b: Sample backbone cable to termination hardware connection record.

Some systems will require individual connection records for each cable pair, while others have a separate data entry screen for creating multiple connection records based on a specified starting pair/pin position and the number of cable pairs to be terminated. Specify the proper cable pair to pin ratio if required. An example might be a ratio of 1 to 2, where one cable pair is connected to two pins on a 66 Block or 110 Block. The screens for specifying the connections will be different for each CMS, however the basic information will normally include the cable name, cable end or location, cable pair(s), the termination device, side, and the connector or position. Some samples of horizontal and backbone connection records are shown in Tables 29.5a and 29.5b

In order to complete the set up of the backbone and network structure, the next step is to connect the controlling equipment such as network hubs, controllers, data switches, etc., to the termination hardware or patch panels where connections will be made to cable pairs or other termination hardware.

Follow the instructions in the Vendor's Reference Manual for connecting a piece of controlling equipment to a termination block or

ITEM	EQUIPMENT			BLOCK / PANEL	
NAME:	LG1D11			SNAPP01	
SIDE:	RIGHT			LEFT	
STARTING PORT/PIN:	1			1	
ENDING PORT/PIN:	4			1	
RATIO:	4	(PORTS)	TO	1	(POSITIONS)

Table 29.6: Sample connection record for connecting a single patch panel connector to the TA ports on the controller.

ITEM	EQUIPMENT			BLOCK / PANEL	
NAME:	LG1D11			SNAPP01	
SIDE:	RIGHT			LEFT	
STARTING PORT/PIN:	1			1	
ENDING PORT/PIN:	4			32	
RATIO:	1	(PORTS)	TO	8	(POSITIONS)

Table 29.7: Sample connection record with 32 patch panel connections to a 3270 controller.

patch panel. This may be a single step that connects all of the device ports to all of the termination block positions at the same time, or could require that each device/port to termination position connection be entered one at a time. Be sure to specify the proper sides of the both the controlling device (top/bottom, left/right, front/back) and the termination block or patch panel. A good rule to apply is to reserve the front, top, or right side of the panel or block for individual circuit connections, and then make all more permanent backbone/network connections to the Back, Bottom or Left (BackBone Link) side of the panel or block.

The screens for specifying the connections will be different for each CMS, however the basic information will normally include the controlling equipment name, side and connector(s), and the termination device name, side and connector(s) or position(s).

A helpful practice when creating connections between two termination blocks, or between two different types of equipment, is to develop a rule for which block or device is entered first, such as always listing the controlling equipment first and then the patch panel or block. This may not seem important since the connections will be established either way, but with some CMS products this will affect how you query the database and how the records will be displayed in reports.

The sample connection record shown in Table 29.6 demonstrates one way a 3270 Controller might be connected to a patch panel with 32 ports. This is a simplified approach where the four Terminal Adapter ports on the controller are connected to the back of the patch panel through a single connection. This approach would not work with a CMS which does not allow multiple connections to a single port or connector.

Another method of connecting the four Terminal Adapter ports to the patch panel would be to connect all 32 panel positions individually to the Terminal Adapter ports, eight positions per Terminal Adapter port, as shown in Table 29.7 Both of these examples are simplified approaches where the TMA (Terminal Multiplexer Adapter) cards are not tracked as separate components.

Connecting a card to an Ethernet hub will be very similar to the equipment to block/panel

ITEM	EQUIPMENT			BLOCK / PANEL	
NAME:	HUB-01			CARD-01	
SIDE:	RIGHT			LEFT	
STARTING PORT/PIN:	S1			T1	
ENDING PORT/PIN:	S1			T1	
RATIO:	1	(PORTS)	TO	1	(POSITIONS)

Table 29.8: Sample connection record for a card to a network hub.

ITEM	EQUIPMENT		BLOCK / PANEL	
NAME:	PC1000		JK101D	
SIDE:	BACK		FRONT	
STARTING PORT/PIN:	2		1	
ENDING PORT/PIN:	2		1	
RATIO:	1	(PORTS) TO	1	(POSITIONS)

Table 29.9: Sample connection record for a PC to an office outlet.

record, in that a card acts as a patch panel where individual user circuit connections are made to the front (right or top) of the card. Table 29.8 shows a one-to-one ratio for connecting the card to the slot on the hub. This uses an example of a CMS where user-defined port names are allowed, such as S1 to indicate Slot 1 and T1 to indicate Tab 1. You will note in the above examples that the front, top, or right side of the patch panel or card is where the individual port connections will be defined.

Creating The Circuits

Once you have a sample base of components which will make up the individual circuits, you can test the ability to link or connect these components together. If your desire is to simply track the location of components and the overall backbone structure, you may choose to simply stop at this point and not track individual circuits and usage against the components. Otherwise, a representative portion of each segment of the network backbone will be created and tested when all cables types, termination hardware, controlling equipment, and patch panels have been defined and connected. The following steps will guide you for setting up connections and individual circuits based on your specific catalog.

If the CMS you have selected has an automatic circuit create function or module, you may want to create your test circuits using that process. We will cover the individual steps involved in creating circuits in order to test and check the connections of separate segments of the circuit, for systems which do not provide automatic circuit creation.

The first step in creating circuits is to determine the point at which the circuit will begin. For demonstration purposes, let's use the desktop equipment as the beginning point of the circuit. Follow the Vendor's Reference Manual for connecting a piece of equipment to an outlet or horizontal cable. The screens for specifying the connections will be different for each CMS, however the basic information will normally include the desktop equipment name, the device side and/or connector/position, and the cable name, cable end/location, cable pair(s), or outlet/jack name and connector of the horizontal

ITEM	CABLE		EQUIPMENT / BLOCK	
NAME:	101V		123-1224	
END/SIDE:	BEGINNING		BACK	
STARTING PAIR/PIN:	2		1	
ENDING PAIR/PIN:	3		1	
RATIO:	2	(PAIRS) TO	1	(POSITIONS)

Table 29.10: Sample connection record for a phone to two horizontal cable pairs.

ITEM	CABLE			EQUIPMENT / BLOCK	
NAME:	101D			LG1D11	
END/SIDE:	ENDING			FRONT	
STARTING PAIR/PIN:	1			12	
ENDING PAIR/PIN:	1			12	
RATIO:	1	(PAIRS)	TO	1	(POSITIONS)

Table 29.11: Connection record for horizontal cable pair directly to controlling equipment port.

cable to which the equipment will be connected.

Connect the appropriate side (front/back, left/right, top/bottom) and connector (RJ45, Slot, 1, 2, etc.) of the desktop equipment (PC, Printer or Terminal) to the appropriate pin/pair or connector of the horizontal cable or outlet.

The connection record in Table 29.9 shows how a PC might be connected to the office outlet. This example reflects a configuration where the outlet contains a single RJ45 connector for connection to desktop equipment. The connection record shown in Table 29.10 shows how a phone might be connected directly to a horizontal cable. This example reflects the configuration where the cable pair connectors on the outlet are attributes of the horizontal cable.

Repeat the above step for each desktop device which is connected to the horizontal cable or outlet in each of the three offices where desktop devices were created. These steps could be combined on a single screen or record, or could require individual connection records, depending on the CMS architecture and functionality.

If connections are directly to the horizontal cables, be sure to specify the proper end of the cable to be connected to the desktop equipment. A look-up table or validation restriction may assist in preventing a user from connecting two items which are not located in the same room. Remember to follow the rule for whether the beginning/originating end, or End 1, is closest to the User's office or to the source of the circuit.

If connections are made to the outlet, be sure to specify the proper side of the outlet to be connected to the equipment. This will usually be the front side of the outlet or jack, since the back of the outlet or jack is most likely connected to the horizontal cable pairs.

The next step in building the circuit is to create the horizontal cable connections in the local wiring closet. First, we will cover the connections to a port on a piece of data controlling equipment or to a patch panel in the local wiring closet. Follow the Vendor's Reference Manual for connecting a cable pair to a piece of equipment, patch panel, or termination block. The screens for specifying the connections will be different for each CMS, however the basic information will normally include the cable name, cable

ITEM	CABLE			EQUIPMENT / BLOCK	
NAME:	101D			SNAPP01	
END/SIDE:	ENDING			FRONT	
STARTING PAIR/PIN:	1			12	
ENDING PAIR/PIN:	1			12	
RATIO:	1	(PAIRS)	TO	1	(POSITIONS)

Table 29.12: Connection record for horizontal cable pair to a port on a patch panel.

ITEM	BLOCK / PANEL			BLOCK / PANEL	
NAME:	SB01			JK101D	
SIDE:	FRONT			FRONT	
STARTING PIN/POS:	3			1	
ENDING PIN/POS:	6			4	
RATIO:	1	(PIN/POS)	TO	1	(PIN/POS)

Table 29.13: Sample connection record for a horizontal cable pair block to an Ethernet port block.

end/location, cable pair(s), termination device, and the device side and/or connector(s)/position(s), similar to the connections for the desktop equipment. Again, the type of connection will depend on whether the termination blocks are tracked as separate component or as attributes of the horizontal cable.

Connect the appropriate cable end and pair (1, 2, 3, or 4) of the horizontal cable to the port on the controlling equipment, the patch panel, or termination hardware.

The type of connection records will depend on the CMS selected and how you have defined your catalog. Table 29.11 shows the simplest approach where the pair on the horizontal cable is connected directly to the port on the controlling equipment, while Table 29.12 shows a connection from the horizontal cable pair to a port on a patch panel. A third option is shown in Table 29.13 where each piece of termination hardware is individually tracked, so the connection is from termination block and position (horizontal cable pairs) to termination block and position (Ethernet ports).

Specify the proper end of the cable and the appropriate side of the controlling equipment, patch panel, or termination hardware. A good rule to follow with respect to sides of equipment, patch panels, and termination blocks is that all permanent backbone or network connections should be made to the back/bottom/left, while patches and individual circuits are connected (patched) to the front/top/right. This may not apply to every piece of equipment or hardware, but it can be applied as a general rule.

For every horizontal cable which has a PC, terminal or printer connected in the office location, create a connection to a port, panel or block in the local wiring closet. If your data controlling equipment and port connections are handled at the Main Equipment Room where a riser cable connection is required, complete the steps outlined in the next few paragraphs on wiring closet cable connections before making connections to the ports, panels, or blocks.

The next step is to create the connections from the horizontal cables to the riser cable in the local wiring closet. This may be represented as a cable and pair cross-connect to another cable and pair, or as a connection between two termination

ITEM	CABLE # 1			CABLE # 2	
NAME:	102V			RC-REN1E	
END/SIDE:	ENDING			BEGINNING	
STARTING PAIR:	1			26	
ENDING PAIR:	1			26	
RATIO:	1	(PAIRS)	TO	1	(PAIRS)

Table 29.14: Connection record for horizontal cable pair to riser cable pair.

ITEM	BLOCK / PANEL			BLOCK / PANEL	
NAME:	SB01			RB02	
SIDE:	FRONT			FRONT	
STARTING PIN/POS:	3			1	
ENDING PIN/POS:	6			4	
RATIO:	1	(PIN/POS)	TO	1	(PIN/POS)

Table 29.15: Sample connection record for a horizontal cable station block to riser cable block.

blocks. First, we will cover the connection as a cable pair to cable pair connection, and then termination block to termination block.

Connect the appropriate cable end and pair (1, 2, 3, or 4) of the horizontal cable to the appropriate cable end and pair of the riser cable. The screens for specifying the cable-to-cable connections will be different for each CMS, however the basic information will normally include the cable names, the cable ends/locations and the cable pair(s) for each cable as shown in Table 29.14.

As indicated with equipment-to-equipment connections, it is important to establish a rule for which cable will be entered as the first cable in the connection record. For example, enter the horizontal cable as the first cable and the riser cable as the second cable, or enter the riser cable as the first cable and then the campus cable. For some CMS products, this will affect how you query the database and how the records will be displayed in reports. This rule should apply whether the connections are cable pair to cable pair or termination block to termination block.

If tracking termination hardware as separate entities, connect the appropriate termination block and position(s) where the horizontal cable is terminated to the termination block and position(s) where the riser cable is terminated. The screens for specifying the block-to-block connections will normally include the block names/IDs, the side of the block, and the pin/connector positions for each block as shown in Table 29.15.

If the circuits continue into another building to connect to a voice switch or outside trunk line, you will need to create the connections of the riser cable and pair(s) to the campus cable and pair(s). This may be represented as a cable and pair cross-connect to another cable and pair, or as a connection between two termination blocks. The connection records will be very similar to those for connecting the horizontal cable pairs to the riser cable pairs.

The next step is to create the connections to the riser cable in the Main Equipment Room, which might be to a port on a voice switch, a piece of data controlling equipment, or to a patch panel. Follow the Vendor's Reference Manual for connecting a cable pair to a piece of equipment, patch panel or termination block. The screens for specifying these connections will be different for each CMS, however the basic information will normally include the cable name, cable end/location, cable pair(s), termination device/equipment, device side, and the connector/port. These connections are similar to those shown in Tables 29.11 though 29.13, except you would use the riser cable name and pair(s) in place of the horizontal cable name and pair(s).

Connect the appropriate cable end and pair (1, 2, 3, or 4) of the riser/campus cable to the port on the controlling equipment, the patch panel, or to the termination hardware. The format for this connection will depend on whether the termination hardware is set up as a separate component or as an attribute of the riser cable.

Tracing The Circuits

The previous sections have covered the steps required for adding individual components to the CMS records, connecting cables, desktop equipment, controlling equipment, and termination blocks, and creating individual circuit connections.

The circuits should be complete from desktop equipment to a Port in the local wiring closet, Equipment Room, or PBX Room.

In order to test the circuit trace or display functionality, the locate the PC or desktop device in the first office. This might be done in the Inventory module, a separate Trace module, a Circuit module, or by selecting an icon on the floor plan drawing. Follow the navigational outline in the Vendors Reference Manual to determine how a trace report is handled.

Once the database has located the record, use the trace key command or menu option to display a trace or report of the circuit connections for the PC. If the connections were created correctly and the catalog is properly defined, the trace should show each component of the circuit. The trace display or report will appear differently in each CMS product. The following illustrations should give you an idea of what the trace should provide in the way of components. The number of components will be determined by the whether termination hardware and patch panels are handled as separate components as shown in Figure 29.2, or as attributes of the cables and controlling equipment ports as shown in Figure 29.3.

Check to see that each component of the circuit is identified with the name/ID of the component, the component type (PC, cable, etc.), the side or end, the pair/port or connector, and the location. This is the minimum amount of information which should be displayed, since it is required to determine the path of the actual circuit from room to room, from termination device to termination device, or from cable to cable.

Also check the circuit display to make sure that information regarding the ports, connectors, or cable pairs is properly displayed, and represents the connections in a way that is readily understandable to the technician and CMS user.

If the circuit display stopped before reaching the controlling equipment port, or did not show the proper sequence of connections, double check each connection record to make sure the correct connectors were used, the correct side of the device was connected, and the correct end of the cable was connected. These are the most common areas where errors or inconsistent connection records may prevent the entire circuit path from being displayed.

If you have checked all your connection records and the circuit display still does not provide the entire path, work with the CMS Vendor to identify which components in the circuit need modification in the to obtain a better circuit display. Catalog adjustments can take many forms, especially if the trace logic relies on certain port configurations, a certain sequence of devices, or the specific use of "front" or "back" sides on a device.

Testing Rules and Restrictions

Once you are satisfied with the way in which the database handles the circuit connections, you should test whether there are any restrictions which prevent certain connections from being created. For example, an individual may attempt

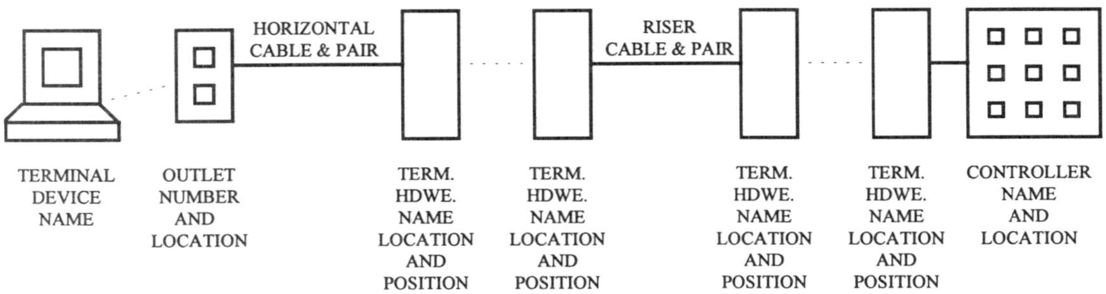

Figure 29.2: Circuit components with termination hardware tracked as separate entities.

to connect the wrong end of a cable to a port on a piece of controlling equipment. Another type of improper connection might be where a user attempts to make a connection between two items which are in different physical locations. This could result from entering the wrong equipment name in the connection record, or from an error in creation of the inventory record which specified the location of the patch panel or controlling equipment. There are many ways in which these errors could be created.

You may want to determine if the CMS has built-in restrictions which prevent a user from connecting items which are not in the same location, or will prevent connections between items from different network systems such as connecting a phone to an Ethernet Port on a Hub. The following tests may not be applicable to all CMS, however they are offered as a guide in testing some of the restrictions which may or may not be built into the CMS you selected.

In an earlier section you created an additional termination block in another room for test purposes. Attempt to create a connection record between a termination block in the local wiring closet and the termination block in the remote wiring closet. You might want to use the termination block where the riser cable is terminated. Complete all the fields on the connection screen. If there are restrictions in the CMS database to prevent items in different locations from being connected together, you should receive a warning or error message. If no warning or error message appears and the record is accepted, you will know there is no restriction on connection of items in different locations. If the record was accepted, you will need to delete the connection record in order to delete the test block from the CMS database.

Some CMS systems have the ability to create rules for connections which will restrict connections and circuits by the circuit type. For example, there may be a rule which would prevent a circuit from starting with a Phone and ending with a port on a Network Hub. Other rules might apply to the number of cable pairs or pin connections required for specific circuit types, which would warn a user when the wrong number or type of connection is specified. To test for connection restrictions, attempt to connect a phone to a port on a Hub or Data Patch Panel. Complete all the fields on the connection screen to see if a warning or error message occurs. If no warning or error message appears and the record is accepted, you will know there is no restriction on circuit type connections. If the record was accepted, you will need to delete it now. You may wish to test other restrictions by randomly creating incorrect connection records to see if any other warning or error messages are displayed.

Finishing the Support Tables

In the beginning section of Chapter 28 you prepared lists of information for the validation tables including buildings, departments, cost codes, etc. We suggested not entering all the items

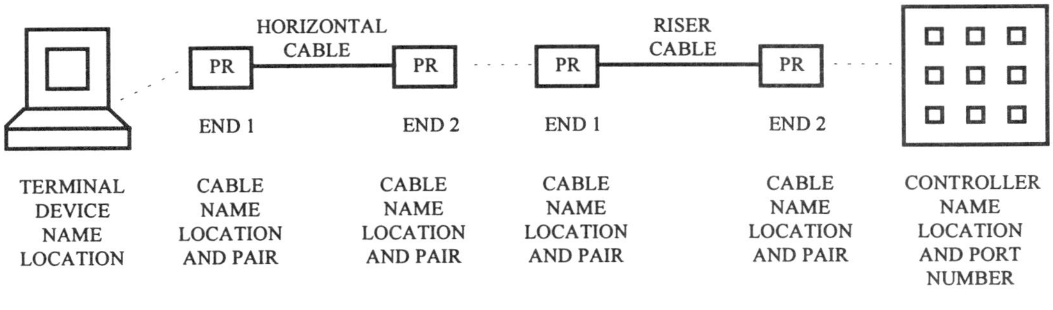

Figure 29.3: Circuit components with termination hardware as attributes of the cable and control equipment.

into the database until final testing could be completed. If you do not plan to utilize the Service Order or Trouble Ticket portions of the CMS, you will want to add the remaining information to those validation tables now.

If you plan to utilize the Service Order and Trouble Ticket functions of the CMS, continue with the testing of the information in the Service Order and Trouble Ticket documentation in Chapter 30 before completing the support tables. You will want to make sure the building, floor, and room locations are properly represented and do not require any modification before adding a lot of information which may need to be changed later.

TESTING SERVICE ORDER, TROUBLE TICKET AND REPORTS

30

By this point you should have set up the catalog of cable and equipment types, and have entered some preliminary data records for testing. We can now discuss the process for testing the Service Order, Trouble Ticket, and report functionality of the CMS. This will make sure they are providing the anticipated information before these functions are used for daily activities. Proper testing of the CMS operation with your sample data will identify areas where information or functionality is lacking, where cable or equipment definitions need some tuning, or where special procedures need to be developed to address record updates and maintenance. You will be in a better position to create specific documentation and training materials for managing and maintaining the CMS records.

If you have selected a CAD-based CMS, the activities outlined below will need to start with the display of the floor plan drawing where a specific item is selected for movement or change, or a room is selected in which new service or components will be added. The steps outlined below are based on using a CMS to manage the daily Moves, Adds, and Changes (MACs) of voice and data connections. Depending on the extent of the records in the CMS, all of the testing activities here may not be appropriate for your individual implementation. Select those areas which apply to your individual recordkeeping needs.

Testing the Service Order

If you are planning to use the Service Order function to manage MACs, you will want to test the entire process for creating the order, printing copies of the order for the technicians, making any changes or modifications to a circuit, and then closing the service order. Some of the items to be tested are listed in Table 30.1.

Create a Service Order to move a User and their equipment from one of the offices you created in Chapter 29. Let's say that the rooms you created with Users and equipment were Rooms 1101, 1102 and 1103, and that rooms 1104 and 1105 were the vacant offices. The Service Order should clearly indicate that User #1, PC1000 (PC), and Phone 123-1234 in Office 1101 are moving to Office 1104. The screen formats for each CMS system will differ, so you will want to complete the information as appropriate. Some systems will have a separate screen or operation for each component to be moved, while others will indicate all affected items on a single screen. You will want to check to see if the existing circuit information and cross-connects for the PC and phone are displayed or accessed.

The Service Order may or may not also provide the opportunity to add, delete, or change

- ADD NEW INVENTORY ITEMS
- TEST ABILITY TO CHANGE CROSS-CONNECTS
- CHECK FOR SERVICE ORDER REFERENCE ON TROUBLE TICKET SCREEN
- TEST FOR ABILITY TO HANDLE CIRCULAR MOVES
- LISTING OF AVAILABLE OFFICE OUTLETS OR HORIZONTAL CABLES

Table 30.1: Service order items to be tested.

the circuit elements and connections. In this case, connections must be broken for the old circuits, equipment moved to the new location, and new cross-connections created based on the horizontal cables or outlets in the new office. If this option is provided, use the test data to request that existing circuits and cross-connects be disconnected from the horizontal cables or outlets in Office 1101, and then make new circuits and cross-connects to the horizontal cables or outlets in Office 1104. These operations should flow smoothly on the Service Order screens and be completely and understandably presented on the printed Service Order document. Check to see if the CMS database will provide a list of available horizontal cables or outlets in the new room.

Once you have completed all the steps for creating the Service Order, request a printed copy to be issued to the technician. This may be in a variety of formats such as an Electronic Mail message, remote printing to the technician's local printer, or to a local printer for pick up by the technician. Review the printed document to determine if the information provided to the technician has all the instructions and circuit information required to complete the move request. This information may be a combination of textual report and printed drawings. If any additional information is needed, determine if it can be provided through minor catalog adaptation, other Service Order fields, or a separate document or report, in that order.

Prior to closing the Service Order created above, create a second Service Order where User #3 in Room 1103 will be moving all of their desktop equipment and services into office 1101, where User #1 is moving out of. This will indicate if you can handle simultaneous multiple moves through the Service Order. In many organizations, it frequently occurs that an office is being vacated by one user and will be occupied by another user during the same time frame. Sometimes two users simply switch offices, frequently referred to as a "circular" move.

You will want to request the existing circuit information and cross-connects for the PC and phone in Office 1103 to be disconnected, and then make new connections to the new horizontal cables or outlets in the Office 1101. The Service Order screen may provide some indication that the horizontal cables or outlets in Office 1101 are currently in use or are pending disconnection. Check to make sure that you can assign new circuits and cross-connects to those components while they still have a pending disconnect status. If connections can only be made to available items, you will likely have to cut and close several Service Orders in sequence or else handle the changes and updates manually. In some CMS systems, you may be able to manage all changes for multiple moves through a single Service Order. While none of these approaches may be a problem, these are points which should be addressed in procedures and training materials for individuals who will be creating Service Orders.

Now create a third Service Order. This time you will add a new User to Office 1105, where a new phone and PC will need to be added to Inventory for use by this employee. Check to see if the Service Order has the capability to add new Users to the database, without having to first create the User record in a separate module or section of the database. Then check to see if a new PC and Phone Extension can be added to the database and scheduled for installation and connection in Office 1105. This will determine if items must first be created in the inventory module prior to creating a Service Order referencing their use or connection. In many of the CAD-based systems, these activities are handled through the selection of an icon, indicating where the item is to be placed, and then completing the record information as to name/ID, type, department, etc. You may wish to see whether the Service Order allows the addition or deletion of control equipment, termination hardware, or pathways as well, but these functions are not nearly as critical.

Before you close any of the Service Orders, go to the inventory tables and do a query on several of the affected items such as the phone the PC, and the horizontal cables to see if there is a status indicator referring to a pending Service Order, Move, or Disconnect. If there are indicators regarding the status, you will want to check these indicators after closing the service orders to make sure they have been properly updated as in-use to reflect the completed status.

Request a report of all open Service Orders. Each of the CMS products will have different options for how the reports can be issued such as by user, technician, building, department, date entered, due date, etc. Run at least the normal set of reports you would find helpful in managing MAC activities to determine if the necessary information is in the reports.

When closing the Service Order, there are several things you will want to check for. These include the ability to make changes in the circuit after the service order has been issued, checking that all status indicators for the cables and equipment have been updated, and checking the old and new circuit and cross-connect information for proper changes.

Close the first Service Order where User #1 moved to the new office. Before closing the Service Order, check to see if a change can be made in the cable pair or port connection for one of the circuits. If there are multiple horizontal cables in each of the offices, you may want to see if you can make a change to the cable which was assigned. Determine whether any changes made by the technician in completing the Move can be addressed in the Service Order. Otherwise, the Service Order may need to be closed as originally issued with any other cross-connect changes done manually through another module of the database.

Most of the CMS products do permit circuit changes after the Service Order has been issued, however some require the closing of the initial order and the creation of another Service Order to make the desired change in the circuit. While this can be handled through procedures and training, it is important to make sure that these type of changes are updated in the database, and not left as specified in the original Service Order.

Now close the second Service Order where User #3 moved into User #1's old office. If the CMS product has a Trouble Ticket module, leave the third Service Order open (where the new User was added to Office 1105). This will provide an opportunity to test the Trouble Ticket to see if there is any indication regarding a pending service request. If there is no Trouble Ticket, then you can close the third Service Order.

Testing the Trouble Ticket

If you are planning to use the Trouble Ticket function to handle trouble calls and issue Trouble Tickets, you will want to test the process for creating tickets, printing copies of the Trouble Tickets for the technicians, making any adjustments in the circuit resulting from the problem resolution, and closing the Trouble Ticket.

Create a Trouble Ticket on the new phone which was installed for the new User added in Room 1105 (the third service order opened previously). Check to see if there is any indication regarding the pending Service Order. This is a valuable feature for use on a Help Desk, since the problem may be related to a Move or circuit changes being performed as a result of scheduled service. Also check to make sure the circuit information is available through the Trouble Ticket. This is very important in assisting with troubleshooting the problem.

Once you have completed the steps for creating the Trouble Ticket, request a printed copy to be issued to the technician. This may be in a variety of formats such as an Electronic Mail message, remote printing to the technician's location, printing locally for pick up by the technician. Review the printed document to determine if the information provided to the technician has the problem description and circuit information required to locate the equipment, circuit components and cross-connects. This information may be a combination of textual report and printed drawings. If any additional information is needed, determine if it can be provided through another Trouble Ticket field or if it must be provided as a separate document or report. Sometimes, the circuit information may have to be obtained from the circuit trace.

Request a report of all open Trouble Tickets. Each of the CMS products will have different options for how the reports can be issued, such as by technician, building, department, date, etc. Run at least the normal reports you would find helpful in evaluating trouble calls, especially if the Trouble Tickets have codes or tables which define trouble or resolution types. These types of reports can be helpful in identifying trends in types of trouble calls, either by location, equipment or service type, or cabling problems. Review the reports to determine if the necessary information is documented in the reports.

The functionality of the Trouble Tickets will vary for each of the CMS products. Some will require an indication as to the problem resolution code or activity, time to repair the trouble, and who handled the Trouble Ticket. The thing to determine is the capability to make changes in the circuit resulting from the problem resolution, such as a change in cable pair, port assignment, or defective equipment. Having the capability to make the change while closing the Trouble Ticket simplifies the update process and ensures circuit changes are reflected in the CMS records. If this function is not available, be sure that the procedures are in place to handle manual updates.

Testing the Reports

The reporting functions of the CMS can be a very important part of managing the cable and connectivity information. If you cannot retrieve the records in a printed format which is useful and provides the required information, it limits the benefits and usefulness of the records. This is the reason many CMS Vendors have added an ad-hoc report writer function either through their own software program or through the use of a third-party report writer.

The types and format of required reports will vary for each organization and will vary from system to system. Often, these reports are based on older manual reports or spreadsheets. In running and printing the desired reports, test the different print options such as printing to a local printer, a remote LAN printer, or to a plotter. Some of the basic reports required by many organizations are listed below. You may wish to run several of these reports now to review their format and the information contained in each report.

- Connections by Cable Pair for a Specific Cable
- Port Connections for a Specific Wiring Closet by Type, by Controlling Device, and by Horizontal Cable
- Equipment by Type, by Location, and by Department
- Remaining Available Capacity by Cable
- Report of all Drawings or Floor Plans

Create a report and specify that it be saved or filed to a disk (or a report directory) in ASCII format for downloading into another database. This should be a report with several columns so you can test the ability to duplicate those columns and records in the desired format. Some systems will allow the user to specify the file name and extension such as CABLERPT.XLS or CABLERPT.WKS for a cable report which might be imported to an Excel or Lotus spreadsheet. Another option might be a report which will be included as part of a text document, where the file name might be EQUIPMNT.TXT or EQUIPMNT.DOC for importing into a word processing application. Such features are growing more common, especially in CMS software designed for a Windows environment.

Other CMS systems will create their own names and extensions for the exported file. This may require renaming the file with an appropriate extension so it will be recognized by the targeted system if the format is compatible. Most word processors (and many spreadsheets) can import text in ASCII (.ASC) format.

Once you have created the file and saved it in ASCII format, proceed to import it into a document or spreadsheet. You can then look at the report and manipulate the data or add comments and titles. This usually requires some special commands or formatting of the file once it has been imported. In Excel, for example, you may be required to use the Data/Parse command to define the appropriate column breaks. The procedure for utilizing the reports which have been created in ASCII format should be provided by the CMS Vendor, or should be covered in the Reference Manual provided with the CMS. The Vendors are usually able to work with you to resolve any problems with ASCII reports or other files generated by the CMS.

If you are satisfied with the results of the Service Order and Trouble Ticket testing, you will need to enter any remaining information in the validation and look-up tables now.

WORKING WITH CAD DRAWINGS

31

If you have selected a CAD-based CMS, you will need to do certain tasks to prepare the CAD drawings for use, and learn some things about working with CAD drawings in general. We have referred to many of the set-up and testing tasks in the last four chapters, but will go over them again here. We will address some of the issues for adding CMS items to the drawing, testing the drawing functionality, and modifying and maintaining the drawings ongoing.

As discussed in Chapters 6 and 17, a CAD-based CMS can either have native CAD functionality (which makes it the primary system interface) or else be built around a database where the drawing interface is optional. Each of the systems which utilize CAD drawings has different functions and methods for placing equipment icons or symbols on the drawings, and for routing of cables and pathways. Depending on the CMS tool selected, some of the activities below may not be appropriate for your individual implementation. Select those areas which apply to your individual CMS and specific recordkeeping requirements.

In the case of a native CAD system, the activities for handling changes to connections and to the location of equipment is usually handled through the CAD floor plan or drawing, since this is where these elements are displayed and referenced. Drawings are therefore required, and must be developed from surveys, paper drawings, or existing CAD drawings. Specific training will be required to manage information through the CAD drawings.

The process for converting paper drawings to CAD can be one of two methods: digitizing or scanning. With digitizing, a human operator places the drawing on a special pad and uses a device to trace over the drawing elements, one line at a time, to recreate the lines and shapes electronically. The scanning process, on the other hand, simply runs the drawing through a scanner which optically scans the drawing in one or more passes and maps it as a bit image called a raster file. Usually these images are then converted from raster to vector format, where lines and objects become defined. Most CAD systems have their own specific vector formats, much like word processing software use different file formats.

While scanning would at first appear to be the faster method, it is always an imperfect process. The original scan has a limited resolution and no way of knowing a real drawing element from a coffee stain or a speck of dirt. The raster-to-vector conversion is also imperfect. It takes a human operator some time to clean up the drawing and create different drawing layers, as is usually done. Scanning works best on fairly basic, architectural paper drawings in good condition.

By digitizing the drawing, the operator has already created a vector format and done the layering, and perhaps has done so in the final CAD format. Obsolete or unnecessary information can be ignored. Complex drawings showing a high level of detail or which include plumbing, HVAC, or electrical information are good candidates for digitizing. If the drawing is now in the CMS CAD system format, no further actions are required and the drawing is ready for use.

Preparing and Importing Drawings

If you had existing CAD drawings in a different CAD system or have just prepared them for each of your buildings and floors, the first step will usually involve creating a DXF file from the existing CAD file. "DXF" stands for Drawing Exchange Format, where different CAD programs will recognize the walls, lines, symbols, and text from the original drawing and be able to convert them to a native format. Some of the CMS products require that specific settings, drawing limits, or a drawing scale/size be identified prior to creating the DXF file, such as an indication as to point of origin with respect to the site layout, or to the different elevations or floors. Check the procedures outlined in the Vendor's Reference Manual for any specific instructions prior to creating the DXF files.

When you create the name of the DXF file, you will want to give it the same name you will use as a reference in the CMS database. Most of the CMS products will retain the current file name when the drawing file is imported. If the current file name is LRP0001A, and you want the database to refer the drawing as REN01 for the first floor of the Research Engineering Building (REN), you should name the DXF file REN01.DXF. This way when it is imported into the CMS database, the drawing reference name will be REN01. Some considerations for the naming and layering of drawings are covered in Chapter 17.

You may be required to specify the file and directory where each DXF file is to be stored. Check the Vendor's Reference Manual for specific instructions on copying the files into a designated directory. After you complete the import process for each drawing, display each individual drawing to make sure the drawing was created in the proper size and that all the components were properly positioned. If the drawing is too small or large, you may need to adjust the limits or scale in the original drawing file and create another DXF file.

Drawing files used in a networked system will usually be located on a single server along with the CMS database. They will likely be in a different directory, and you must be sure that each CMS user can access the specific directory where the drawings are. The CMS should allow only one user to access a specific drawing at a time, so changes are not made simultaneously by two users. Keeping a single set of drawing files will make it easier to maintain them ongoing. They should be backed-up routinely along with the database itself.

Placing and Moving Icons

When first populating each of the drawings with icons and symbols for your cables, pathways, and equipment it is best to follow a systematic approach. Each of the CAD systems work a little differently in this manner. Some systems are set up to enter a sequence of components for a specific room or location, while others may be set up so that items of a similar type are entered consecutively. For example, after inserting a phone the CMS will prompt the user whether they wish to insert another phone. In this case, you would want to organize your records so that all phones would be created together, then PCs, printers, etc. Test the CAD functionality to determine if it prompts the user to add another item of the same type or whether it prompts to place another item in the same location. If the CMS prompts for the next item to be placed in the same room, it may be best to organize your records so that all items are added based on their location.

Begin by placing the termination hardware on the drawing so that as cables are created or located, they can be connected to the appropriate termination devices such as outlets or blocks. Most CMS products have a process for controlling the routing of cables so they follow a straight line, such as holding down the shift key while routing the cable, or activating an Ortho-type (Orthogonal or right-angle) command. This feature will keep the cable routings neat and following a designated path. If you know the actual routing of a pathway, it should be indicated on the drawing. Once the termination devices and cables have been located on the drawing, then it is easier to place icons for desktop and controlling equipment at the proper location for connection to the outlets or blocks and panels. If icons are used on a plan drawing to represent termination hardware or control equipment, be prepared for the drawing to get very busy as numerous icons are stacked on each wiring closet.

The process for moving items and changing connections on the drawings will be different for each CMS. The CAD-based systems are easier to manage in an environment where each office is permanently and identically wired for voice and data connections, and desktop equipment is tracked against a room location or jack. Buildings that use dedicated wiring are a good example, where patch cords or jumpers are permanently in place between horizontal cable and backbone cable, or between horizontal cable and equipment ports. This simplifies changes to connections which must be made between cables, termination equipment, patch panels, or controlling equipment, since a typical Move only requires relocation of the desktop devices. If an item is selected on the drawing and moved from one office to another, the only updates required on the equipment record are to the location and the outlet to which the equipment is now connected. The CAD-based CMS will do one or both of these updates automatically.

Individual circuits associated with each desktop device may require changes in the wiring closet cross-connects when the desktop equipment is moved or relocated. There are several steps required to maintain both the database connection records and the drawing floor plan where the icons for the desktop devices are placed. If an item is selected on the drawing to be moved from one location to another, the location record must be updated and the connections to the jack or outlet, the horizontal cable, and the port or riser cable pairs also need to be updated. Some CMS products will prompt the user, while in CAD mode, whether they wish to update these records and connections in the database. Other systems will leave any required cross-connect changes entirely up to the user, while there are even systems which do not provide any indication as to the connections or circuits associated with moving the item from one office to another. In these systems, the floorplan drawings often become out of synch with the database records, since there is no link or process for updating both the connection and drawing records from the same segment of the database.

There are graphic systems available which provide illustrations of patch panels and control equipment. Changes in patch cord connections between panel and wiring can be made on the picture of the device itself by "clicking" on the appropriate connection points, with the database updated accordingly. Such a system will require the development of graphic pictures for each panel or device.

Some CMS products have strong links between the database and CAD modules such that the drawing "knows" the room number of a given space, or "knows" the outlets at each location. Some systems therefore have the ability to automatically move an icon on the drawing if the room number or outlet connection for that icon (item) is changed in the database. Of course, moving the icon automatically updates the database location field. Check both the database and drawing sections of the CMS whenever a change is made regarding the location or connection of an item on a floor plan drawing.

Test the database from both the CAD module and the database connection module to determine what impact a Move from either module will have. The first test should be from the CAD portion, where an item is selected and moved to another office in the same floor or wing. Observe any prompts or options for changing the circuit connections and database records. Then check the records from the database portion to see if the item record and connections reflect the changes made from the drawing section.

The second test should be done from the database portion of the CMS tool, where the connection of a desktop device is disconnected from the current outlet or jack and reconnected to a new outlet or jack. Make any necessary changes in the desktop device location record and any required circuit connections in the wiring closets to data ports or riser cables. Then check the records from the CAD portion to see if the icon for the item has been relocated to the new office, or if it still appears in the old room.

These activities will help to identify the procedures and steps which must be completed with each move or change, in order to keep both the database records and connections in synch with the CAD portion of the CMS. As mentioned earlier, CAD systems can provide additional benefits for planning and change management, but they can require additional time and effort to maintain the records properly. This is the reason some organizations only track the more permanent portions of their communications structure on CAD drawings - items such as the backbone cables, horizontal cables, outlets or jacks, termination hardware and controlling equipment that remain fixed and do not require extensive drawing maintenance.

Modifying or Changing Plan Drawings

Each CMS will have different levels of built-in CAD functions. Depending on whether the CMS you have selected is a CAD-based system or a database linked to CAD drawings, the process for making changes to the base floor plan drawings will vary.

A CMS which utilizes the CAD drawing only as a reference point for locating icons will have limited CAD functionality such as the ability to zoom in on a particular area, pan to another portion of the drawing, and plot or print a drawing. These systems require that modifications to the base drawing be made in the original CAD program, and then the drawing is re-imported into the CMS. There is no loss of cable and device information overlaid on the drawing, but this information may then have to be modified through the CMS. With these systems, it is important that all the settings of the revised drawing file be the same such as size, scale, limits, etc. Otherwise, this will impact the accuracy with which the location of icons, symbols and cable routings are overlaid on the re-imported drawing.

The CAD-based systems usually have full CAD capability where the user can move or add walls, doors, text, etc. and change the actual floor plan layout. With these systems there is no need to re-import another version of the drawing, since all required modifications to walls, doors, text, etc., can be done through the CAD functions of the CMS database.

Printing and Plotting the Drawings

The ability to designate printing or plotting of drawings in the CMS systems will vary from system to system. CAD-based systems usually offer more flexibility in designating the size of the area to be printed and whether the file should be sent to a laser printer or a plotter.

You should test the plotting and printing capabilities of your CAD drawings early in the implementation stage so that you identify any limitations in obtaining printed copies of the drawing files. You should display all components of a specific type, such as all PCs or horizontal cables, and request a printed copy of the displayed drawing. This will show how the information is displayed and whether icon names or IDs are indicated on the printed drawing along with the icon itself. Some laser-type printers have the ability to emulate a plotter using an add-on cartridge or software, but the text resolution of a D (2' X 3') or E (3' X 4') size drawing printed at 8 1/2 X 11 inches is very limited.

Zoom in on a small area of the drawing, such as 5 to 10 offices near the wiring closet, and then request a printed copy of that area. This will identify whether you can specify only a portion of the drawing to be printed, and whether the printing is done at the right scale.

You will want to determine how you can export drawing files from the CMS database. This will be required for systems which do not provide a high level of CAD functionality, and where you may wish to add text, comments, or individual item names to the drawing icons through separate CAD software.

You will also want to determine where the export (DXF) files are stored when they are created, and if the CMS database gives you the option of naming the DXF file it creates for export. This will allow you to locate the file and copy it to a disk or separate CAD directory where you can work with the file. In order to work with the DXF file in the CAD system, you will most likely have to use the DXF-In command or utility, which then regenerates the drawing as it was displayed in the CMS software.

WORKING WITH A SQL REPORT GENERATOR

32

Many of the major CMS systems use third-party database software such as Oracle, Informix, or Sybase. The CMS software operates on top of the database software, so to speak, after a certain set of structured tables are created in the database. The CMS uses the specific command language of the database to communicate with it and perform operations on it.

For systems such as this, the database software and tables can be accessed directly (not going through the CMS) by using the Structured Query Language (SQL, pronounced like "sequel") tools available for that database. Other software known as Report Generators use a similar approach that may be applied to a number of different database types. Some CMS software comes with a built-in or third-party report generator for creating ad-hoc reports, while other software will require purchase of the SQL software specific to that database.

In this chapter, we will discuss some of the functions performed using SQL tools or report generators to retrieve information from various tables in the CMS database and to structure the information into a useful report format. Often, there are specialized requests or reports which can help managers and supervisor to monitor activities and capacity utilization on cables and network controlling equipment.

The use of SQL tools or report generators will require the user to understand the query language and report formatting commands required to accomplish a particular result. The SQL commands are usually created as a batch file which when executed operates on the particular database tables, accumulates the proper records into a file, and then can print out the file in a specified format. The specific SQL commands and batch file structure must be learned from the SQL documentation itself. In some organizations, the CMS uses a "company-standard" database that is supported by a specific person or persons familiar with the database structure and the SQL language. In such cases, this person or group can provide the necessary expertise for generating custom reports or queries.

Creating Custom Queries and Reports

The built-in reporting capabilities of each CMS vary greatly, and are usually a reflection of the CMS' own data structure. Many so-called "canned" reports in a CMS are generally able to query and provide information from a particular table in the database. An example might be to list all the analog telephones on a particular floor. Reports concerning the connections of things are usually limited to a specific type of connection query, such as showing what phones are connected to which outlets on a given floor.

The use of SQL tools can greatly enhance these capabilities by creating queries across several tables, by looking at connected or associated information in different ways, and by allowing the user to specify the final report format. The SQL commands allow for queries over a range of values, rather than just matching a particular value or string. While a canned report might be able to list all PC-to-outlet connections on a floor, a SQL report might be able to limit this to PCs with two connections or more. This type of report would utilize the SQL commands which select only those records where a particular PC appears more than one time in the connection table. Another variation through SQL would be to list only those PC-to-outlet connections which are carrying a specific network service.

A custom report might include a list of all PCs which have internal or external modems. This must be determined by looking at connection records where an equipment type matching the type "Internal Modem" or "External Modem" and an equipment type matching "PC" are contained in the same connection record.

A very sophisticated, hypothetical query might be to identify the phone numbers of all users with more than one PC connection to the network. In this example, the database must first be queried with respect to PC-to-outlet connections, identifying the right PCs. Then a second table is queried to identify the user of each of those PCs. A third table is queried to determine the phone number of each of these users. As each of these queries is performed, the information is gathered into a single file and formatted so that each relevant PC, user, and phone number is displayed.

SQL tools can be used for requesting special information from the database which would not normally be in a table or report format. The requested information is based on matching certain field criteria from multiple tables, and can apply to fields that cannot be queried through the CMS itself. An example of this type of special query might include a report of all Service Orders requesting new service over a specified period of time, for a specific department, and listing who authorized the request. Another example might be a report of Trouble Tickets for a specific building, where the problems involved a defective cable pair which were handled by a specific technician over a specific time period.

Making Global Record Changes

Another very useful function which can be handled through SQL tools is the correction or revision of certain information in a database table. These would be global changes which affect multiple records in the database, normally requiring hours of revisions on a one-by-one basis. Character strings in multiple tables can be changed at the same time.

An example of this type of change might be a change in department or division names, where two departments may have merged into one and the records now need to be revised so that all the records referring to either of the two old departments now reflects the new department name. Through the use of the SQL tools, a command can be issued which looks for records which refer to either of the two old departments, and replaces the department name for those records with the new department name.

If a network system address is used for naming control equipment such as Hubs, Cards and Ports, SQL can be used to make the required table modifications where the network address has been changed. Through an SQL tool, a global command can be issued which looks for all device names matching a certain character string in the Name or Network Address field, and changes the current series of characters to a new series of characters. An example of this would be where a 3270 controller had a Network Address or Name of LG1D11, and the patch panel and ports all carried these 6 characters in their Name or Network Address field. If the new controller ID became P01A31, the SQL command would look for any device name (Hub, Card, Patch Panel, or Port ID) beginning with LG1D11 and replace those 6 characters with P01A31. Such revisions must be approached carefully and systematically so that all affected records are properly updated and any validation information is properly maintained.

Another useful function handled through SQL tools is the creation of specialized report formats and menu screens which can be created outside of the actual CMS software. These can be used by managers, supervisors, and analysts for obtaining routine information from the database without having to log into the system and request the report. Through the use of SQL tools, special report formats can be created and saved on file, with a menu screen developed for use in requesting those particular reports to be run to retrieve information from the database. These reports can be used for monitoring cable capacity, Service Order activity, trends in trouble calls and problems, and numerous other routine requests.

DOCUMENTATION

33

Having completed the final catalog and made preparations to load the database, it will be useful to prepare a complete set of documentation regarding the CMS and its associated procedures. This documentation provides a basis for training additional users, guides the administrator with further implementation, and serves as a general user's manual ongoing. This documentation includes, and is in addition to, the Reference Manual provided to you by the CMS Vendor.

In the course of your project development and testing, you have prepared a number of tables and documents that can now assist you in preparing this documentation. The information which might be contained in this additional documentation would be specific to your implementation of the CMS, including a description of the purpose and goals for the CMS recordkeeping, your company-specific catalog covering the classes and types of equipment and cables, naming and labeling conventions, procedures for working with Service Orders and Trouble Tickets, workflow diagrams, and roles and responsibilities. Additional information might include training exercises, reference tables for sites, buildings, and departments, and the names and telephone numbers of individuals to contact for CMS questions and support, both within your organization and the Vendor's Company. A list of potential CMS documentation items is given in Table 33.1.

It is always beneficial to provide users with general information, in the form of a CMS or Project Overview, on the reasons for implementing the CMS and the goals and benefits anticipated from maintaining accurate records. This helps to ensure users understand the purpose behind learning and using new tools, especially if it changes the normal routines for performing their job. The overview can also describe the type of information contained in your specialized documentation and how it can be used in conjunction with the manuals provided by the CMS Vendor to make the most effective use of the CMS functions and records.

It is helpful to provide users with information as to who to contact in case of problems with the CMS software, or for questions regarding use and operation of the database. Include both the name, location and telephone number of the local CMS Administrator for your organization and the primary contact and telephone number for the CMS Vendor. This avoids frustration on the part of users when they are having problems working with the database, get locked up in the system, or receive an error message which they do not understand.

This is a good place to cover any special rules or procedures for working with the CMS database which may not be covered in the Reference Manual, or which need to be reinforced to improve the accuracy and consistency of how information is entered into the CMS database. If the CMS does not validate against the room number, you should emphasize how all room numbers are to be entered especially if room numbers are a combination of letters and numbers, such as A101, or 122E, or 105-A. State the exact format which is to be followed when entering room numbers, including examples, so there is no confusion on when to use hyphens, dashes, spaces, slash marks, etc.

• GOALS AND OBJECTIVES
• KEY CONTACTS AND REMINDERS
• VENDOR REFERENCE MANUAL
• ERROR MESSAGES AND WARNINGS
• SUPPORT TABLE LISTS
• SAMPLE SCREENS AND REPORTS
• CABLE CONFIGURATION DIAGRAMS
• DEVICE CONFIGURATION DIAGRAMS
• LISTS OF MAJOR BACKBONE COMPONENTS
• CABLE AND DEVICE NOMENCLATURE
• LIST OF CAD DRAWINGS
• DEVICE ENTRY PROCEDURES
• SERVICE ORDER PROCEDURES
• TROUBLE TICKET PROCEDURES
• WORKFLOW DIAGRAMS

Table 33.1: General list of user-specific CMS documentation.

Another item to cover in the overview is the proper use of upper and lower case letters, especially if you are working with a database that is case sensitive. With case sensitive products, it is best to enter all information in CAPS to maintain consistency and assist in retrieving records for queries or reports. If there are special function keys used for short cuts or routine actions, it may be good to list those here. While they are most likely covered in the Vendor's Reference Manual, it's a good policy to review those with users so they remember to use them whenever possible.

The Vendor documentation, in the form of a Reference Manual or User's Guide, provides basic information about structure and navigation in the CMS. Menus, commands, function keys, etc. should be described in detail. The steps in creating each type of record should be detailed, as well as instructions for the Service Order, Trouble Ticket, and canned reports. If the CMS includes a basic catalog of cables and/or equipment, these types should be listed and described, and their relationship to one another and common usage as circuit components should also be discussed. Overall administration tasks for the CMS such as table maintenance, user profiles and privileges, drawing imports, network operation, and system troubleshooting should also be discussed in detail. Your additional documentation must show how all this applies to your catalog, your procedures, and your environment.

If it is not addressed in the Reference Manual, include instructions for users in how to handle error messages or warnings. This might involve what to do when receiving an error or warning message which does not deal with the normal routing functions of the database. Instructions might include requesting a "print screen" of the error message and making a note of what function they were performing at the time the error message appeared. Sometimes, this additional information can assist the Vendor or system administrator in resolving the problem or preventing it from happening again in the future.

Reference Tables and Catalog Diagrams

Ongoing, your documentation should provide an easy reference for understanding, adding, or modifying information in the CMS. You should include copies of reports from the CMS support tables which list the various designations for sites, buildings, floors, etc. There are probably similar reports listing all connector types, port types, connection configurations, cable types, and equipment types which should be included. As these tables are modified over time, individual user documentation should be updated. Lists of department codes/abbreviations, cost codes, etc. could also be maintained, but are not critical. These will help users to gain better understanding and easily refer to the different elements in the CMS database.

You may want to include some copies of "print screens" to show how some of the more commonly used database screens look, and how information needs to be entered. This is especially helpful for screens where connections are entered between cables and devices, or between network components such as hubs, cards, servers, and cables. This is also a place where required fields can be identified so users can become familiar with what information must be completed on each record before it will be accepted by the CMS. It becomes a helpful reference for users to determine how to add new types of devices or cables to the database as new models/types of communications equipment and components are added to the network.

A copy of each canned report should also be included. This will provide a quick reference for each report format, indicating the related information available on each. While some reports are generated by a query from a standard screen, that does not mean all of the fields of that record are automatically provided on the report. This way, the user does not have to generate a report just to see its format.

You should include a diagram or sketch of the connection configuration for each major type of cable and device, including termination hardware, desktop devices, and control devices. You probably have already done so as part of your catalog development. These diagrams will be invaluable in explaining the configuration of each component and how interconnection of these components is handled, such as how desktop devices are connected to outlets or horizontal cables, how network hubs are treated with respect to cards and ports, and how patch panels are treated in connection with controlling equipment. This will help users when making circuit connections or adding new controlling equipment, patch panels, termination hardware, or cables.

A general description for a cable configuration diagram and a device configuration diagram are given in Figures 33.1 and 33.2. They can be used as a "key" or legend for describing or

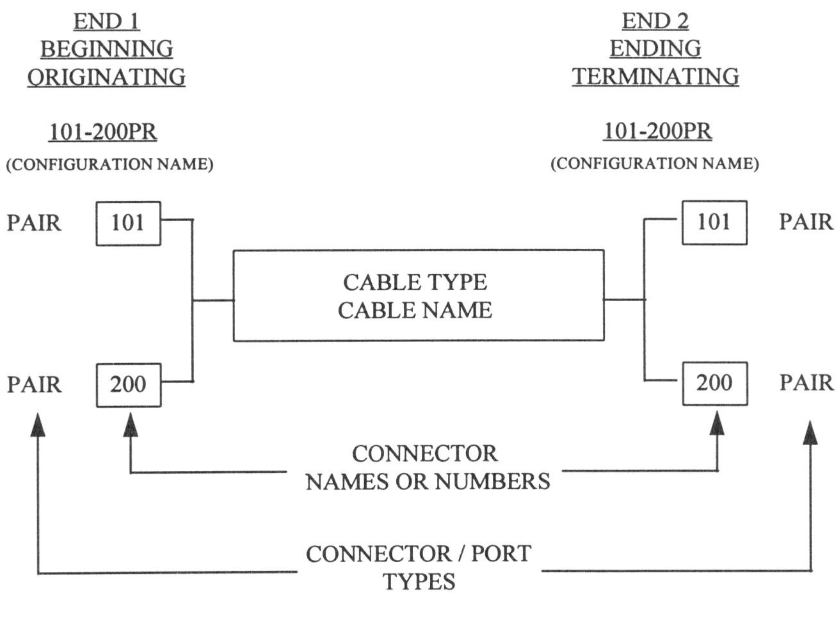

Figure 33.1: Key cable configuration diagram.

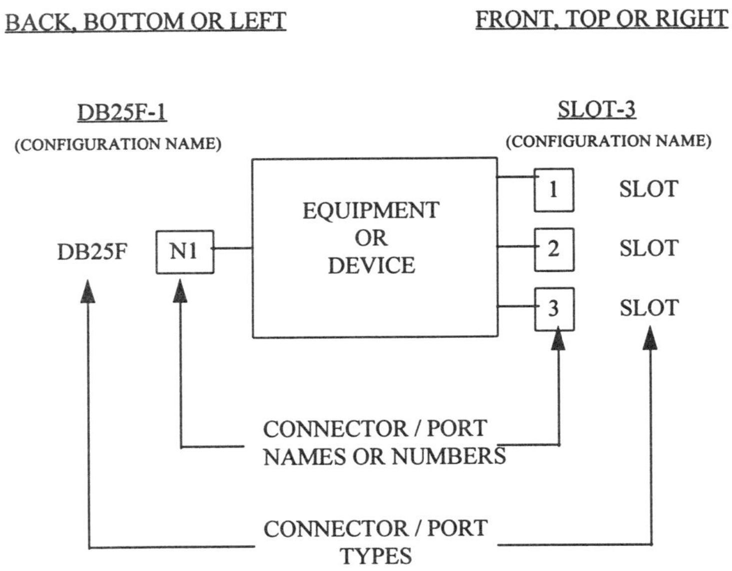

Figure 33.2: Key device configuration diagram.

understanding the specific configurations. An example of a possible device configuration diagram for an expandable Ethernet hub is shown in Figure 33.3. The use of configuration names is optional, depending on the CMS.

Once the major components of the backbone cable and controlling equipment have been added to the database, it may be beneficial to print out reports of all the riser and campus backbone cables indicating their to and from locations, pair counts and cable types, (voice, fiber, data, tie, etc.), and the locations of wiring closets and data centers or computer rooms. This information will be a helpful reference to users when making circuit connections in the database, in selecting the right cable, and determining cable location. This also provides the proper room numbers for the wiring closets to insure new cables or equipment have the correct room designations.

It is beneficial to have a complete source for the naming conventions developed for identifying each component. A description or diagram explaining the naming convention for each type of cable or equipment should be included. This will help users when adding new equipment or cables and keep them following the same format. This will avoid new equipment being added under different names. Some examples of how you might document your naming conventions are provided in Table 33.2 and Figure 33.4. Your naming conventions may not follow these formats, but these illustrations should provide a reference for how you might approach the documentation.

If you are using a CMS with CAD capabilities, you may want to include printed copies of the floor plan drawings for quick reference when planning multiple moves or when providing instructions to local vendors for installing new cables or equipment. They provide a means for documenting the exact cable routing or the intended location of new equipment. They also provide a quick reference for naming or numbering conventions of rooms, outlets, or wiring closet locations for each building or floor. Drawings are particularly useful in environments where the floor plans for each building and floor differ widely from area to area.

If you do not make copies of the drawings for each user's manual, try to place a full-size copy of the appropriate floor plan drawing in each Telecommunications Closet. The user's manual should then include a list of the available drawing files, the database drawing name, and the areas that each drawing covers.

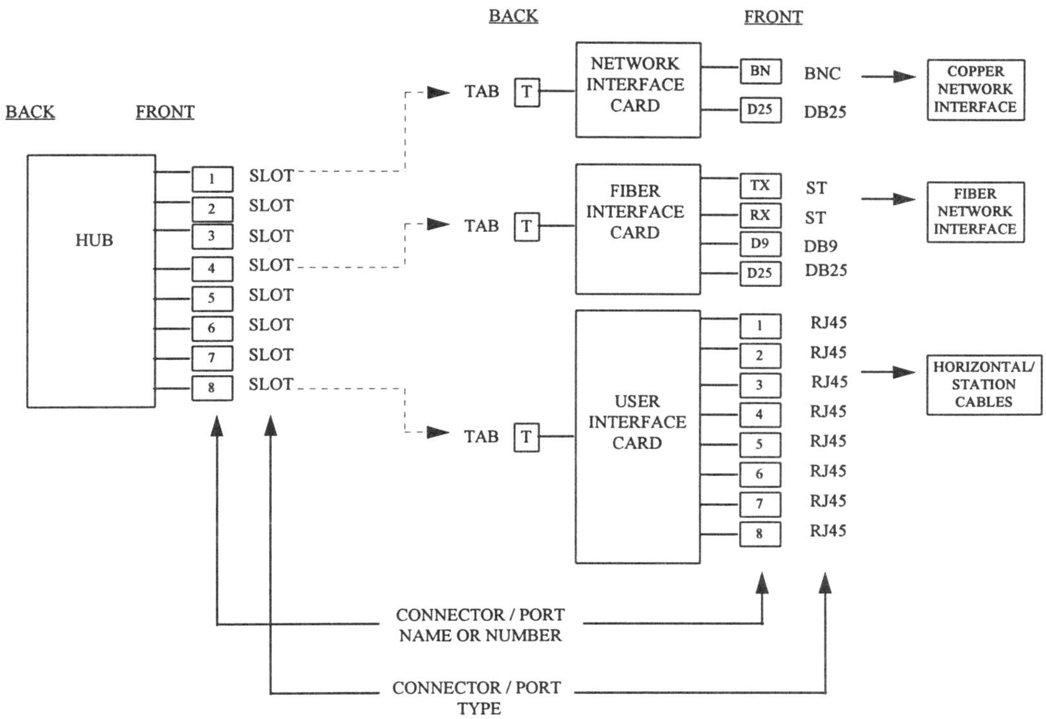

Figure 33.3: Possible configuration diagram for an expandable Ethernet hub.

For certain devices, you may consider developing step-by-step instructions for creating the device and connecting it in the CMS. This may apply to very complex devices with many connection options, or to simple devices created frequently. Preparing documentation in this detail can be time-consuming, so it should be done only when necessary.

Procedures

You should include any special instructions to be followed when creating and closing Service Orders or Trouble Tickets, such as whether new circuits are to be made when the

DATABASE COMPONENT	NAMING CONVENTION
HORIZONTAL CABLE	BLDG + FLOOR/WING + OUTLET NUMBER EXAMPLE: REN1E101
RISER CABLE	RC - BLDG + FLOOR/WING EXAMPLE: RC-REN1E
NETWORK HUB	NETWORK TYPE + H + BLDG + FLOOR/WING + SEQ # EXAMPLE: EHREN1E01 E = ETHERNET; T = TOKEN RING

Table 33.2: List showing documentation of possible nomenclature for database components.

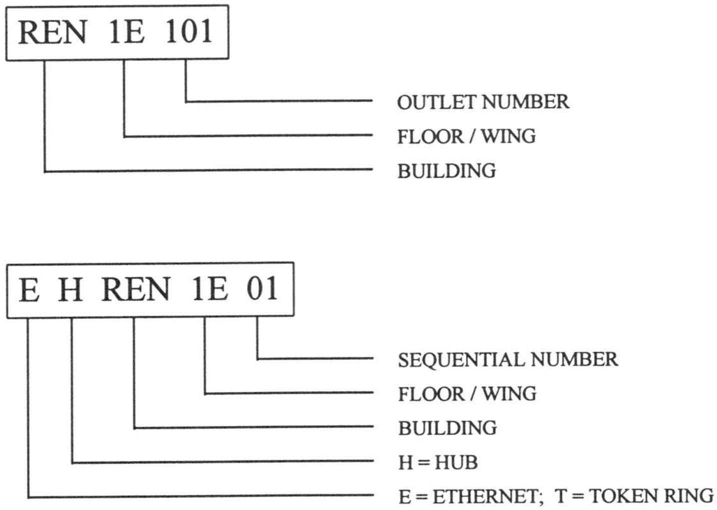

Figure 33.4: Diagram showing possible nomenclature for database components.

Service Order is opened or after the technician has completed the work and provided the circuit connections. Indicate how the document will be sent to the technician, and how finished orders will be returned. Specify any additional documents or reports which need to be provided with the Service Orders as they are issued to the technicians, such as additional circuit information. Include any specific information which must be added when the Service Order or Trouble Ticket is closed, such as technician time, resolution, technician name/initials, and how the circuit changes resulting from trouble resolution are to be handled. Some fields are optional and will not be required to close the order or ticket.

Copies of workflow diagrams should be included which document the flow of information and activities required to maintain and support the CMS database. These workflows should clearly identify the roles and responsibilities of each individual or group which is a part of the process for making Moves, Adds and Changes to the components which are tracked in the CMS records. This will help to insure individuals understand their responsibilities with respect to updating the database, and whether they are individually responsible for the updates or for passing the information to others who will make the necessary changes in the CMS records.

The last item to consider for your documentation might include exercises which can be used for training or orientation with the database. These exercises should cover simple tasks such as adding new components to the database. The exercises should be specific with reference to the names or IDs for the equipment or cables, and how to connect these components to form a circuit. The exercises will help new users begin to navigate around the system and become familiar with the screens and naming conventions. Basic training exercises will be provided by the CMS Vendor on the operation and navigation in the CMS, but their training programs are not tailored to your specific catalog and naming conventions. Creating exercises to help users get familiar with your specific implementation of the database will be beneficial.

TRAINING

34

After you have your documentation underway, you must consider the formal training required for each individual affected by the connectivity management process. In this chapter, we will discuss some of the items to include in the training sessions for new users, as well as some additional training which can complement the training materials and program provided by the CMS Vendor.

Prior to the initial Vendor training, review with the Vendor any specific information you want them to address during the introduction portion of the training. This information will help the Vendor focus on the areas which will be most beneficial to your users. Be sure they know which segments/modules of the database you are planning to use in the initial stages. Some organizations prefer to implement different stages of the software over a period of time, in which case the training should be phased to cover only those functions which will be used immediately. Training on functions that cannot be used immediately is often forgotten and wasted.

The standard training that most Vendors provide is oriented towards the database administrator or to persons who will routinely and directly use the CMS. This training is set up to train users in how to navigate around the database and the CAD interface, how to add, change and delete inventory items, how to connect different components together, and how to make changes in the circuit connections. If the Vendor does training shortly after installation it is unlikely that the database will be populated in its final form, so users cannot see their own catalog and procedures in action. The optimum approach is to allow the Vendor to provide basic functional training to a core of users including the project manager and CMS Administrator. This core group will be involved in project development and can take responsibility for training additional users according to their specific needs after the CMS has been set up to your specific requirements. In other words, the Vendor's role may be on a "Train the Trainer" basis.

The core team and others working directly with the CMS will require training exercises for creating and modifying the catalog and drawings. These exercises might include simple tasks like defining new types of cables and devices, and adding individual cables and devices such as telephones, PCs, outlets, and horizontal cables. Additional exercises are recommended for connecting various components into a circuit. The core group, and particularly the project manager and the CMS Administrator, must become thoroughly familiar with the entire catalog and naming conventions.

Training should be scheduled in groups with individuals who have similar job functions, so the class is not spending too much time on an area which is not applicable to all users. Data technicians and network support groups will have different interests and needs than voice technicians, the Help Desk, or the Customer Service Center. One approach would cover all the basic functionality of the software, such as adding and deleting inventory items and adding and removing connections and database components, in the early part of the training sessions. The Service Order and Trouble Ticket functions, which are of more value and interest to the Help Desk and Customer Service Center, can be covered on a separate day. This would provide the opportunity for technicians and other network services individuals to skip training on creating and issuing Service Orders or Trouble Tickets.

A training session should be scheduled with the technicians on understanding and responding to Service Orders and Trouble Tickets according to the procedures developed. Here, the emphasis must be placed on understanding the instructions, acting accordingly, and providing information back to the CMS Administrator or Group Leader in the proper format rather than interfacing directly with the CMS itself. A good understanding of the catalog and nomenclature is required by all parties.

Users should only receive specialized training (the specific catalog and procedures) in the areas of the database which they will use. While training in the other modules or sections of the database may be covered briefly, spending too much time on these areas may cause users to become confused or frustrated because the training is not applicable to their use and involvement with the CMS.

When an individual user is scheduled for training, make sure their workstation has been set up for access to the CMS so they can begin accessing the system as soon as they have completed the training. If users are trained but cannot put their training to use, they will quickly forget what they learned in the class. Schedule the training for users just prior to their need to work with the records in the database.

Training sessions should be limited to half-day sessions if possible. Full-day sessions tend to overwhelm users and cover more than they may be able to absorb. This also provides the opportunity to do training in split shifts where half of the individuals (technicians, network support, Help Desk, Customer Service, etc.) can attend training in the morning, with the remainder attending training in the afternoon. This allows for coverage to be maintained in each area, since only half of the individuals will be in training at any one time. If users must travel to the Vendor's facility for the training, this may not be feasible. Initial training for each role could take two to four half-day sessions and should allow for plenty of assistance for the first few weeks of system operation. Remember that familiarity will come with repetition and continued use of the system.

If the initial software training is provided by the CMS Vendor, you may wish to have the CMS Administrator provide additional training at your facilities after users have completed the Vendor's training program. This additional training could be on a one-on-one basis with each user at their own workstation. This can be beneficial in covering areas which were confusing or unclear to the user, or which emphasize the activities they will perform on a routine basis. All personnel will ultimately need some ongoing training on the catalog and nomenclature, since they must be able to recognize and use configuration information with confidence.

Ask the Vendor if they can set up a duplicate copy of the CMS database in another directory so it can be used for training purposes. You would want to do this after you have set up the core catalog, including the validation and look-up tables, and have entered enough data for users to learn the cable and equipment configurations and naming conventions. This normally does not involve a lot of work, and can be very beneficial. This test database gives users a chance to work with actual database records without the having to restore the records to their original status. The live database should not be used for training exercises if it can be avoided.

The following sections provide some guidance on the types of training which might be appropriate for the different individuals or groups who will be working with the CMS records.

CMS Administrator and Project Manager

The CMS Administrator and/or Project Manager are key individuals who will be responsible for the initial set-up and implementation of the CMS database. These persons should receive comprehensive training in every aspect of the database operation. Training should include setting up new users, passwords and security levels, creating and modifying catalog items, maintaining tables, and creating custom reports (either through the CMS software, an ad-hoc report writer, or SQL software).

The CMS Administrator or Project Manager should receive training on handling major revisions or changes to tables in the catalog resulting from major building renovations, organizational restructuring (merging of departments or divisions), or reconfiguring of network components. They should have a basic understanding of the tests to perform when new releases, enhancements or customizations are installed in the database. The administrator must understand the database functions which might be affected by changes to the network or CPU operating systems. Training should be required for database file management such as archiving Service Order or Trouble Ticket records and routine database backup and restoration.

Customer Service Center Users

Personnel from the Customer Service Center who will be creating Service Orders for routine Moves, Adds and Changes (MACs) should be thoroughly trained in the nomenclature for cables, equipment and ports, so they can identify the types of equipment and components in the database. They should also receive comprehensive training in the Service Order portion of the database, with specific training in how to retrieve current circuit connections and how to query for available outlets or horizontal cabling at the office or workstation locations. Training should include adding new inventory items, such as PCs, phones, printers, etc., and adding and deleting circuit connections.

Individuals who will be making and breaking cross-connects and pre-assigning the cable pair and port connections should receive specific instructions (with written documentation) on the rules for making those connections, such as preferences for selecting available cable pairs or ports and selecting the appropriate office outlet. If personnel are not familiar with the network equipment located in the wiring closets, they may need some additional orientation in network configurations and circuit connectivity. This training could be provided by a network technician who could take them through a few wiring closets to gain an understanding of what the records in the CMS database represent. They should have an understanding of what actions are taken by the technicians in making circuit connections and the function of the individual network components. This will help them use the CMS records and help make them more knowledgeable in dealing with their clients.

Help Desk Personnel

Help Desk personnel who will be creating Trouble Tickets for handling problem calls should be thoroughly trained in the nomenclature for cables, equipment, and ports so they can identify the types of equipment and components in the database. They should also receive comprehensive training in the Trouble Ticket portion of the database including any codes or tables used in identifying problem types or trouble resolution. Training should include how to retrieve and trace circuit connections based on several types of information such as user name, office number, equipment ID or outlet number.

If the Help Desk will be responsible for updating the CMS based on changes made during the trouble repair, they should be trained in how to change these connections (or in how to delete current connections and add new ones). As with Customer Service personnel, they may need some additional orientation in the area of circuit connectivity. This training could be provided by a network technician who could take them through a few wiring closets to gain an understanding of what the records in the CMS database represent. They should have an understanding of what actions are taken by the technicians in making circuit connections and the function of individual network components. This will help them use the CMS records and make them more helpful to users who call the Help Desk with a problem. If the CMS is tied in to a network management system, then Help Desk and technical personnel will need additional training on the links between these systems, how to respond to specific alarms, and how to obtain physical location and configuration information from the CMS.

Voice Technicians

Technicians who will be making cross-connect changes for voice circuits should be fully trained in the nomenclature established for the CMS records, especially riser, campus and horizontal cables, as well as the procedures for completing Service Orders and Trouble Tickets issued by the CMS database. They should receive training on the information contained in the Service Order and Trouble Ticket documents and in reading the voice circuit assignments. They must have a clear understanding of how changes are to be reported back to the Customer Service Center or Group Leader. Training should include information on reporting any changes they make or identify which were not handled through the normal Service Order or Trouble Ticket process. If the ports on the PBX are tracked in the CMS database, the technicians should be instructed to report any changes in port assignments.

Data Technicians

Technicians who will be making cross-connect and configuration changes for data circuits should be fully trained in the nomenclature established for the CMS records, especially network equipment and port naming, as well as procedures for completing Service Orders and Trouble Tickets issued by the CMS database. They should receive training on the information contained in the Service Order and Trouble Ticket documents and in reading the data circuit assignments. They must have a clear understanding of how changes are to be reported back to the Customer Service Center or Group Leader. Training should include information on reporting any changes they make or identify which were not handled through the normal Service Order or Trouble Ticket process. They may require additional training in the specific configurations established for each type of network equipment so they can query the CMS records or request reports of data connections to assist them in troubleshooting, reconfiguring, or replacing a piece of network equipment.

PART 4: PHYSICAL INVENTORY

PLANNING AND SCHEDULING

35

Once you have established a connectivity management plan and determined a method for recordkeeping, you will need to plan and schedule the physical inventory of cable and equipment components which will be tracked as part of the CMS records. You will need to determine the sequence of inventory activities, estimate the time required to complete the physical inventory tasks, and set up a schedule for completing the full implementation. The general activities for a physical inventory are discussed in Chapter 1 and are given in the flow chart (Figure 1.6) covering the third phase of CMS implementation, Figure 35.1 below. The extent of these activities were originally defined in the primary planning tables of Chapter 13 in terms of buildings, floors, cable types, etc. These charts will be useful now as a reference for the information which needs to be recorded, and in calculating the total number of wiring closets and office areas to be inventoried. A preliminary estimate of project activity durations was prepared in Chapter 20, and you may wish to refer to these estimates at this time.

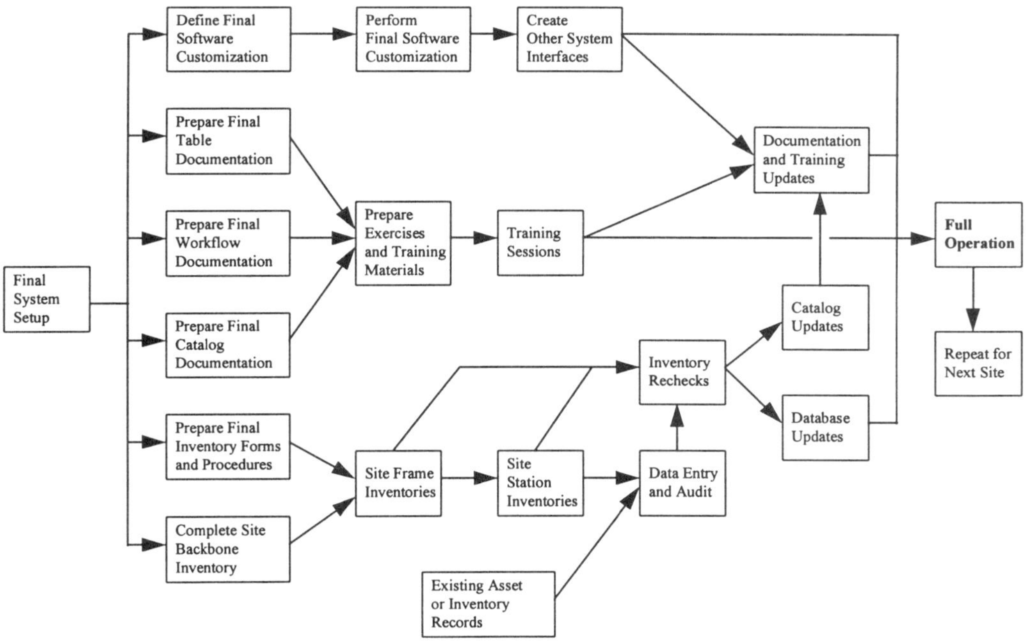

Figure 35.1: Flow chart of the third phase of CMS implementation, including physical inventory activities.

ACTIVITY TO BE COMPLETED	ESTIMATED TIME	INDIVIDUAL / GROUP RESPONSIBLE
RECORD/LABEL BACKBONE CABLES	2-4 HR./CLOSET	LEAD TECHNICIAN
RECORD/LABEL TERM. HDWE.	2-4 HR./CLOSET	LEAD TECHNICIAN
RECORD/LABEL NETWORK CONTROL EQUIPMENT	2-4 HR./CLOSET	NETWORK SVCS
RECORD NETWORK CONTROL EQUIP. BACKBONE CONNECTIONS	2 HR./CLOSET	NETWORK SVCS
RECORD/LABEL HORIZONTAL CABLES	2-4 HR./CLOSET	CMS ADMINISTRATOR
INVENTORY INDIVIDUAL BACKBONE CABLE PAIR CONNECTIONS	4-8 HR./CLOSET	INVENTORY CREW
INVENTORY INDIVIDUAL CONTROL EQUIPMENT PORT CONNECTIONS	2-4 HR./CLOSET	INVENTORY CREW
INVENTORY/LABEL OFFICE COMPONENTS AND CONNECTIONS	10-15 MIN./OFFICE	INVENTORY CREW
AUDIT INVENTORY RECORDS	1 DAY/5 INV. DAYS	PROJECT LEADER
DATA ENTRY OF INVENTORY RECORDS	2 DAYS/5 INV. DAYS	DATA ENTRY CLERK
AUDIT CMS DATABASE RECORDS	1 DAY/5 DATA ENTRY DAYS	CMS ADMINISTRATOR

Table 35.1: Sample list of activities to be included in physical inventory.

Defining Tasks and Required Information

In order to begin planning for physical inventory work, your first task is to determine what activities will be involved in the inventory process. These activities might begin with recording and labeling of all backbone cables and equipment located in each of the wiring closets or administration points, including the recording of all cable pairs and control equipment ports. Since these are the major components which will make up individual circuits, they usually represent key elements of the inventory. Following the inventory and labeling of these components, the next step will include recording the actual physical connections to each of these items which link them with the network or system backbone. Once the backbone structure and connections are completed, you can record the individual cross-connects between each of these components and the horizontal wiring (outlets) in the wiring closets. The final steps will include the inventory of any equipment and connections to horizontal cables or outlets located in each of the office areas. Additional information might include the employee information and the user's department.

You will need to review the catalog of cable and equipment types to determine the actual cables, devices, and equipment that must be inventoried at each administration point. For example, individual backbone cables, termination hardware, and control equipment may need to be inventoried and labeled in each Telecommunications Closet and Equipment Room. If so, the connections between these cables and devices must also be recorded at the pair and port level. Make a chart similar to the one in Table 35.1 listing each of the activities which need to be completed for your specific implementation. On this chart, you may also want to provide columns for the estimated time required to complete each task, and which individual or group will be responsible for performing the task.

CABLE ELEMENT	REQUIRED FIELDS
EXISTING NAME	HC-1
NEW CMS NAME	RC-REN1E
PAIR COUNTS	1-200
CABLE TYPE	VOICE RISER
MFR.	AT&T
SPEC./PART #	3631-1
TO BUILDING	REN
TO FLOOR	1E
TO ROOM	1E ER
FROM BUILDING	REN
FROM FLOOR	B
FROM ROOM	B22E

Table 35.2: Example list of information to be recorded for backbone cables.

We will cover how to estimate the required inventory time later in this chapter, which will determine whether current staff should be responsible for completing each task or whether an inventory crew will be required. While existing staff may be the best qualified to handle the inventory and labeling of the individual components and circuits (from the standpoint of familiarity), this often creates a conflict with the normal daily job functions which must be completed at the same time. In order to move the project along at a reasonable pace, designated inventory crews are often used to handle the inventory tasks.

Any existing records can be used to assist the inventory process for establishing preliminary routing of backbone cables and locations of control equipment. If existing naming conventions will be used, then this information must be available during inventory. If current cable names will be used, for example, then the proper cable name and pair count for each closet should be determined before inventory, if it is not already labeled. If current network device names will be used, then this information must also be prepared and available during inventory.

Your chart may be shorter than the one shown in Table 35.1, depending on how much information you plan to track in your recordkeeping system and whether you plan to use existing data. It may require auditing of existing records for a portion of the project, rather than a full inventory.

The estimated times given in Table 35.1 are general estimates, and may vary greatly from location to location. How well things are currently labeled will have a major impact. Tracking down a number of backbone or tie cables in a single wiring closet may require inventory of multiple closets, and then using a process of elimination to determine which cables run between which closets. There will likely be a few cables or outlets which cannot be tracked down in a reasonable amount of time and should therefore be ignored, or handled as a future task when time allows for further investigation.

Using the chart you just created, make a list of the specific information which needs to be recorded for each type of component during each of these activities. Using the backbone cables as an example, your list might be similar to the one shown in Table 35.2. A similar list should be prepared for desktop equipment, for example, as shown in Table 35.3. You will want to include everything you plan to track for each item, so that you have all the information in a single place or on a single sheet. You should refer to the lists you developed when defining the project scope, covered in Chapter 13, and the ultimate description of each component defined in your catalog in Chapters 21 and 28. You can then use

DESKTOP ELEMENT	REQUIRED FIELDS
NAME OR ID	PC1000
TYPE	PC
MODEL	PS2/70
SERIAL NUMBER	0021134570
USER NAME	JOHNSON, M.L.
BUILDING	REN
FLOOR	2E
ROOM	240E
DEPARTMENT	ENGG

Table 35.3: Example list of information to be recorded for desktop equipment.

these lists as a basis for creating inventory forms for the actual recording of information. The steps for developing inventory forms will be covered in the next chapter. Note that you will likely come across some components during inventory which have not yet been defined in the catalog, and therefore a device type and configuration must be created for these items.

Estimating the Time to Complete Tasks

In Chapter 20, we offered some guidelines were offered for estimating the time required for backbone inventory, office and frame inventory, and data entry. As mentioned, these estimates can very greatly depending on different circumstances. To get a realistic estimate of timing for these tasks in your facilities, it will be necessary to walk through these activities in a typical wiring closet on a typical floor. Prior to conducting the sample inventory below, refer to the next chapter for creating inventory forms which can be used for these activities.

In order to determine the estimated time required to handle each task, work with an individual from the group who has primary responsibility for managing the cables or equipment at that location. Use the lists for each task developed above to record the necessary information. Select a typical wiring closet and complete each of the inventory/labeling tasks involving the cable, equipment, and connections in that wiring closet.

Begin by first taking an inventory of all the cables and equipment at that location and then recording the connections for each item to the backbone. After the backbone connections have been recorded, the final task in the wiring closet should be to inventory and record the individual circuit connections to each of the cables and equipment. Keep track of the approximate time required to complete each task, and add it to the chart. You can then multiply the estimated time by the number of closets in a specific building to determine the total number of labor hours required in that facility. This time estimate should help to determine whether the existing staff has the time available for these tasks, whether the tasks need to be completed on an overtime basis, or whether designated inventory personnel should be used.

The next task to consider is the inventory of the individual offices. Review your list of information to be recorded in each office and any labeling required, such as for PCs, terminals, printers, outlets, etc. Select an area of 15-20 offices or workstations associated with the wiring closet you just worked in, and complete each of the inventory and labeling tasks involved in the office inventory. Keep track of the time required to complete the 15-20 offices, and develop an estimated time requirement for each office, then add it to the chart. As with the tasks involving the wiring closet components and connections, the time estimate should help determine whether the existing staff can handle the office inventory, or if a separate inventory crew will be required.

You may wish to increase this estimate by 20 to 50 percent to allow for problems and slowdowns that will naturally occur. For example, moving furniture to determine an outlet number will take additional time, and perhaps require a second person. Some number of offices on a floor will be locked and must be revisited at a later time. Performing inventory after-hours will probably increase this problem rather than reduce it. Another delay occurs if the occupant of an office is on the phone or having a meeting, since it is courteous to return later to inventory that office. Mislabeling and misrecording will take time to be rechecked. There is a law of diminishing returns, and if a problem cannot be resolved within an hour or two, it is probably best to let it go or follow up at a later date.

Following the inventory activities, the information recorded on the inventory forms should be audited for completeness, accuracy and consistency prior to data entry of the records into the CMS database. This helps to correct errors before they get into the CMS records. The Project Leader or CMS Administrator should review each of the information sheets, and make a note of the time required to audit the information. This may be indicated on the chart in terms of hours or days in relation to the number of inventory hours/days.

The last two areas where you need to develop time estimates are the actual data entry process and the auditing of the records after they have been entered into the CMS database. Record the time required to enter all of the cables, equipment, termination hardware, desktop equipment, and the individual connections from

BLDG	FLOOR OR WING	NO. OF WIRING CLOSETS	NO. OF OFFICES	EST. DAYS PER CLOSET	EST. DAYS OFFICE INVENTORY	DATES SCHEDULED
ADM	B	1	20	3	1	WKS 5/8-5/22
ADM	1	1	50	2	2	"
ADM	2	1	50	2	2	"
FAC	1	1	40	1-1/2	2	WK 5/29
FAC	2	0	20	0	1	"
REN	BE	1	25	3	1	WK 6/5
REN	BW	0	25	0	1	"
REN	1E	1	50	2	2	WK 6/12
REN	1W	1	50	2	2	"
REN	2E	1	50	2	2	WK 6/19
REN	2W	1	50	2	2	"
REN	3E	1	50	2	2	WK 6/26
REN	3W	1	50	2	2	"

Table 35.4: Sample schedule for physical inventory.

the inventory forms for the areas you have covered. It is usually most efficient to add all items of the same type at one time, such as adding all equipment, then cables, then connections. Record the time required in terms of the number of data entry days or hours required per day of inventory. This is also a good time to see where data omissions and conflicts become evident, so that the information can be rechecked. You will need a process for handling specific rechecks and making sure the updated information gets into the database before an area is considered complete.

The final step is to run the necessary reports or queries in the CMS database in order to review the data which has been entered. The Project Leader or CMS Administrator should review the data for completeness, accuracy and consistency, and correct any errors.

Developing a Schedule

After you have developed the list of tasks and the estimated time to complete them, you can develop a schedule for each of the sites, buildings, and floors to be included in the CMS project. Remember to break areas down into manageable sections where all inventory and data entry tasks can be completed in one or two weeks.

Referring to the Primary Planning Table (Table 13.5) developed in Chapter 13, make a chart listing the sites, buildings, floors and/or wings to be inventoried. Prioritize the list in the order in which they will be inventoried. As shown in Table 35.4, add the following columns along the top of the chart: Number of Wiring Closets, Number of Offices, Estimated Time Required, and Scheduled Dates. The table here has been filled in with estimated times based on the LRP Site discussed in Chapter 2. It is best to leave extra days available in the schedule for areas which require a little extra time, or for any areas which need to be revisited to clarify data or to record any information which was missing on the initial inventory records. Of course, the overall schedule will depend on the number of persons involved in the inventory.

Leaving some extra time in the schedule also allows for unexpected delays which might result from sick days, vacation days, or other problems which impact the ability of the inventory crew to complete the tasks. This provides time to make any necessary revisions or modifications to the inventory forms and procedures, and helps to make sure the data entry process can keep up with the flow of inventory records being generated.

Main cross-connect locations will take additional time to inventory, perhaps twice as much time as a regular Telecommunications Closet.

As an example, consider the basement Equipment Room in the Administration Building, which houses the site PBX and the Main Distribution Frame for the PBX. It is assumed that the machine side of the frame is already labeled with the PBX port designations, but that recording and relabeling of the 1200 backbone cable pairs will take approximately 4 hours. The tracing of jumpers between backbone cable and PBX ports will take approximately 12 hours, and the inventory of incoming horizontal cables and connections will take an additional 4 hours. Inventory of data control equipment and connections and for rechecks on all connections requires 4 hours, bringing the total inventory time to 24 hours or approximately three days. Note that this is an estimate based on using trained personnel. Your timing may vary significantly based on existing conditions, the extent of additional equipment, and the efficiency of the inventory personnel. For the office inventory in the ADM basement, 20 offices at ten minutes per office requires about 3 and 1/3 hours to complete. Allowing for minor delays and rechecks, a full day should be allocated to this task.

A second example is the inventory of the 2nd Floor East in the REN building. There are 200 backbone cable pairs which will take about two hours to record and relabel. There are approximately 100 4-pair horizontal cables and 50 IBM Type 1 data cables which must be recorded and labeled, taking another four hours. Inventory of the control equipment, optical fiber cable, and control equipment connections will take an additional two hours, and inventory of the horizontal cable connections will take about four hours. Total inventory time is therefore 12 hours. At least two days should be allocated for this closet. For the office inventory, 50 offices at twelve minutes per office requires about 10 hours to complete. Again allowing for minor delays and rechecks, two full days should be allocated for the REN 2E office inventory.

Based on the inventory schedule, you can then determine how much data entry effort will be required to support the inventory activities. Data entry should be scheduled to immediately follow the completion of each section (floor, wiring closet, etc.) so that any conflicts or problems can be addressed as quickly as possible following the inventory. This also provides for immediate use of the records upon the completion of each area.

DEVELOPING INVENTORY PROCEDURES AND FORMS

36

For each type of area which must be inventoried, such as backbone, frame, and office, we will discuss some methods for developing specific procedures and forms for use during the physical inventory. Spending some additional time up front in preparing the proper inventory forms and procedures can save inventory time, save data entry time, and help avoid additional site visits to verify or correct bad information. The inventory procedure for each area, along with the data entry process, should be designed to reduce missing and conflicting information and make these errors or omissions very apparent so that rechecks can be scheduled.

To begin, refer to the list of tasks to be completed during the inventory process developed in Table 35.1 in the previous chapter. This will identify the areas for which procedures and forms must be developed. Your procedures should include specific step-by-step instructions for what information is to be recorded, in what format, and how any labeling or tagging is to be handled. The procedures should give examples of how each of the items is to be recorded, as well as providing a list of options or choices for indicating equipment or cable types. This helps to achieve consistency in how the forms are completed and how the information is entered into the database. Samples of completed forms for each type of equipment or cable should be included for reference.

Other information which may be helpful to include in the procedures, especially when assigning numbers to termination hardware or to controlling equipment in wiring closets, is to indicate where the numbering scheme for labeling is to begin. For example, you can adopt a convention such as beginning to the left as you enter the wiring closet and continue working around to the right. For patch panels or equipment racks, be specific as to numbering from top to bottom, and whether to skip a number if there is an empty space in the rack where a piece of equipment might be installed at a later date.

Preparing detailed inventory procedures avoids confusion and provides clear direction for the individuals conducting the inventory. In the early phases of the inventory, as individuals become familiar with the procedures, they act as a check list which can be referenced to verify that all required steps have been completed.

The process of inventory includes five basic components which can be completed in three steps, as listed in Table 36.1. The first component is the recording and labeling of all backbone and tie cables running between wiring closets, along with the termination devices and termination positions of each cable pair, if desired. As cables are inventoried, pathways should also be named, labeled, and recorded. The second component is recording and labeling of all local control equipment in the wiring closet such as LAN hubs, 3270 controllers, and multiplexers. Termination device and position for each port must also be completed, if desired, where ports are cross-connected on termination hardware rather than at the control device itself. Third, the horizontal cables and their termination devices and positions (if desired) must be recorded and labeled in the wiring closet. Taken together, these three elements create a complete inventory of all the "things" in the wiring closet, along with their terminations. All three tasks can be completed in the same time frame and represent the first step in physical inventory.

STEP 1: WIRING CLOSET CABLE AND EQUIPMENT INVENTORY
BACKBONE AND TIE CABLES AND PATHWAYS
LOCAL CONTROL EQUIPMENT
HORIZONTAL CABLES
STEP 2: FRAME INVENTORY
PORT-TO-HORIZONTAL CABLE CONNECTIONS
PORT-TO-BACKBONE CABLE CONNECTIONS
BACKBONE CABLE-TO-HORIZONTAL CABLE CONNECTIONS
STEP 3: WORKSTATION INVENTORY
OUTLET LOCATIONS
DESKTOP EQUIPMENT
EQUIPMENT-TO-OUTLET CONNECTIONS

Table 36.1: Summary of the steps in a complete physical inventory.

The fourth component involves the recording of connections between these three components: backbone cables, ports, and horizontal cables. We have previously referred to this as a frame inventory. It requires that the first three steps be completed beforehand, so all items are already labeled and connections can easily be determined and noted. A complete frame inventory in the wiring closet is the second step of the overall physical inventory. It can be an extensive process which we will discuss more thoroughly in a moment.

The fifth component and the third step in a physical inventory is the recording and labeling of horizontal cables (outlets), desktop devices, and connections at the workstation or office area itself. This will likely be the final step, since it can serve to audit services that are actually in use versus those that are simply connected without being used. It confirms the location of each outlet and provides any desired user information which will be tracked in the CMS.

Having covered each component and each step in the inventory process, you can break each down into the detailed procedures and forms necessary to make each task go as quickly and accurately as possible. You may have to adapt these procedures to reflect your specific recordkeeping and cataloging requirements. Some adaptation may also be desired to best fit the process of data entry required by a specific CMS. The following procedures are offered as a general case, providing for the recording, labeling, and auditing of all physical elements of the telecommunications infrastructure.

Inventory Procedures for Cables and Equipment

Beginning with Step 1, the procedures for backbone cable inventory might look something like those listed in Table 36.2. Note that the naming conventions are indicated on the procedures. If an extensive list of cable types is being maintained, attach this list to the procedures.

For each procedure, include clear instructions on how labeling is to be done, including designations for each different type of cable and equipment, termination hardware, or patch panel. Also include any special instructions with regard to labeling such as color coding for cable type or equipment type, and where the labels are to be placed on each type of component (under make/model number, top/center, bottom left hand side, etc.). Make sure the inventory crew has a complete supply of the required labels or can make them up as they go. If a sequential numbering scheme is to be used for labeling cables and equipment, it is beneficial to create a list of individual names or numbers which can be used as a checklist with each number checked off as it

ITEM NO.	ACTIVITIES TO BE COMPLETED
1	Fill out a separate sheet for each Backbone Cable.
2	Record Wiring Closet Location - Building, Floor and Room.
3	Record existing Cable Name.
4	Record new CMS Cable Name (Example: RC-REN1E) Cable Type + BLD + Floor/Wing RC = Riser Cable ADM B 1 2 FC = Fiber Cable REN B 1 2 3 E W CC = Campus Cable FAC 1 2 TC = Tie Cable
5	Label Cable with new CMS Name - ALL CAPS using Brother Label Maker, with 1" White Tape (or follow TIA/EIA 606 color codes).
6	Record Cable Type. (Riser, Campus, Fiber, Tie)
7	Record Cable Pair Counts. (Example: 1-200)
8	Record Termination Device Names and Positions where cable pairs are terminated.
9	Label Cable Pair Counts over each Termination Block, using same type label as for Cable Name. Otherwise, use appropriate designation strip or write pair numbers (by five's) on the wiring block with indelible marker.
10	Record Cable Manufacturer. (Belden, AT&T, General Cable, Mohawk)
11	Record Manufacturer's Spec./Part Number. (Example: 3161-C)
12	Record the Conduit or Cable Tray in which the cable is contained. (Example: CWREN01) Note: If the Pathway has not yet been named and labeled, assign an appropriate name and label the pathway with the new name. Then, record pathway name and information on a pathway inventory form.
13	Make any comments regarding special conditions or concerns.
14	When Inventory Sheet is completed, review sheet for completeness. Date and initial the sheet at the bottom where Indicated.
15	Return completed sheets to: James Jensen, ITG, ADM-127

Table 36.2: Sample procedures for inventory of backbone cables.

used. This provides continuity from day to day and among different inventory teams.

An example might be where each network hub in a wiring closet is to be labeled with a building and equipment type prefix, followed by a sequential number for each hub located in that wiring closet. The list might include a list of available hub numbers such as RENHUB01, RENHUB02, RENHUB03, etc. It may be possible to use labels prepared ahead of time.

The inventory of control equipment may require that a device chassis may be inventoried on one sheet, while the cards or modules plugged into the chassis are each recorded on separate sheets. In this fashion, their particular port types and connectors can be described and the overall capacity of the chassis managed. Recording of termination position and any cable connections will be on the device sheet for a non-expandable device (such as an 8228 token ring MAU) or on the sheet for each card for an expandable device

ITEM NO.	ACTIVITIES TO BE COMPLETED
1	Fill out separate sheets for each piece of Equipment in the Wiring Closet. Begin at the left as you enter the Wiring Closet and work around to the right. Numbers should be assigned in racks from top to bottom.
2	Record the new CMS Equipment Name. (Example EHREN1E01) Network Type + H + BLDG + Floor/Wing + Sequential Number E - Ethernet H ADM B 1 2 01 T - TokenRing H FAC 1 2 02 S - SNA H REN B 1 2 3 E / W 03
3	Label Equipment with new CMS Name in ALL CAPS, using Brother Label Maker with 1/2" Green Tape. Place label on front of control equipment, in upper left hand corner.
4	If there is a space in the rack where a new piece of control equipment could be installed at a later date, record the hub sequence number for that space on a blank inventory sheet and note in the comments section - "Spare location in rack." Do not use that sequence number for any existing equipment.
5	Circle the Type of control equipment on the sheet. If the type is not listed, write in the equipment type next to the list provided on the sheet.
6	Record the Serial Number of the control equipment, and any existing labels or names on the equipment.
7	Record the Wiring Closet Location - Building, Floor/Wing and Room Number.
8	For expandable hubs, record the number of cards located in the equipment.
9	Record the Termination Device and Positions where equipment port connections are located.
10	Record Network Connections to control equipment by recording the type of connector and connector name, and the termination device and positions or cable and pair to which the network connections are made.
11	For expandable hubs, fill out a separate inventory sheet for each card. (Use Form CARD01.)
12	Record new CMS Name for Card. (Example: AB21/A01) Network Segment + Slot Position on Hub AB21/A 01, 02, 03.....12
13	Circle the Card Type. If the type is not listed, write in the card type next to the list provided on the sheet.
14	Record the Hub ID in which the card is installed, and the Wiring Closet Location - Building, Floor and Room.
15	Label all Cards with the new CMS Name, using Brother Label Maker, with 1/2" Green Tape. Place label on the front of the card at the bottom (or far left for cards located in 3030 hubs which are installed sideways.)
16	Record the Termination Device and Positions where equipment port connections are located.
17	Make any comments regarding special conditions or concerns regarding the cards. Staple the inventory sheets for the cards to the inventory sheet for the hub to which they are installed.
18	Review inventory sheets for completeness, then date and initial the sheets at the bottom, where indicated.
19	Return completed inventory sheets to: James Jensen, ITG, ADM-127.

Table 36.3: Sample procedures for inventory of control equipment.

(such as a Bay Networks 3030 hub). Sample procedures for inventory of local control equipment, whether expandable or non-expandable, are given in Table 36.3.

Also, during the recording of control devices, any connection to the network backbone (on the "back side" of an active device, so to speak) should also be recorded. This connection

ITEM NO.	ACTIVITIES TO BE COMPLETED
1	Record the Wiring Closet Location - Building, Floor/Wing and Room Number.
2	Record the different types of horizontal cables located in the wiring closet, such as Voice, Data, Category 3, Category 4, Category 5, Type-1, Type-2, Coax, etc.
3	For each different type of horizontal cable, record the first cable name/number, the last cable name/number and the cable type. If the cable names utilize a sequence of 101A, 101B, 101C, 101D, be sure to note there are 4 cables for each cable number.
4	For each different type of horizontal cable, record the termination device and position for the first cable of that type, and the termination device and position for the last cable of that type.
5	If termination devices are not labeled, label the termination devices using the Brother Label maker with 1" blue tape, indicating the beginning and ending horizontal cable names on the wall or rack over the set of termination devices.
6	Make any special notes or comments for each cable type.
7	Review inventory sheets for completeness, then date and initial the sheets at the bottom, where indicated.
8	Return completed inventory sheets to: James Jensen, ITG, ADM-127.

Table 36.4: Sample procedures for inventory of horizontal cables in the Telecommunications Closet.

could be to a cable pair or to another control device, such as Ring-In and Ring-Out connections on a token ring MAU, for example, or the network interface of an Ethernet hub. Many of these connections may be on optical fiber cable. If the connection cannot yet be identified, mark the form or list this connection as an item to recheck later. Note that a network interface card may be treated as a separate card or as part of the chassis itself.

Procedures for recording and labeling horizontal cables in the wiring closet are given in Table 36.4. If outlets will be numbered by frame location (termination position), proceed directly to this step. If the frame position of each outlet is not known, then each outlet must be toned from the office to identify its frame location. This can be a very time-consuming process, and will have a significant impact on the pace of inventory work.

Note that if tracking of termination hardware is desired, then the specific termination device should be named, labeled, and recorded at the same time the cable pairs and/or control devices are inventoried. For example, as backbone or horizontal cable pairs are recorded on a sheet, the block or panel on which they are terminated should be named and labeled, and the termination position for each pair noted on the cable inventory sheet. Ports on control equipment may or may not be terminated on a separate wiring block or patch panel since the port on many devices uses a connector (RJ45, BNC, etc.) that can be directly connected to the horizontal cable termination. For ports that are terminated on a separate device, the termination device should be named, labeled, and recorded at the same time the control equipment and ports are inventoried. The termination position for each port should then be noted on the control equipment inventory sheet.

Inventory Procedures for Connections

The process of recording connections can be very straightforward in the case of smaller, well-organized wiring closets, or fairly complex as in the case of older, larger wiring closets and equipment rooms. Considering that backbone cables, horizontal cables, and ports are what is available to be connected, there are essentially three types of connections which can be made, as shown in Figure 36.1. These are the backbone cable-to-port, port-to-horizontal cable, and backbone cable-to-horizontal cable connections noted under Step 2 of Table 36.1. Network

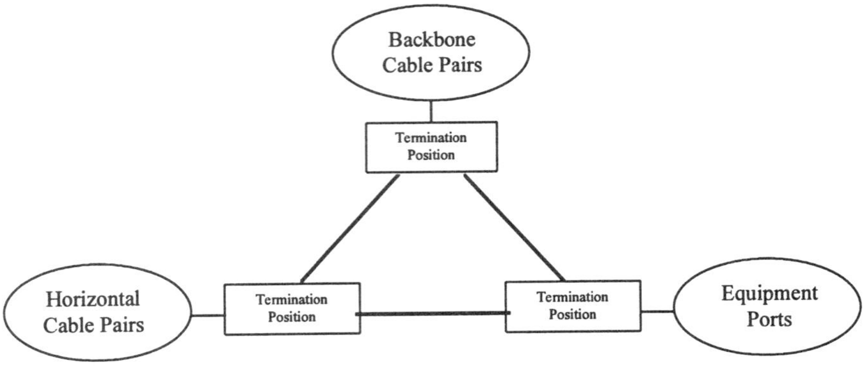

Figure 36.1: Diagram of the three types of connections to record during frame inventory.

backbone connections to control equipment have already been considered, and we will ignore for the moment that any cable or device can be connected to itself since the procedures provided here will pick up such connections anyway.

In order to capture all three sets of connections, it is only necessary to inventory two of the three connection fields, such as at the backbone cable terminations and at the port terminations, or at the backbone cable terminations and the horizontal cable terminations. A list of the termination positions of the third termination field can be used as to audit the recorded data.

You are free to choose which two of the three fields you use, but the control equipment ports and the backbone cable pairs are recommended. Duplicate entries can be corrected immediately by inspection. After these frames are complete, the port and backbone cable pair lists can be cross-referenced against one another to check for any conflicts and omissions. Further, the appropriate port and backbone connections can be written on the list of horizontal cables, which will reveal any additional duplications or conflicts with respect to the horizontal cable connections. Finally, the list of horizontal cables is rechecked against each termination position on the frame to be sure that no horizontal cable connections were missed, to pick up any connections between horizontal cables, and to note any missing horizontal cable/jack numbers.

There are special cases where simpler methods can be applied. For example, there are certain situations where only LAN connections are being tracked. If the LAN ports are connected to dedicated horizontal cables used exclusively for the LAN, then only the port terminations need to have the connections noted. Upon completion, the connections can be cross-referenced and written on the list of dedicated horizontal cables to find any duplicate entries. The list of horizontal cables can then be rechecked against each termination position on the frame, as before.

A similar situation exists for inventory of only voice connections and dedicated voice outlets. In this case, the backbone cable list can be used to record connections, followed by cross-reference to the horizontal cable list and recheck against the horizontal cable frame.

In yet other cases there are no connections between backbone cable and local ports, and all connections involve the horizontal cable. The list of horizontal cables can be used for recording connections, with the appropriate port and backbone cable connections cross-referenced and written on their respective lists to find any duplicate or conflicting entries. The port and backbone cable lists can be quickly rechecked against their respective termination fields to locate connections which may have been missed.

Sample inventory procedures are given in Tables 36.5a, 36.5b and 36.5c. The procedures are

ITEM NO.	ACTIVITIES TO BE COMPLETED
1	Check to make sure all backbone cables are labeled.
2	Check to make sure all control equipment and patch panels are labeled.
3	Start with Port 01 on each of the patch panels for the 3270 SNA Controllers. Write the controller name and patch panel ID on the inventory sheet. Trace each connection to the appropriate cable pair, or termination position. Record the cable name and pair or termination device and position connected to each port next to the port number on the controller inventory sheet. If there is no connection, indicate "NC" next to the port number on the inventory sheet. Use a separate inventory sheet for each patch panel.
4	Record the connections to the cards on the network hubs. Start with the non-expandable hubs by recording the port connections on the existing hub inventory sheet. Start with Port 01 for each hub. Then inventory the connections to the cards on the expandable hubs. Start with Port 01 on Card 01 of each network hub. Write the Hub ID and the Card ID on the inventory sheet. Trace each connection to the appropriate cable pair or termination position. Record the cable name and pair or termination device and position connected to each port next to the port number on the card inventory sheet. If there is no connection, indicate "NC" next to the Port number on the inventory sheet. Use a separate inventory sheet for each card on each hub.
5	After all the connections for the controlling equipment, patch panels and termination devices have been recorded, compare the inventory sheets with the controlling equipment patch panels or hubs to make sure a connection was recorded for every port which has a patch cord or connection.
6	When inventory sheets are completed, review sheets for completeness. Date and initial the sheets at the bottom where indicated.
7	Continue with procedures for inventory of backbone cable connections.

Table 36.5a: Sample procedures for inventory of wiring closet connections - control equipment.

divided into three phases: inventory of control equipment port connections; inventory of backbone cable pair connections; and a cross-reference of control equipment and backbone cable connections with the list of horizontal cables. The last phase allows for the auditing of connections and identifies any duplications.

The procedures for frame inventories should clearly define how connections are to be recorded, with forms that list each pair and provide columns for recording the connections in terms of cable name and pair, termination block and position, or patch panel and port. If termination hardware is not tracked, record each connection as the port or cable pair. On the other hand, if terminations are being tracked, record each connection as the termination device and position where the cable pair or port is terminated.

All connection fields should be filled in. For example, in recording the connections to ports on a patch panel, start with Port 1 and continue through to the last port in the sequence. If there is no connection for a port, indicate N/C for No Connection. This helps to make sure each port connection has been traced and recorded, and clearly identifies that there is no connection rather than a connection which was missed.

If splitters are commonly used at the outlet for multiple connections to a port or cable pair, it is beneficial to indicate how those splitters are to be recorded on the connections. This avoids confusion on the part of the inventory crew and the data entry personnel. By clearly identifying the use of a splitter during the inventory process, it avoids a question or revisit to verify the dual connection at a later date.

After the inventory of connections has been completed, the procedures require the inventory crew to compare the connections noted on each list against the actual connections on the frame. They should check to make sure a connection has been recorded for each port or

ITEM NO.	ACTIVITIES TO BE COMPLETED
1	Check to make sure all Backbone Cables and Termination Devices are labeled.
2	Use the Inventory Sheet which has the Proper Cable Pair Count listed, such as 1-200, or 201-600. Record the CMS Cable Name on the Sheet. (Example RC-REN1E)
3	Record the Wiring Closet Location - Building, Floor/Wing and Room Number.
4	If termination hardware is being tracked as separate components, record the termination devices and positions next to each of the cable pairs on the inventory sheet.
5	Starting with the first cable pair, trace each connection (jumper or patch cord) to the appropriate horizontal cable pair, controlling device port, or termination device and position. Record the connection next to the Cable Pair on the Inventory Sheet. If there is no connection, indicate "NC" next to the cable pair on the inventory sheet.
6	After all connections have been recorded, compare the inventory sheets with the cable frame to make sure a connection was recorded for every pair which has a jumper or patch cord. Trace any missed connections and make the necessary corrections to the inventory sheet.
7	Make any comments regarding special conditions or concerns regarding the cable and connections.
8	Review Inventory Sheets for completeness, then date and initial the sheets at the bottom, where indicated.
9	Repeat Steps 2 through 8 for each backbone cable located in the Wiring Closet.
10	Continue with procedures for recording connections to horizontal cables and auditing inventory sheets.

Table 36.5b: Sample procedures for inventory of wiring closet connections - backbone cables.

termination position or cable pair which has a jumper or patch cord. This will identify any connections which were missed, or were recorded where none actually exists. Any discrepancies should be resolved by retracing the connection to identify where the error was made. Procedures should include instructions for personnel to review all inventory sheets and forms for completeness before to leaving the wiring closet or going onto the next item to be inventoried.

Inventory Procedures for Office Areas

Inventory procedures for the office area equipment and connections should be very specific as to how information is to be recorded, how labeling is to be performed, and how the locations of outlets will be shown on a floor plan drawing or sketch. Instructions should require that areas with no equipment or connections be noted as "N/C" so there is no question as to whether the office area was inventoried or an outlet located.

A list of outlets should be provided to the inventory crew so they can check off the jacks as they are located. Depending on the method used to approach the office inventory, this information may already be listed on the inventory forms. This helps to quickly identify missing outlets when an area has been completed, and also helps to identify incorrectly recorded outlet numbers if a jack is checked off twice.

Procedures should include instructions to make sure all offices are left in the same order and condition in which they found them, to review all inventory sheets for completeness upon finishing an area, and to place the inventory sheets in order by jack number or by office number depending on how the forms are organized. This will help to identify any missing areas, makes it easier for auditing of the information, and enhances the flow of the data entry process. A sample set of instructions for inventory of office equipment and connections is shown in Table 36.6.

Prior to conducting the inventory of the office areas, you can issue a notice or letter to employees in the affected areas to inform them of

ITEM NO.	ACTIVITIES TO BE COMPLETED
1	After all connections to the controlling equipment, patch panels and termination device positions have been recorded, use the inventory sheets where the connections were recorded to cross reference the connections onto the list of horizontal cables.
2	If you do not have a list with all the horizontal cables for the wiring closet, prepare a list of the cable names/numbers, in numerical order, with columns for Cable Name, Pair, Connection and Comments.
3	For each horizontal cable connection recorded on the control equipment inventory sheets, record the port or termination device and position next to the cable name. If two connections were recorded for the same horizontal cable, double check the connections to identify which one was recorded incorrectly. Make any necessary connections to the inventory sheets for the controlling equipment port connections.
4	For each horizontal cable connection recorded on the backbone cable inventory sheets, record the cable and pair or termination device and position next to the cable name. If two connections were recorded for the same horizontal cable, double check the connections to identify which one was recorded incorrectly. Make any necessary corrections to the inventory sheets for the backbone cable connections.
5	Compare the list of connections which were recorded onto the list of horizontal cables against the actual horizontal cable frame to make sure a connection was recorded for every horizontal cable which has a jumper or patch cord. Trace any connections which did not appear on the list and make any necessary corrections to the list of horizontal cable connections.
6	If any horizontal cables were missing and not located on the horizontal cable frame, make a note next to the cable name/number of the list of horizontal cables indicating that particular cable is missing.
7	Make any special notes or comments for regarding the horizontal cable connections.
8	Review inventory sheets for completeness, then date and initial the sheets at the bottom, where indicated.
9	Return completed inventory sheets to: James Jensen, ITG, ADM-127.

Table 36.5c: Sample procedures for inventory of wiring closet connections - horizontal cables and audit.

the scheduled inventory. This letter should inform them of the purpose of the inventory and ask their cooperation in allowing the inventory crew to accomplish their required tasks as quickly as possible. The notice or letter should include the scheduled dates for the inventory and the approximate time it should require the inventory crew to complete their tasks in each office. This will help to speed up the inventory, especially if the inventory is conducted during office hours. Without prior notice to employees, unnecessary time is spent explaining to users the reasons for the inventory and the possible disruption it causes.

Developing Inventory Forms

Inventory forms should be in a format which is easy to fill out, leaving as little chance as possible for error and omission of data. They should flow in a way that follows the inventory pattern as well as the data entry process. A good starting point is to use the data entry screen in the CMS software as a basis for creating the inventory form. This screen lists each of the fields which must be completed in the database and the order in which they will be entered.

Make a first draft of your inventory form listing all the fields on the CMS data entry screen. Using a Cable Inventory Form as an example, you could simply prepare an inventory form for backbone cables which would list all of these fields, with a line after each one where the information would be entered similar to the one shown in Table 36.7. However, the inventory crew would most likely not know the length of the cable, the date it was installed, or possibly the performance rating, so this information would not be necessary on the inventory form. Perhaps all the cables belong to one department, so that this

ITEM NO.	ACTIVITIES TO BE COMPLETED
1	Begin in the Basement or on the First Floor of each Building. Be sure you have inventory sheets with the list of horizontal cables/outlets for each floor/wing of the building.
2	Go to each Office or Workstation in the building and locate each outlet in the office or workstation.
3	Complete the outlet inventory sheet fields for each outlet as it is located..
4	Record the Room Number or Workstation ID next to the outlet/jack number on the Jack Inventory Sheet.
5	Record the User's Last Name and Initials on the inventory sheet.
6	Record the Company Tag Number on the office equipment which is attached to the outlet/jack.
7	Record the Equipment Class on the inventory sheet. Options: PC, Phone, Printer, Terminal, Fax, Ans. Mach., Modem
8	Record the Equipment Type. See reverse side for list of options.
9	Record the Serial Number for the equipment. If there is no Serial Number, note "No #" on the inventory sheet.
10	Check the Floor Plan Drawing to see if the outlet is shown on the drawing. If not, make a note of the outlet number on the drawing in the appropriate location. If office furniture is being tracked, mark the position of the desk or workstation on the floor plan drawing.
11	If an outlet is located which is not labeled, mark it on the Floor Plan Drawing as an "Unmarked Outlet", and assign it a Temporary Number. Record the connections to the unmarked outlet on a separate inventory sheet. Temporary Outlet Numbers should be assigned from the "Temporary Outlet Number Sheet" included in your Inventory Manual.
12	When inventory sheets are completed, put the inventory sheets in numerical order by outlet number and review the sheets for completeness. Date and initial the inventory sheets at the bottom where indicated.
13	Return completed inventory sheets to: James Jensen, ITG, ADM-127

Table 36.6: Sample procedures for inventory of office equipment and connections.

field might be listed on the form and can be filled in before or after the inventory so the form is complete for data entry. If you also want to track fire performance rating on the cable, you might indicate this on the comments line. Enter this on the form as "Fire/UL Rating (Comments Field)," so that the inventory crew knows what to record and the data entry personnel knows where to enter the information on the cable record screen.

The next thing to consider is the flow of the information coming from the inventory crew. Check to see if any modifications are needed to insure that all the information is recorded, and that filling out the form flows easily. A review of the list indicates that the Fire/UL Rating is usually listed with the Manufacturer and Part/Specification Number, so it might be more appropriate to move that field to just below Part/Specification Number to be sure the inventory crew remembers to record the information while they are recording the Manufacturer and part number. This saves the inventory crew from referring back to the same area two different times. Data entry personnel must temporarily skip over this field until they can enter it in the Comments field, but it makes more efficient use of the inventory time.

Another way to improve the ease of use and accuracy of the inventory forms is to provide the choices to select for certain fields, such as the cable type, on the form itself. The individual only has to circle the appropriate type rather than refer to a separate list of options and write in an entry. This will save time, provides consistency in how the information is provided for data entry, and avoids incorrect entries. An example might be to list the choices for the Cable Type field in parenthesis following the field name, as is done in the revised cable inventory form shown in Table 36.8. This would prevent an individual from entering Copper, House, or Voice as the Cable Type, instead of Riser.

Selection lists should be used on the inventory forms wherever validation tables support the information to be recorded, such as desktop equipment types (PC / Phone / Printer / Fax / Modem, etc.). Providing the inventory crew with the acceptable entries for each field helps to ensure accuracy and consistency. If there are only a small number of buildings, you may wish to provide the appropriate building abbreviations for fields requiring a building name, or the same for departments. If space does not permit listing all options, you might want to consider using a 2-sided form, where the options could be listed on the back side for quick reference.

When developing the inventory forms for recording connections to individual backbone cable pairs, it can be beneficial to record the connections for both ends of the cable on a single form. This will help to identify any areas where dead jumpers have been left in place that will need to be disconnected. A quick review of the forms at the end of the inventory process will readily identify cable pairs which have a connection on one end but not on the other. Prior to disconnecting the jumper, a verification of the end showing no connection should be made to check if an error was made in recording the cable pair connections. This provides a method for cleaning up the wiring closet frames and ensuring the information tracked in the CMS properly reflects the actual connections in the wiring closets. Two sample forms are shown in Tables 36.9 and 36.10. The form shown in Table 36.9 would apply where termination hardware is tracked as a separate component in the CMS database, while Table 36.10 applies to a situation where termination hardware is treated as an attribute of the cable pair.

When developing the inventory forms for network control equipment, it may be helpful to include a list of the ports or pin positions on the form with a space provided for recording the connection to each position. This provides all the information about the equipment or hardware type, location, and connections on a single form. Set up the inventory form to closely resemble the CMS data entry screen where the connections will be entered into the database.

It is useful to distinguish between non-expandable and expandable devices, as mentioned earlier and discussed in Chapters 3 and 21. A non-expandable device and its ports (connection points) can be described on a single sheet, such as the example given in Table 36.11 at the end of this chapter. Expandable devices require a sheet to describe the chassis and the chassis positions, while each card installed in a chassis position will require a separate sheet for describing its ports (connection points). An example of an inventory form for the chassis itself is given in Table 36.12, while an example for an individual card or module

DATA FIELD	INVENTORY
CABLE NAME:	___________
CABLE TYPE:	___________
PAIR COUNT:	___________
CABLE LENGTH:	___________
CONDUIT:	___________
DEPARTMENT:	___________
INSTALLED DATE:	___________
MANUFACTURER:	___________
PART NO./SPEC #:	___________
TO BUILDING:	___________
TO FLOOR:	___________
TO ROOM:	___________
FROM BUILDING:	___________
FROM FLOOR:	___________
FROM ROOM:	___________
COMMENTS:	___________

Table 36.7: Sample fields from the CMS cable screen.

DATA FIELD	INVENTORY	
CABLE NAME:	______________________________	(Example: RC-REN1E)
CABLE TYPE:	RISER/CAMPUS/TIE/HORIZONTAL	(Circle One)
PAIR COUNT:	______________________________	
CONDUIT:	______________________________	
TERMIN. DEVICE:	FIRST PAIR:_____________	LAST PAIR_____________
TERMIN. POS'N::	FIRST PAIR:_____________	LAST PAIR_____________
DEPARTMENT:	_______________	
MANUFACTURER:	AT&T/BELDEN/COMM SCOPE/MOHAWK/_____________ * (Circle One)	
PART NO./SPEC #:	______________________________	
FIRE/UL RATING	______________________________	(Comments Field)
TO BUILDING:	REN / ADM / FAC	(Circle One)
TO FLOOR:	B 1 2 3 (Circle One)	E / W (Circle one for REN)
TO ROOM:	______________________________	
FROM BUILDING:	REN / ADM / FAC	(Circle One)
FROM FLOOR:	B 1 2 3 (Circle One)	E / W (Circle one for REN)
FROM ROOM:	______________________________	
COMMENTS:	______________________________	

Table 36.8: Sample inventory form for backbone cables.

is given in Table 36.13, both at the end of this chapter.

When recording the list of horizontal cables located in each of the wiring closets, prepare an inventory form similar to the one shown in Table 36.14. This type of form ensure that all cable types, naming conventions, and termination devices and positions are captured. This information will be used for preparing lists of all horizontal cables or outlets by wiring closet, which can then be used for recording horizontal cable connections or as a checklist for auditing inventory information.

There are different methods for developing inventory forms for office outlet connections. One method is to prepare the inventory forms based on a list of the horizontal cables which are terminated in the wiring closet, and then recording the information against the appropriate outlet as each one is located. This is usually most beneficial where the same type of cabling and outlets are used throughout all offices. It also provides an easy way to identify which outlets were not located during the office area inventory and captures the information for each outlet on a uniform basis. A sample office inventory form based on outlet numbers is provided in Table 36.15 at the end of this chapter.

Another way to approach the office area inventory form is to set up the form on an office by office basis, where outlet numbers and connections are recorded for each office. This method is preferred where outlet numbers vary widely from office to office or even within an office, such as where a given office might include Outlets #102 and #214. This is typical of facilities where additional outlets are added over time. Inventory forms arranged by office number also work well where the types of horizontal cables or outlets differ from office to office. This type of approach allows the inventory crew to record the type of outlet during the inventory. A sample office inventory form based on office room numbers is provided in Table 36.16 at the end of this chapter.

You can use the connections recorded against the horizontal cables or outlets in the wiring closet to verify the active use of those connections at the office location. This is done by adding a column to the office inventory sheet, where the wiring closet connection to the horizontal cable can be indicated. It provides a way to verify that there is a connection in the office for every connection found in the wiring closet. Often, during the Move, Add, and Change process, connections are left active in the wiring closet which are not being used at the office location. During this inventory they can be disconnected and later reused. If you are using a list of outlets for recording office equipment and connections, this same form could be used for the initial recording of connections in the wiring closet, as described earlier in this chapter.

Once you have finished preparing the forms, use them to inventory the first wiring closet and the office areas associated with that wiring closet. This will test out the forms and determine whether they are easy to fill out, capture all the required information, flow along with the inventory process, and whether any modifications are needed. Actually entering the data into the CMS may reveal further modifications to speed up the process and make the forms easier to audit.

Review the final procedures and inventory forms with the inventory crew, the data entry personnel, and anyone who will be auditing the inventory records after they have been entered into the CMS database. This way all individuals who will be involved in the recording, entering and auditing of the information will have an understanding of what is required.

During the process of inventory, data entry, and auditing, some arrangement should be made for capturing any changes to circuit cross-connects which are made between the time the connections were recorded and the time at which

CABLE NAME:___________________________

CABLE PAIR	LOCATION -- END 1 BLDG: _________ FLOOR: _________ ROOM: _________ DEVICE / POSITION	CONNECTED TO: PORT CABLE / PAIR DEVICE / POSITION	LOCATION -- END 2 BLDG: _________ FLOOR: _________ ROOM: _________ DEVICE / POSITION	CONNECTED TO: PORT CABLE / PAIR DEVICE / POSITION
1	/	/	/	/
2	/	/	/	/
3	/	/	/	/
4	/	/	/	/
5	/	/	/	/
6	/	/	/	/
7	/	/	/	/
8	/	/	/	/
9	/	/	/	/
10	/	/	/	/
11	/	/	/	/
12	/	/	/	/
13	/	/	/	/
14	/	/	/	/
15	/	/	/	/
16	/	/	/	/
17	/	/	/	/
18	/	/	/	/
19	/	/	/	/
20	/	/	/	/
21	/	/	/	/
22	/	/	/	/
23	/	/	/	/
24	/	/	/	/
25	/	/	/	/

Table 36.9: Sample inventory form for backbone connections with termination hardware tracked separately.

CABLE NAME:____________________________

CABLE PAIR	LOCATION - END 1 BLDG: ________ FLOOR: ________ ROOM: ________ CONNECTED TO: CABLE/PAIR (DEVICE /POS.)	LOCATION - END 2 BLDG: ________ FLOOR: ________ ROOM: ________ CONNECTED TO: CABLE/PAIR (DEVICE /POS.)	CIRCUIT NUMBER/ COMMENTS
1	/	/	
2	/	/	
3	/	/	
4	/	/	
5	/	/	
6	/	/	
7	/	/	
8	/	/	
9	/	/	
10	/	/	
11	/	/	
12	/	/	
13	/	/	
14	/	/	
15	/	/	
16	/	/	
17	/	/	
18	/	/	
19	/	/	
20	/	/	
21	/	/	
22	/	/	
23	/	/	
24	/	/	
25	/	/	

Table 36.10: Sample inventory form for backbone cable connections where termination hardware is an attribute of the cable.

the information is being used for managing circuit changes. This may be done in a number of ways, depending on the activity levels of the areas which are being inventoried.

Some organizations have made use of wiring closet change forms which can be posted on the door of the wiring closet to capture any changes which are made following the inventory. These change sheets can be collected after the data has been entered into the CMS records so that revisions to the circuit connection records can be made. Other organizations have put a temporary freeze on changes for each area as it is inventoried and the records are entered into the CMS. This avoids any circuit changes being made which would impact the accuracy of the information which has been recorded. Choose the method which works best for your organization, but be sure you capture any changes which are made prior to full utilization of the CMS records.

NETWORK HUB ID/DEVICE NAME: ____________________

(TH + BLDG + FLR/WNG + Seq. #) - Token Ring

(EH + BLDG + FLR/WNG + Seq. #) - Ethernet

TYPE: IBM-8228 / DEC200 / ________ (Circle One)

CLASS: HUB-NE

SERIAL #: ____________________

STATUS: I

NO. OF PORTS: ____________________

BUILDING: REN/FAC/ADM (Circle One)

FLOOR: B 1 2 3 E / W (Circle One)

ROOM: ____________________

DEPT.: ITG/ADM/ENG/OPR/SLS/MFG (Circle One)

PORT CONNECTIONS: **CONNECTED TO:**

PORT TYPE	SIDE	PORT NAME	TERM. DEVICE & POSITION	CABLE NAME OR DEVICE NAME	CABLE END OR DEVICE SIDE	CABLE PAIR OR DEVICE POS'N
________	F	____	________	________	____	________
________	F	____	________	________	____	________
________	F	____	________	________	____	________
________	F	____	________	________	____	________
________	F	____	________	________	____	________
________	F	____	________	________	____	________
________	F	____	________	________	____	________
________	F	____	________	________	____	________
________	F	____	________	________	____	________
________	F	____	________	________	____	________
________	F	____	________	________	____	________
________	F	____	________	________	____	________

NETWORK CONNECTIONS:

CONNECTOR TYPE:	______	______	______	______
CONNECTOR NAME:	______	______	______	______
CONNECTED TO:	______	______	______	______
	______	______	______	______
	______	______	______	______
	______	______	______	______

COMMENTS: ____

____

____

Completed By: ____________________ Date: ____________________

Table 36.11: Sample inventory form for non-expandable control equipment.

PART 4: PHYSICAL INVENTORY

NETWORK HUB ID/DEVICE NAME: ___________________________

(TH + BLDG + FLR/WNG + Seq. #) - Token Ring

(EH + BLDG + FLR/WNG + Seq. #) - Ethernet

TYPE:	3000-4 / 3000-5 / 3030 / 6000	(Circle One)
CLASS:	HUB-E	
SERIAL #:	____________	
STATUS:	I	
OLD/CURRENT LABEL:	____________	
NO. OF SLOTS:	____________	
NO. OF CARDS INSTALLED:	____________	
BUILDING:	REN / FAC / ADM	(Circle One)
FLOOR:	B 1 2 3 E / W	(Circle One)
ROOM:	____________	
DEPT.:	ITG/ADM/ENG/OPR/SLS/MFG	(Circle One)

NETWORK CONNECTIONS:

CARD NAME / ID: ______________________

CONNECTOR TYPE	CONNECTOR NAME	TERMINATION DEVICE / POSITION OR CABLE / PAIR
_____	_______	___________/_______
_____	_______	___________/_______
_____	_______	___________/_______
_____	_______	___________/_______

COMMENTS: ____

____

____

Completed By:__________________________ Date:___________________

Table 36.12: Sample inventory form for expandable control equipment.

NETWORK CARD ID / NAME: ____________________________

(Network Segment + Slot on Hub)

(Example: RA23/A01)

TYPE: 3205 / 3514 / 3534 (Circle One)

CLASS: CARD

HUB ID: ____________

STATUS: I

NO. OF PORTS: ____________

BUILDING: REN / FAC / ADM (Circle One)

FLOOR: B 1 2 3 E / W (Circle One)

ROOM: ____________

PORT CONNECTIONS: **CONNECTED TO:**

CARD ID	SIDE	PORT	TERM. DEVICE & POSITION	CABLE NAME OR DEVICE NAME	CABLE END OR DEVICE SIDE	CABLE PAIR OR DEVICE POS'N
________	F	01				
"	F	02				
"	F	03				
"	F	04				
"	F	05				
"	F	06				
"	F	07				
"	F	08				
"	F	09				
"	F	10				
"	F	11				
"	F	12				

NETWORK CONNECTIONS (For Network Interface Cards)

CONNECTOR TYPE: ________ ________ ________ ________

CONNECTOR NAME: ________ ________ ________ ________

CONNECTED TO: ________ ________ ________ ________

COMMENTS: ____

Completed By: ____________________ Date: ____________________

Table 36.13: Sample inventory form for cards or modules in expandable control devices.

WIRING CLOSET LOCATION:

BUILDING: REN / FAC / ADM ______ (Circle One)
FLOOR: B 1 2 3 E / W ______ (Circle One)
ROOM: ______________________

HORIZONTAL CABLE TYPES:
CIRCLE EACH TYPE OF HORIZONTAL CABLE IN THE WIRING CLOSET LOCATION:

CATEGORY 3	COAX	(Circle all types of Horizontal Cable in TC Room.)
CATEGORY 4	TYPE-1	
CATEGORY 5	TYPE-2	
VOICE	DATA	

FIRST CABLE TYPE:

		TERMINATION DEVICE	POSITION
CABLE TYPE:	________		
FIRST CABLE NAME/NUMBER	________	____________	______
LAST CABLE NAME/NUMBER	________	____________	______

COMMENTS: ____

SECOND CABLE TYPE:

		TERMINATION DEVICE	POSITION
CABLE TYPE:	________		
FIRST CABLE NAME/NUMBER	________	____________	______
LAST CABLE NAME/NUMBER	________	____________	______

COMMENTS: ____

THIRD CABLE TYPE:

		TERMINATION DEVICE	POSITION
CABLE TYPE:	________		
FIRST CABLE NAME/NUMBER	________	____________	______
LAST CABLE NAME/NUMBER	________	____________	______

COMMENTS: ____

FOURTH CABLE TYPE:

		TERMINATION DEVICE	POSITION
CABLE TYPE:	________		
FIRST CABLE NAME/NUMBER	________	____________	______
LAST CABLE NAME/NUMBER	________	____________	______

COMMENTS: ____

Completed By:__________________________ Date:________________

Table 36.14: Sample inventory form for recording horizontal cable names by wiring closet.

BLDG: REN / ADM / FAC (Circle One)   FLOOR: B / 1 / 2 / 3 E/W (Circle One)

JACK ID #	JACK PAIRS	TERM. DEVICE & POS.	ROOM NO.	USER NAME LAST+ 2 INIT'LS	DEPT.	OFFICE EQUIP ID	EQUIP. CLASS	TYPE / MODEL	SERIAL NO.	HOUSE PRS. / DATA PORT
101	123									
101	4									
102	123									
102	4									
103	123									
103	4									
104	123									
104	4									
105	123									
105	4									
106	123									
106	4									
107	123									
107	4									

Completed By: ______________________   Date: ______________

Table 36.15: Sample inventory form for office inventory, arranged by outlet number.

BUILDING:	REN / ADM / FAC	(Circle One)	USER LAST NAME: ______
FLOOR:	B / 1 / 2 / 3 E / W	(Circle Floor/Wing)	USER INITIALS: ______
ROOM	______		EXTENSION: ______
DEPARTMENT:	ITG/ADM/ENG/OPR/SLS/MFG	(Circle One)	

JACK / OUTLET	TERM. DEVICE & POSITION	JACK TYPE	CN #	OFFICE EQUIP EXT. / TAG ID	EQUIPMENT CLASS (Circle One)	TYPE / MODEL (Record Make/Model)	SERIAL NUMBER	CN #	HOUSE PAIRS / DATA PORT
			V1		ANALOG/DIGITAL			1	
			V2		ANALOG/DIGITAL			1	
			D1		PC/TERM/PRINT				
			D2		PC/TERM/PRINT				
			V1		ANALOG/DIGITAL			1	
			V2		ANALOG/DIGITAL			1	
			D1		PC/TERM/PRINT				
			D2		PC/TERM/PRINT				
			V1		ANALOG/DIGITAL			1	
			V2		ANALOG/DIGITAL			1	
			D1		PC/TERM/PRINT				
			D2		PC/TERM/PRINT				
			V1		ANALOG/DIGITAL			1	
			V2		ANALOG/DIGITAL			1	
			D1		PC/TERM/PRINT				
			D2		PC/TERM/PRINT				
			V1		ANALOG/DIGITAL			1	
			V2		ANALOG/DIGITAL			1	
			D1		PC/TERM/PRINT				
			D2		PC/TERM/PRINT				

Completed By: ______ Date: ______

Table 36.16: Sample inventory form for office inventory, arranged by office number.

LABELING

37

During the process of inventory it is essential to properly label all cables and devices for future reference. Included below are some guidelines for labeling of the cables, equipment and termination hardware which will be tracked in the CMS records. The labels will become the reference point and connection between the record and the item itself that the technicians and network support personnel will depend on when working with connections and equipment. If the equipment and cables are not labeled to correspond with the database records, the records provide little benefit when specifying which items are to be moved, changed, or connected.

Labeling should be consistent for each location and for each type of component, such as labeling all network hubs with the same type and color label, and should be located in the same place on each device. Labeling methods will vary based on the types of hardware and equipment which is being tracked. Procedures should be established for labeling each type of component including the type of label to be used, the specific color coding, and where the labels are to be placed. This consistency will make it easier for individuals to go from one area to another and be able to quickly identify and locate the desired component.

There are several options for labeling cable and connectivity components. There are label software products which can generate printed labels for a range of cable names or equipment components using a sequential numbering scheme combined with a fixed prefix. Some software, with the appropriate printer, can produce bar-coded labels as well. There are several types of label makers which can be used to create simple, single-strip labels for immediate application while in the field, such as the Brother or Dymo label makers. Other labelmakers can automatically produce sequential labels in a variety of styles, such as are available from Brady USA Inc. There are numerous types of self-adhesive labels which

- USE LABELS WITH A PERMANENT ADHESIVE
- USE LARGE LABELS WHICH ARE EASY TO LOCATE
- LABELS SHOULD BE NEAT AND EASY TO READ
- LABELS SHOULD BE UNIFORM IN SIZE AND TYPE
- USE THE SAME COLOR SCHEMES FOR ALL LOCATIONS
- LABELS SHOULD BE IN THE SAME PLACE ON SIMILAR COMPONENTS

Table 37.1: Elements for selecting and using equipment and cable labels.

come in blank sheets where names or IDs can be created through the computer, a typewriter, or may even be hand-written on the labels. Components may be labeled by using a fine-tip permanent ink pen to write the name or ID directly on the individual item. A few points to keep in mind when selecting the method you will use for labeling cables and equipment are provided in Table 37.1.

Remember that labeling must be approached with precision. Handwritten labels must be legible and in the proper case. A method should be adopted to distinguish the number "1" from the letter "l", the number "0" from the letter "o", the number "2" from the letter "z", and the number "5" from the letter "s". All labels should be in a highly visible location. You may wish to refer to Chapter 3 and Chapter 4 for additional information on labeling techniques.

Labeling in the Wiring Closets

Once you have decided on the naming conventions for cables, equipment, and termination hardware, each component should be labeled accordingly, especially those located in wiring closets. Start by labeling the cables and equipment is usually at the wiring closets, main distribution frames, and computer rooms where the major backbone components are located. Labeling in the wiring closets will assure that individual circuit connections recorded during the physical inventory will refer to the proper components in the CMS records.

If the components have existing labels, review this labeling to determine if it is consistent with the naming conventions you have developed for use with the CMS. If new labels are required, be sure to remove old labels as the new ones are installed or develop a method for distinguishing the old labels from the new ones. In some circumstances, the old labels may need to be left in place if they are referenced in other recordkeeping systems.

If you have a list of the components to be labeled with their new names, it is most efficient to prepare the labels in advance and install them prior to the inventory of individual circuit connections. This saves time and helps to ensure the labels are created accurately and consistently. Segregate the labels by equipment types, cable types, and wiring closets and make a separate folder or envelope for each. This will make it easier to locate and attach the proper labels when you are in the wiring closet.

If you do not have a list of the components to be labeled but plan to develop this

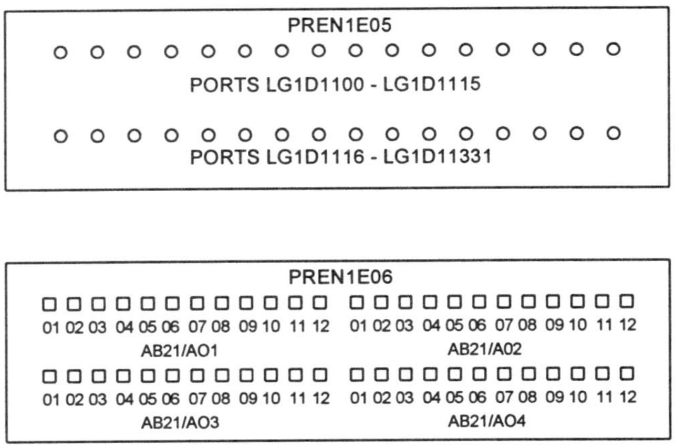

Figure 37.1: Recommended labeling for a patch panel where network ports are presented.

list through an inventory process, it may be best to take a portable label-maker with you so the labels can be created as each item is inventoried and assigned a name or ID. This accomplishes both the inventory and labeling tasks at the same time. Be sure to prepare a reference list for each type of equipment and cable indicating the label format, color, and where the labels are to be placed on the individual components.

Backbone cables should be labeled at each end, with individual cable pairs labeled at their point of termination. Horizontal cables should be labeled at each end and at their point of termination. For termination hardware, each block should have its own label so individuals do not have to count the blocks to be sure they have the correct one. Rows or columns may be labeled at the top or side indicating the cable names and counts, or the equipment and ports represented by the group of blocks. If space permits, the actual pin positions/port counts or cable pairs represented by each block should be indicated on the block.

Control devices should be labeled with a logical network ID or a physical asset ID according to the naming conventions developed in the CMS catalog. For connection tracking, the port termination should be labeled on the wiring block or patch panel, and on the control device itself if each port is a separate connector on the device. For some devices, this can be met if the device itself is labeled and the individual port connectors are identified with a sequence number such as 01, 02, 03, etc. Labels for patch panels should indicate their CMS name/ID as well as the port names (or control device ID and port sequence numbers) which are represented on the patch panel. An example would be patch panel PREN1E05, which represents ports LG1D1100 - LG1D1131. The panel should be marked with PREN1E05, LG1D11, and a sequence number for each connection point (port) on the panel, as shown in Figure 37.1.

Each pathway should be labeled at each end, at each space where pathways come together, and at intermediate points if the pathway is accessible. Splice enclosures should be labeled on the cover, and it is helpful to label the cable ends within the splice enclosure.

If you are using a CMS which assigns its own names or IDs for the cables and equipment, you will need to do the labeling after the items have been created in the CMS database. This may require two visits to the wiring closet - one to inventory the items so they can be created in the database, and a second trip to install the labels once the identifiers have been assigned by the

TERMINATION POINT	COLOR	PANTONE	COMMENTS
DEMARCATION POINT	ORANGE	150C	CENTRAL OFFICE TERM'S
NETWORK CONNECTIONS	GREEN	353C	NETWORK CONNECTIONS OR AUXILIARY CIRCUIT TERM'S
COMMON EQUIPMENT PBX, HOST, LAN, MUX	PURPLE	264C	MAJOR SWITCHING AND DATA EQUIPMENT TERM'S
FIRST LEVEL BACKBONE	WHITE		MC-IC CABLE TERM'S
SECOND LEVEL BACKBONE	GRAY	422C	IC-TC CABLE TERM'S
STATION	BLUE	291C	HORIZONTAL CABLE TERM'S
INTERBUILDING BACKBONE	BROWN	465C	CAMPUS CABLE TERM'S
MISCELLANEOUS	YELLOW	101C	AUXILIARY, MAINTENANCE, ALARMS, SECURITY, ETC.
KEY TELEPHONE SYSTEMS	RED	184C	

Table: 37.2: TIA/EIA 606 Color Codes for termination positions.

CMS and noted on the inventory sheets.

If you plan to color code the labels to distinguish the different components but have not established a specific color code, you may want to use the color code in the TIA/EIA 606 Standard shown in Table 37.2. Consider placing a reference sheet showing the color coding in each wiring closet. Alternative color codes can be developed to differentiate among network services.

Labeling Office Outlets

If the outlets have existing labels, review these labels to determine if they are consistent with the naming conventions you have developed for use with the CMS records. If the CMS outlet name includes a Building/Floor prefix or identifier, it may not be necessary for that information to be shown on the actual outlet, since the user or technician would already know the building and floor. If new labels are required, be sure to remove old labels as the new ones are installed, or develop a method for distinguishing the old labels from the new ones. In some circumstances, the old labels may need to be left in place if they are referenced in other recordkeeping systems, or until new labels have been installed in both the office and wiring closet.

If the labeling of the office outlets is to be done during the physical inventory of the office equipment and connections, it would be beneficial to have these labels made up in advance so the inventory crew does not have to create an individual label for each jack during the inventory activity. Preparing labels in advance also avoids having to keep track of the outlet numbers as they are used, and also avoids the possibility of the same name/number to be assigned to two different jacks. If office outlet labels are to follow the termination sequence in the wiring closet, toning of the individual jacks may be required to determine the proper number.

Labels for office outlets should be easy to read but not objectionable to the individual user. If labeled on a removable cover, the horizontal cables or the internal outlet fixture should also be labeled in case the cover is removed. Often overlooked is the need for specific work area names and labels. Each office or cubicle should be labeled near its entrance.

Labeling Desktop Equipment

If new labels need to be installed on desktop equipment such as PCs, printers, terminals, etc. these should be installed during the physical inventory of the office outlets and connections. This ensures the new name corresponds to the right equipment and connections. Depending on the naming conventions developed, the new name may include a portion of the serial number, equipment type, or some other information as an identifier in the label in which case labels cannot be prepared in advance.

If the naming convention involves the use of some prefix for the equipment type (such as PC, PR, TM, etc.) followed by a sequential number, it may be most efficient to prepare these labels in advance. This way the inventory crew simply has to remove the next available label and attach it to the desktop device, then record the office location and connections for the device. This avoids the need for the inventory crew to keep a list of the desktop device names/labels which have been used. A set of labels for PC's numbered PC0001 to PC0200, Printers PR0001 to PR0200 and Terminals TM0001 to TM0200, can be made up in advance for the first office area to be inventoried. This gives the inventory crew an opportunity to work with the pre-made labels and to determine if they meet the needs for labeling desktop equipment.

Another option might be to have an individual take the pre-made labels and go to each office and install the labels on the desktop devices prior to the inventory work. This would make the inventory process go a little faster since the labels would already be attached, and the inventory crew would simply have to record the equipment type, location, and connections. The inventory crew could have some spare labels with them for labeling any equipment which did not already have a label.

Selecting the Labels to Use

Before you begin any labeling, it is best to select the type of labels which will be used for

• COMPUTER-GENERATED LABELS DOT MATRIX LASER JET STRIP PRINTER • PORTABLE LABELMAKERS ELECTRONIC MANUAL • OFFICE SUPPLY LABELS TYPE-WRITTEN HAND-WRITTEN

Table 37.3: Types of labels and label-making devices.

labeling each of your components. This will provide for uniform labeling throughout your facilities, and will help technicians become familiar with the labels on the cable and connectivity components. A variety of labels and label-making methods is summarized in Table 37.3.

- Computer-Printed Labels: One option is to utilize a PC-based label software program. These software packages are designed to create a range of labels utilizing a specified format (with or without a prefix) and numbering scheme. This type of software-programmed label is most efficient for labeling desktop devices, termination hardware, and horizontal cables or office outlets which are based on a sequential numbering scheme. These label software programs, designed for creating labels, are offered by a number of software supply manufacturers.

  Labels are available for dot matrix printers, laser jet printers, and strip printers such as the Seiko Smart Label Printer and the CoStar LabelWriter XL. Strip printers come with printer drivers or macros to interface with common word processing software.

- Portable Electronic Label Makers: A less expensive and more flexible option for field labeling is to use a portable label maker which has the ability to create pre-programmed labels utilizing a prefix or specific numbering scheme. This can be can be beneficial for creating labels as needed in the field and for areas where a large number of items will have similar labels with sequential numbering, such as horizontal cables or termination blocks. Portable electronic label makers for cable and equipment are manufactured by Brady USA Inc. Brady also makes a wide variety of labels for labeling telecommunications components and distributes a labeling guide specifically tailored to the TIA/EIA 606 Standards. Other general purpose label makers are made by Brother International Corporation and Casio.

- Manual Label Makers: Another alternative is to utilize a hand-held label maker which can create labels one at a time for attachment to the cables or equipment. These label makers can produce labels of different colors and sizes, depending on the size tape which is used. These tools and the label tapes can usually be purchased at local office supply stores or might be purchased through your company's stationery supply section. These labels have a good adhesive and stay in place. They also offer flexibility in the size of the label which is created. However, these types of labels may not be the most efficient for labeling office outlets since they require each label to be individually created, which may be time-consuming when creating a large number of labels of the same type. Brother International Corporation and Dymo manufacture these types of label makers.

- o Office Supply Labels: A simple, inexpensive method of labeling is to select a set of labels available from the local office supply store or a label manufacturer such as Avery, Dennison, Maco, etc. These labels come in a wide variety of sizes, shapes and colors, and can then be used in a number of ways. The labels must have a permanent adhesive. Many labels are now available which can be used with laser or dot-matrix printers, so the actual labels can be typed into a word processing or spreadsheet software package and printed to a local printer. Labels can also be typed on a typewriter or hand-written.

Any of the above approaches will work, so chose the type of labeling method which best meets your needs. The key is to select the type of label to be used and check to make sure they adhere well and fit into the desired location on the cable or equipment. Again, decide on the proper colors and sizes required and be sure to order an adequate supply. You may wish to use a simple, inexpensive type of label to do your initial labeling, and then after all components have been identified and labeled, replacing these with more permanent labels at a future date.

WORKING WITH EXISTING DATA

38

If you are planning to use existing data to load your CMS database, there are certain considerations that will help assure complete and accurate loading. Some of the items we will cover include auditing the data for accuracy, transforming the records into the proper format for entering the information into the database, and checking the records for proper format and performance once they have been entered into the CMS.

The first step in preparing your existing data for entry into the CMS database is to audit the data for accuracy and completeness. You will not want to enter incorrect or incomplete records, since this will impact the effectiveness of using the information which you will retrieve from the CMS. Data should be as accurate as possible when it is entered into the database.

Take a sample portion of your existing records and audit the data in order to determine the level of accuracy. You will need to go out to the wiring closets and offices to double-check this information. Use a sample that represents approximately 10 to 15% of the overall records. While this may seem like extra effort, it will greatly help to identify incorrect data before it appears in the database. If the information audited is accurate, the records can most likely be utilized. If errors are identified in this small sample, you may wish to audit a larger sample or assign an individual to audit and correct the information prior to using it in the CMS database. If the data is seriously in error, you should consider performing a full inventory instead, as this might actually take less time than a full audit. Collecting fresh information is often quicker than sorting through data full of conflicts and omissions.

Organizing the Data

Once you have determined that the existing data can be used in the CMS, the next step is to organize the data into the proper format for most efficient transfer of the records. If you are planning to transfer the data directly from the existing database to the CMS database, you will need to take several steps to make the data

- REVIEW AND MODIFY FIELD LENGTHS
- REVIEW AND MODIFY FIELD NAMES
- ARRANGE DATA FIELDS IN PROPER SEQUENCE
- IDENTIFY AND ADD “REQUIRED” DATA FIELDS

Table 38.1: Initial steps required to prepare existing records for transfer to the CMS database.

compatible with the new system.

Table 38.1 shows a list of some of the activities which may be required in order to prepare your existing database records for loading into the CMS. Note that you may have to manually create tables of supporting information such as connector types or connection point configurations prior to using the existing data. In fact, it is likely that the entire catalog of cable and equipment types must be manually loaded, with existing records used for item information only. Of course, the new catalog must be developed with the existing data in mind.

Review the field lengths for the existing data to make sure the field lengths are the same size or smaller than those in the CMS database. If any of the field lengths are larger than those in the CMS database, you will need to revise the information contained in those fields so that it will fit properly, or else have the CMS fields modified. This may require printing some reports to identify those records which need modification.

Field names should also be modified, where possible, to match those in the CMS database. If the existing records are in a word processing or spreadsheet software package, the file may be revised to reflect the new headings/field names. If these cannot be revised, the transfer command or program will need to specify that the data from the existing field name is to be transferred to the corresponding CMS field name. If the records are to be entered manually, the individual doing the data entry should be provided with a reference list indicating corresponding field names for the existing system and the new CMS.

If possible, arrange the fields in the order in which they appear on the CMS record screen. With some existing database formats this may not be possible, however most word processing or spreadsheets allow for the rearrangement of columns or fields. This will be beneficial with both electronic transfer and manual data entry, and in auditing the CMS records.

Review the screens into which the existing data will be transferred. You will want to identify any fields on these screens which are "required" fields. These are fields which must be completed in order for the record to be accepted. If there are any required fields for which you do not have a field in your existing records, you may need to add a field and enter the "required" field information as a separate step, prior to transferring the information to the CMS database.

Transferring the Data

If the existing data is to be electronically transferred through a batch program or file transfer command, review the procedures and specific instructions for indicating where the existing data is to be placed in the CMS database. This information should be defined in the Vendor's Reference Manual. If no instructions are provided, or they are not detailed enough for your specific data, contact the Vendor to determine the exact procedures for electronic transfer of information into the CMS software. It is often very difficult to delete or modify the information once it has been used to populate the CMS

- REVIEW VENDOR'S PROCEDURES
- CREATE OR LOAD CONNECTOR TYPES OR PORT TYPES
- CREATE OR LOAD ANY EQUIPMENT AND CABLE TYPES OR CONFIGURATIONS
- CREATE OR LOAD INDIVIDUAL ITEM RECORDS
- CREATE OR LOAD CONNECTIONS BETWEEN INDIVIDUAL ITEMS

Table 38.2: Individual steps involved in transferring cross connect records to the CMS database.

records, so it is important to make sure the records are properly set up and transferred in the appropriate format. The Vendors will be more than happy to assist you in setting up the proper format for utilizing your existing data, wherever possible.

If any circuit cross-connects are to be electronically transferred where two or more components are linked together, this usually requires several steps as shown in Table 38.2. Any connector or port types may first need to be created or populated in a specific CMS table, followed by creating the component types and configurations for each of the cable and equipment types. A third step may then be required to create the individual item records before finally loading or creating the connections between the components. This is often a complicated task to accomplish through electronic transfer of information, and should be tested on a small scale before transferring any large amounts of data.

If the existing data will be manually entered into the database through the use of data entry personnel, the existing records should be clearly marked indicating the order in which they are to be entered into the CMS. For example, the process may begin with entry of the component types and configurations, then the individual components, followed by the connections. If the existing field names do not match the CMS database, specify the CMS field names for each type of information to be entered. Also, clearly indicate the order in which the records are to be created, especially where all horizontal cables or termination hardware must be created in sequential or numerical order. Some CMS products will list items in reports and queries based on how they are entered into the system rather than alphabetically or numerically. For these types of systems, it is beneficial to use a systematic approach to entering the data.

When manually entering existing data, be sure to identify any additional "required" field information in creating the CMS records. This might include a Department, Site or Building name which might not appear on the existing data but which is required on the CMS record screen in order to complete the record. This might be noted as a comment on the existing data record for reference by the data entry personnel.

Verifying the Data

After the existing data has been transferred to the CMS database, it is important to audit and verify the records. You will want to make sure the data was properly loaded and that the records can be retrieved and displayed as desired. A list of some items to consider when auditing the records is provided in Table 38.3.

Issue several queries and run several reports which utilize the data to make sure it was properly loaded into the database, such as a list of the cables or equipment, and then a list of any connections which were added for each of the items. Review the reports for consistency in the number of characters in each item name, the proper use of upper and lower case letters, correct ends of cables, correct sides of devices, and for the

- CONSISTENCY IN ITEM NAMES
- CONSISTENCY IN USE OF UPPER AND LOWER CASE LETTERS
- CONSISTENCY IN CABLE ENDS AND EQUIPMENT SIDES
- ACCURACY OF INFORMATION
- COMPLETENESS OF ALL FIELDS
- INCLUSION OF ALL RECORDS TRANSFERRED

Table 38.3: Elements to consider when auditing the CMS records.

proper connectors, ports or pairs. Check for accuracy to make sure the fields were populated with the correct information from the existing data, such as the proper cable and equipment types, locations and departments, etc.. Look for any records which might need modification due to incomplete or missing information. You will also want to make sure that all records were retrieved and that none are missing.

For cables, controlling equipment, or termination hardware, test the ability to make connections to these items to be sure they were created with the proper number of ports or connectors. If any circuit cross-connects were electronically transferred where two or more components were linked together, test those connections by selecting one or more of the circuits and tracing the connections to make sure all the components were properly linked and the appropriate connectors were used.

TRAINING

39

In this chapter, we will discuss the training which should be provided for the inventory team This training is a very important part of the inventory process, since it directly impacts the quality of the records which will be gathered and entered into the CMS. Proper training will result in better inventory records, less data entry errors, fewer rechecks, and an overall smoother implementation. Some elements which should be considered in order to provide adequate training for inventory personnel are provided in Table 39.1.

The Inventory Process

The first step in training your inventory personnel is to provide them with an understanding of the inventory process. This should include a short background on the CMS project and why the inventory is being conducted, how the information will be used, and the importance of accuracy. They should understand how their role in the project fits into the whole picture, so it takes on more meaning than just recording names and numbers.

It is beneficial to provide hands-on training for the portions of the inventory that each person will be assigned to handle. This may include doing an actual office inventory of several offices, recording the cross-connects to horizontal cables in the wiring closets, or taking an inventory of all the connections for a specific type of network control equipment. The more hands-on training they receive, the less questions they will have, resulting in fewer errors and less missing information. This also provides first-hand experience on what is expected, and

- OVERVIEW OF INVENTORY PROCESS
- IMPORTANCE OF ACCURACY
- HANDS-ON EXPERIENCE
- PROPER COMPLETION OF INVENTORY FORMS
- LIST OF ACCEPTABLE ENTRIES
- DETAILED WRITTEN INSTRUCTIONS
- EMPHASIS ON CLEARLY LEGIBLE HANDWRITING
- PROPER TRAINING IN IDENTIFYING DIFFERENT COMPONENT TYPES
- HOW TO AUDIT FORMS FOR COMPLETENESS

Table: 39.1: Elements of training for physical inventory.

approximately how long it should take to complete each portion of the inventory. Personnel will develop an understanding of the time-frame in which tasks should be completed, and time is not wasted before going on to the next section or task.

Filling Out the Inventory Forms

Specific training should be provided on how the inventory forms are to be completed. This will include the information that is required for each field or line on the form, how the information is to be recorded, and a review of the acceptable entry options. These options should also be listed on the back of the inventory forms, or on a quick-reference sheet which they can refer to. This will help to avoid unacceptable entries. Inventory personnel should receive explicit instructions on whether any lines can be left blank on the forms or when blank lines are permissible.

If information is to be recorded on the manufacturer and model number of desktop equipment, specific guidelines should be provided for what information is to be recorded, such as IBM PS70/386 or IBM PS/70. This will avoid wasting time by recording too much information, or in having to go back for additional information. All information must be recorded in the precise form developed in the catalog.

Neat and clearly legible handwriting should be emphasized in filling out the inventory forms. This will help to avoid rechecks and insure the information is correctly entered into the database. If the inventory information will contain a combination of letters and numbers, provide a specific format for distinguishing the number "0" from the letter "O" and the number "2" from the letter "Z". The use of a slash through the zero and the letter "Z" will help to distinguish these characters. Using only capital letters will distinguish the letter "L" from the number "1". Legible handwriting must be used to differentiate the letter "S" from the number "5". Proper use of spaces, dashes, hyphens, underscores, and slashes is also mandatory.

If inventory personnel will be responsible for determining different types of outlets (RJ11, RJ45, BNC, etc.) or equipment types (analog, digital, etc.), be sure they receive the proper training so they can distinguish the different components from one another. Personnel should be given some way to identify the wiring configuration of an RJ45 jack such as 568A, 568B, or USOC if required. Provide them with examples of each type of component which they will encounter during the inventory process. Again, this will help avoid errors and rechecks.

Auditing the Inventory Forms

Inventory personnel should be trained to audit their inventory forms for completeness. They should review all forms to make sure all room numbers, outlets, desktop equipment, connections, etc. have been recorded. Forms should be reviewed to make sure all offices were inventoried, all outlets were located, and all connections were recorded. These steps should be performed as each area is completed prior to moving onto the next area or task. All forms should be reviewed before they are turned in for data entry.

Guidelines should be established for dealing with problem areas and rechecks where the inventory personnel need additional help or guidance from the Project Leader. This will avoid wasting too much time in an area which needs special attention.

DATA ENTRY AND AUDITING

40

In this chapter, we will discuss the training which should be provided for personnel who will be handling the data entry and auditing of the CMS records. This training is very important in order to ensure the accuracy and quality of the records which are entered into the CMS. Proper training will result in a higher quality of data, more accurate records, and early identification of incomplete or incorrect data. If individuals understand what they are entering into the database, they will make fewer errors, will be able to recognize incorrect data, and will initiate the proper rechecks. Some elements which should be considered in order to provide adequate training for data entry personnel are provided in Table 40.1.

The Data Entry Process

The first part of the data entry training should include an overview of the CMS project, including a short background on the purpose of the project, why the inventory records were recorded, how the information in the CMS database will be used, and the importance of accuracy. This will help the data entry personnel to understand how their role in the project fits into the whole picture, so they have a good sense of what the records represent rather than just completing a series of screens.

If the data entry personnel are not familiar with wiring closets and communications equipment, training should include a visit to a

- OVERVIEW OF INVENTORY PROCESS
- IMPORTANCE OF ACCURACY
- ORIENTATION ON EQUIPMENT AND CABLE COMPONENTS
- HANDS-ON EXPERIENCE WITH EACH CMS SCREEN
- PRACTICE USING ACTUAL INVENTORY FORMS
- STEP-BY-STEP WRITTEN INSTRUCTIONS
- PROPER SEQUENCE FOR ENTERING RECORDS
- SAMPLE COPIES OF COMPLETED SCREENS
- HANDLING WARNING AND ERROR MESSAGES
- NOTING DUPLICATE CABLE PAIR OR PORT CONNECTIONS
- LOOKING FOR INCORRECT OR INCOMPLETE DATA

Table 40.1: Elements of training for data entry personnel.

wiring closet and to an office where they can view the different components and their connections. This will provide the data entry individual with an understanding of what each of the records represents and how actual connections are made. The individual will better understand what they are creating in the database and recognize potential errors in the information which was recorded on the inventory sheets. The data entry process is the last line of defense, so to speak, for maintaining the quality, accuracy, and completeness of the CMS records.

Training should include actual hands-on work in the database, with each person completing several entire records for each task performed during the data entry process. An example of each type of connection should be given. This should be done using actual inventory forms completed during the physical inventory. If possible, establish a separate "Test" or "Training" database to perform these activities so live records will not be at risk. The more experience that each person can get, the better the end results will be.

Written procedures should provide step-by-step instructions for logging into the database and navigating to the appropriate modules or screens. There should be an established order for entering the different types of records into the database. Establishing this entry sequence will help to insure that all components and their connections are completely entered into the database, and that no circuit connections are omitted. The instructions for indicating the data entry sequence might look something like the example shown in Table 40.2.

Samples copies of completed screens should be provided whenever possible, with instructions as to where each item of information from the inventory form is to be entered on the screen. This is particularly useful for entering cross-connect records. Having a reference screen which shows the connector types, sides of equipment, ends of cables, etc., will assist the data entry individual in correctly entering the records. If the use of templates, copy commands or copy buffers are to be used, instructions should also

ITEM	ACTIVITIES TO BE COMPLETED
1	ENTER ALL WIRING CLOSET EQUIPMENT
2	ENTER ALL WIRING CLOSET TERMINATION HARDWARE
3	ENTER ALL WIRING CLOSET RISER CABLES
4	ENTER ALL CONNECTIONS BETWEEN WIRING CLOSET EQUIPMENT AND TERMINATION HARDWARE
5	ENTER ALL CONNECTIONS BETWEEN WIRING CLOSET EQUIPMENT AND RISER CABLES
6	ENTER ALL CONNECTIONS BETWEEN TERMINATION HARDWARE AND RISER CABLES
7	ENTER ALL HORIZONTAL CABLES
8	ENTER ALL DESKTOP EQUIPMENT
9	ENTER ALL CONNECTIONS BETWEEN DESKTOP EQUIPMENT AND HORIZONTAL CABLES
10	ENTER ALL CONNECTIONS BETWEEN HORIZONTAL CABLES AND WIRING CLOSET TERMINATION HARDWARE
11	ENTER ALL CONNECTIONS BETWEEN HORIZONTAL CABLES AND RISER CABLES
12	ENTER ALL CONNECTIONS BETWEEN HORIZONTAL CABLES AND CONTROLLING EQUIPMENT PORTS.

Table 40.2: Sample instructions for the order in which inventory records are entered into CMS database.

include how these functions work for duplicating data from record to record. Use of a copy buffer will speed up the data entry process.

If data entry personnel will be entering backbone cable information through a bulk connection process, they should be given a complete understanding of how to specify the cable, the range of cable pairs, the termination device, and the range of termination positions which will be connected in a single step.

Handling Error/Warning Messages

Data entry personnel should be trained in how to handle error or warning messages which appear during the data entry process. These errors or warnings may involve a duplicate name for an inventory item, duplicate assignment of a cable pair or equipment port, an incorrect entry in a validated field, an incomplete field, etc. Written instructions should be provided for how to respond to each type of warning or error message. This will avoid confusion and wasted effort attempting to identify the problem causing the error or warning.

In the case of a duplicate cable pair or port assignment, the data entry individual should be trained in how to determine the other component which was assigned to the same equipment port or cable pair. This will help the inventory individual be able to quickly identify the correct information. All conflicting information should be checked on the inventory form, since either piece of recorded information could be in error. Checking only one half of the problem can result in creating another duplicate cable pair or equipment port assignment when later connections are made. If the problem cannot be resolved on the inventory form, the information must be rechecked in the field by the Project Manager or the inventory team.

Identifying Rechecks

The data entry process must somehow spot and trap missing or incorrect data. Training for personnel should include a specific process for generating rechecks, such as how to flag each of these items by attaching a removable sticker on the inventory sheet, noting any additional information which will help the inventory personnel in correcting the information, and providing this sheet to the proper individual. While most missing or incorrect data should be identified during the initial audit of the inventory records, it is possible that an equipment type or user name might have been missed. Any errors should be addressed as quickly as possible so they do not affect other data entry steps or become involved in a circuit change before the information is checked.

Auditing the CMS Records

Upon the completion of each floor/wing or section, the records in the CMS database should be audited for accuracy, completeness, and consistency. Several different queries or reports should be created which can be used to review the records and check for errors, duplicate assignment

- HORIZONTAL CABLES AND OUTLETS BY WIRING CLOSET
- TERMINATION HARDWARE AND/OR PATCH PANELS
- RISER OR CAMPUS CABLE PAIR CONNECTIONS BY CABLE NAME
- DESKTOP EQUIPMENT BY LOCATION
- DESKTOP EQUIPMENT CONNECTIONS
- HORIZONTAL CABLE CONNECTIONS BY WIRING CLOSET

Table 40.3: Sample reports to be used in auditing CMS records.

- CONSISTENCY IN ITEM NAMES
- CONSISTENCY IN USE OF UPPER AND LOWER CASE LETTERS
- CONSISTENCY IN CABLE ENDS AND EQUIPMENT SIDES
- ACCURACY OF INFORMATION
- COMPLETENESS OF ALL FIELDS
- INCLUSION OF ALL INVENTORY RECORDS

Table 40.4: Elements to consider when auditing the CMS records.

of cable pairs/equipment ports or missing information. Some standard reports which you may want to create for each area are listed in Table 40.3. It is recommended that you run a complete report from each table listing all entries made that day, or the previous day, for review.

As you review each of the reports, you will want to keep in mind the checklist shown in Table 40.4. Review the reports for consistency in the number of characters in each item name, the proper use of upper and lower case letters, correct ends of cables, correct sides of devices, and for the proper connectors, ports or pairs. Check for accuracy to make sure the fields were populated with the correct information from the inventory sheets, such as the proper cable and equipment types, locations and departments, etc.. Look for any records which might need modification due to incomplete or missing information. You will also want to make sure that all records were entered and that none are missing from the reports.

Make a note of any incorrect records, and make arrangements for the information to be verified and corrected in the database as quickly as possible. Once the data has been audited and corrected, you are ready to begin utilizing the records for that section for management of your Move, Add and Change activities.

The CMS records should be used for ongoing auditing of the connections in the wiring closets and office areas. Connection reports for horizontal cables, backbone cables, control equipment, and patch panels should be run periodically and compared to actual connections in the field. This can pick up changes that were made outside of the connectivity management process which would not be identified otherwise. Audits should be done at least quarterly for a certain number of wiring closets on a random or rotating basis. This procedure can be done by the CMS Administrator or their designate.

While the accuracy of the CMS will depend on the soundness of the management process, ongoing audits are a good quality check and can help identify areas where there are problems or breakdowns.

GLOSSARY

GLOSSARY

Alphanumeric	A field or label consisting of both letters and numbers.
Analog	A communications signal based on the transmission of frequencies or tones.
ARMM	An exchange cable designation for a common construction of riser cable.
ASCII	American Standard Code for Information Interchange.
Backbone	The components which form the main telecommunications structure which connects buildings, floors and wiring closets together.
Balun	A specially configured cable adapter to convert from coaxial cable to twisted pair cable and vice versa.
BICSI	Building Industry Consulting Service International.
Bonding Conductor	A conductor that connects noncurrent-carrying parts of electrical equipment, raceways or enclosures to the system grounding conductor.
CAD	Computer Aided Design.
CAFM	Computer Aided Facilities Management designed to track and report on space occupancy and square footage.
Call Accounting	The process of tracking telephone usage and billing the charges for each line back to a user, client or cost center.
Catalog	The list of communications cable and equipment types including descriptions of the type and number of connection points on those items which are used to link to other components.
Circuit	A series of cable and equipment connections which carry information between two points.
CMIP	Common Management Information Protocol language.
CMS	Cable Management System.

GLOSSARY

Coaxial	Cable with an insulated center conductor wrapped with foil and/or braid for the second outer conductor.
Configuration	Defining the name and number of connection points for a particular type of cable or device.
Connectivity	The tracking of connections or associations connections between different components which make up individual circuits and the backbone structure.
Controlling Equipment	A device on the network containing ports or connection points which connect to other components to form a circuit. Some examples include LAN hubs, controllers, network servers, bridges, routers, a voice PBX, etc.
CPU	Central Processing Unit.
CSA	Canadian Standards Association.
Customization	Special enhancements or modifications to a standard product to better meet the specific requirements of a client's environment.
Dataset Structure	The architecture of each database table and its records.
Desktop Device	User equipment located at individual workstations, such as phones, PCs, terminals, fax machines, etc., also referred to as CPE (Customer Premises Equipment).
Digital	A communications signal based on the transmission of two or three signal states representing binary (zero or one) information.
DXF	Drawing Exchange Format.
EIA	Electronic Industries Association.
ELL	Equipment Line Location on a voice switch.
ER	Equipment Room.
FAX	Facsimile transmission device.
FCC	Federal Trade Commission.
FEP	Front End Processor.
Field	A single string of characters which represents one piece of information, such as your last name or a numerical value, which has a specific name and format assigned by the software.
Frame	Wiring blocks, racks, and patch panels where cable pairs and ports are terminated for connection to one another.

Functionality	The operation of a database taking into consideration the structure of the information fields, table structures, indexing, look-up tables, dataset structures, query and reporting formats, component configurations, and connections between different components within the database.
GES	Grounding Electrode System - A set of buried electrodes designed to achieve good ground potential and continuity in the local environment.
Grounding System	A set of buried electrodes or power service ground connection designed to achieve good ground potential and continuity in the local environment, connected to a main ground busbar and then to bonding conductors to each telecommunications closet and individual devices and ground terminations in each closet.
HC	Horizontal Cross-Connect.
Horizontal Wiring	The wiring between the local telecommunications closet and each office or workstation area.
Hub	An active device at a central point which is managing control of the network.
HVAC	Heating, Ventilation and Air Conditioning.
IC	Intermediate Cross-Connect.
Icon	A symbol used on a drawing or screen to represent a specific type of device or circuit component.
Implementation	The process of selecting, installing and setting up a system ready for operation.
Information Outlet	The wall plate or connection point at the local office or workstation where the horizontal cable is terminated, often referred to as a jack or drop.
Interbuilding	Cables or pathways running between two or more buildings.
Interconnections	The connections between two items, such as cable pairs, block positions, ports, etc.
Intrabuilding	Cables or pathways running within the same building.
IP	Internet Protocol.
IPX	Internet Protocol Exchange.
IT	Information Technologies.
LAN	Local Area Network intended for local transport of data, video or voice.
LAT	Local Access Transport protocol.

Look-Up Table	A support table where commonly used information is stored for the convenience of look-up and control of validation.
LU	Logical unit number in an IBM SNA Network.
MAC	Moves, Adds, and Changes.
MAU	Multistation Access Unit (Token Ring) or Media Access Unit (Ethernet).
MC	Main Cross-Connect.
MDF	Main Distribution Frame.
MER	Main Equipment Room.
MIS	Management Information Systems.
MMJ	Modified Modular Jack, a 6-position modular connector with an offset locking tab used for certain data connections.
Modem	A device used to convert a signal from digital to analog format for transmission over extended distances, usually on the public telephone network.
Network Address	An ID assigned to or derived from a device by the local network.
Nomenclature	Naming conventions for cables, pathways and devices.
Pathways	Enclosures of various types in which cable is pulled or otherwise installed, such as conduits, cable trays, etc.
PBX	Private Branch Exchange or local voice switch.
Percent Fill	The percentage of usable area that is already used in a pathway, such as a duct or cable tray.
Percent Loading	The weight of the included cable as a percentage of the total weight allowed in a cable tray.
PVC	The abbreviation for polyvinyl chloride, a material used in manufacturing a type of cable jacketing material.
Required Field	A field where the software requires that an entry be filled in before the record can be saved.
RFP	Request For Proposal.
RFQ	Request For Quotation.

RJ11	Registered jack or plug wiring configuration for a particular type of service based on a six-position modular jack or plug, recognized by the Code of Federal Regulations, Title 47, Part 68. Commonly used to refer to the 6-position modular connector.
RJ45	Registered jack or plug wiring configuration for a particular type of service based on an eight-position modular jack or plug, recognized by the Code of Federal Regulations, Title 47, Part 68. Commonly used to refer to the 8-position modular connector.
Schematic	A block-type diagram often used to represent the components and connections contained in a circuit.
ScTP	Screened Twisted Pair Cable.
SMDR	Station Message Detail Report generated from a PBX.
SNMP	Simple Network Management Protocol language.
Space	A point or location where access and administration of cables and pathways is required.
Splice	A point where one or more cables are joined together by fusing or mechanical connectors.
SQL	Structured Query Language software.
STP	Shielded Twisted Pair Cable.
TA	Terminal Adapter used on 3270-type controllers.
TC	Telecommunications Closet - a space in a building set aside to provide safe, secure and environmentally suitable area for installation and administration of cables and telecommunications equipment.
TCP/IP	Transmission Control Protocol/Internet Protocol.
Termination Hardware	Punchdown blocks or patch panels where cables and ports are terminated to which individual circuit connections are made.
Termination Position	The name of the pair or conductor on a termination device where individual ports or cable pairs are connected (terminated).
Terminations	The end point where cable pairs or ports are punched down for individual circuit connections.
TGB	Telecommunications Grounding Busbar located in each Equipment Room and Telecommunications Closet.
TIA	Telecommunications Industry Association

TMA	Terminal Multiplexer Adapter used on 3270-type controllers.
TMGB	Telecommunications Main Grounding Busbar, usually located near the Main Cross-Connect.
Topology	Specific wiring configuration for connecting individual user workstations to a network.
Transport System	The cables and equipment over which voice and data communications are carried.
USS Code	Uniform Symbology Specification Code developed and maintained by Automated Identification Manufacturers, Inc.
UTP	Unshielded Twisted Pair Cable.
Validation Table	Database tables for specified fields which prevent the entry of information into that field which does not match a pre-defined set of acceptable entries.
Wildcard	The use of a designated character (similar to the asterisk (*) in DOS) used in the criteria string during a query to represent any character or character string within a given criterion field.

INDEX

INDEX

A

Access, levels, 127-130, 211, 235, 291
Access, options, 27-28, 127-130
Administration, 81-86
  Workflows, 83-84, 149-156
  Maintenance, 85-86
  Plan, 151
  Tasks, 86
Advantages, comparison, 277-278
Answering machines
  Defining, 172-173, 175-176
  Labeling, 180
  Naming, 179-180
Archiving data, 212, 257-258
Asset management, 101-104
Asset records, existing, 131-134
Auditing
  Data Entry, 375-376
  Connections, 347-348, 375-376
  Ongoing, 10, 376
Authorizations, obtaining, 157-168

B

Backbone cable, 13, 28, 31, 112-113, 118-121, 247, 292, 296-297, 302-303, 362-363
  Defining, 188-191
  Labeling, 193, 362-363
  Naming, 190-193
Backbone connections, 292, 304-308
Backup, system, 212, 234, 257-258
Barriers, to analysis, 57-58
Benefits, 53-62, 158-162
  CAD, 135-136,
  Indirect, 161
Billing interface, 90, 212, 259
BICSI, 13
Bonding conductor, 16, 40, 46-48, 50-52
Bus network, 12

C

Cable management, definition, 1-2
Cable plant, 11-20. 118-121, 188-197, 296-297, 362-363
  Architecture, 11-20
  Catalog, defining, 188-197
  Dataset tables, 118-121
  Evaluation, 112-114
  Labeling, 191, 195, 362-363
  Naming, 191-197
  RFP, 241-242, 247-251
  Types, 112-115, 118-120, 124-126, 296-297
  Records, 45-46
CAD (Computer Aided Design), 135-140, 299-300, 319-322
  Recordkeeping, 22
  Functionality, 74-76
  Working with, 135-140
  Benefits, 135-136
  Drawings, working with, 293-294, 299-301, 319-322
  Maintaining, 166-167, 139-140
  Reports, 143, 147-148
  RFP, 212, 251-253
  Testing, 291, 293-294, 301, 303-304
CAFM (Computer Aided Facilities Management), 91-92
Call accounting, 88-90
Canadian Standards Association, T527 & T530 13, 20, 50
Case sensitivity, 291, 326

Catalog, defining, 169-208
  Control equipment, 180-188
    Bridges, 186-187
    Cards, 185-186
    Controllers, 180-183
    Hubs, 180-186
    Routers, 186-187
    Servers, 186-187
  Desktop equipment, 172-180
    Answering machines, 172, 175-180
    Facsimile machines, 172, 175-179
    Modems, 172-173, 177-180
    PCs, 172-173, 176-180
    Phones, 172-174, 179-180
    Printers, 172-173, 176, 179-180
    Terminals, 172-175, 179-180
  Diagrams, 175, 177-178, 181-186, 193-194 196, 201, 203-205, 327-330
  Pathways, 206-208
  Setting Up, 169-208, 295-300
  Support tables, 169-170, 291-296, 313-314
  Termination hardware, 195-206
    Blocks, 195-203
    Outlets (jacks), 192-195
    Patch panels, 203-206
  Testing, 295-314
Circuit
  Components, 4, 119, 212, 241-248, 302-308
  Creating, 308-312
  Functionality, 212, 248-251
  Tracing, 311-312
Coaxial cable, 16-17, 30, 118
Composite cable, 16-17
Connections, recordkeeping, 30-32
Connectivity management, definition, 1-2
Controlling equipment, 29, 123-125, 243, 247 291-292, 297-298
  Defining, 180-188
  Labeling, 188, 362-364
  Naming, 187-188
Costs, 53-68, 165-168
  Comparisons, systems, 264, 272-273, 283-284
  Data entry, 65, 68, 164-165
  Hardware, 62-63, 68, 165, 168
  Inventory, 64-66, 68, 163-166
  Software, 63-64, 68, 165, 167-168
  Project management, 165-166, 168
Creating circuits, 308-312
Custom queries, 323-324
Customization, software, 231, 279-282

D

Daisy-chain wiring, 12-13, 35, 56
Data Entry
  Costs, 65, 164-165
  Requirements, 212, 256-257, 340
  Training, 373-376
Dataset tables, 117-126
Desktop equipment, 29, 123-124, 247, 291-292, 297-299
  Defining, 172-179
  Labeling, 180, 364
  Naming, 179-180
Disadvantages, comparisons, 277-278
Documentation, 230, 325-330
Documents
  Administration Plan, 151
  Authorization Request, 167
  License Agreement, 231-232, 287-288
  Project Costs, 62-68, 165-168,
  Project Scope, 116
  Purchase Order, 285-287
Drawings
  Modifying, 322
  Printing, 320-322
  Testing, 299-300
  Working with, 293-294, 299-300, 319-322
Drop (outlet), 26-27, 111, 122, 193-200, 302, 354, 363

E

Employee directory, 131-132
Equipment room, 15
Error messages, 375
Escrow, source code, 286-287
Evaluation matrix, 265-273
Evaluation sheet, 274-276
Evaluations
  CMS demonstrations, 262-263, 274-276
  CMS systems, 261-278
  RFP responses, 261-273
Existing records
  Use of, 131-134
  Working with, 367-370

F

Facilities management, 91-92
Facsimile machines
  Defining, 175
  Labeling, 180
  Naming, 179-180
Fiber optic cable, 16-17, 118
File locking, 235, 291, 294
Formulas, 59-62
  Cable inventory avoidance, 61-62, 161
  MAC savings, 59-60, 159
  Trouble calls, 60-61, 159-160
  Unused circuits, 61, 160-161
Functionality, 69-80
  CAD, 74-76
  Relational databases, 70-74
  RFP, 248-251

G

Global changes, 324
Graphics, see CAD
Gray cable, 12 (also see backbone cable)
Grid location, 75
Grounding busbar, 16, 39-40, 45-47, 50-52
Grounding system, 16, 39-40, 45-47, 50-52

H

Hardware requirements, 62-63, 68, 165-168, 211
Help function, 235
Horizontal cable, 13, 28, 113-114, 118-121, 247, 292, 296-297, 302, 362-363
  Defining, 191-199
  Labeling, 195, 362-364
  Naming, 194-195
Horizontal cross-connect, 15
Hybrid cable, 16-17, 31
Hypothetical site, 17-18

I

Icons, creating, 300, 303-304
  Moving, 320-322
Implementation, phases, 5-7
Index list, 70-71
Information outlet, see outlets
Installation, software, 289-294
Interfaces, other systems, 87-92, 212, 258-259
  Billing, 90, 212, 259
  Call accounting, 88-89, 212
  Facilities management, 91-92
  Network management, 90-91, 214, 258
  PBX, 87-88, 212
Intermediate cross-connect, 15
Inventory, 64-68
  Costs, 165-168
  Forms, 341-360
  Procedures, 341-360, 371-374
  Staffing, 64-66, 166, 338-339
  Timing, 338-340

J

Jacks (outlets), 26-27, 111, 122, 302, 352-353, 359-360, 364-365
  Defining, 192, 196-199
  Labeling, 199, 364-365
  Naming, 198-199
  Recordkeeping, 26-27, 111, 122
Jargon, 12
Junctions, see splice

K

Key, help, 235
Key contacts, 325-326

L

Labeling, basics of, 21, 361-366
  Desktop equipment, 179-180, 364
  Label types, 37-38, 364-366
  Methods, 37-38, 361-364
  Office outlets, 202, 364
  TIA/EIA, 48-52, 363
  Wiring Closets, 362-364
Levels, security, 127-129, 291
Licensing agreements, 231-232, 287-288
Linkages, 41
Look-up tables, 255-256, 291-293

M

MAC, savings, 59-60, 159
Main cross-connect, 15
Maintenance fees, 167, 230, 287-288
Matrix, RFP evaluation, 261-273
Milestones, 107-108
Modems
Defining, 172-173, 177-179
Labeling, 180
Naming, 179-180

N

Naming, basics of, 21, 328-330
Controlling equipment, 187-188
Conventions, 32-36
Desktop equipment 179-180
Diagrams, 33, 330
Outlets (jacks), 198-199, 364,
Patch panels, 205-206
Pathways, 207-208
Termination hardware, 202-203, 206
Network management, 90-91
Nomenclature, 32-36, 328-330

O

Offices, 348-353, 359-360, 364
Inventory, 348-353
Labeling, 364
Operating system, 129-130, 234
Optional information, 118-124
Organizing data, 367-368
OSI model, 1
Outlets, 26-27, 111, 122, 246, 302, 352-353, 359-360, 364-365
Defining, 192-196-199
Labeling, 199, 352-353, 364-365
Naming, 198-199, 364
Recordkeeping, 26-27, 111, 122

P

Pan feature, 74, 140
Passwords, 127-129, 235, 291
Patch panels, 203-206
Pathways, 13-14, 29, 37, 40-42, 48-50, 113-114, 121, 206-208, 242, 299, 301 303-304
Dataset table, 121
Defining, 206-208
Labeling, 208
Naming, 207-208
Planning chart, 113-114
Recordkeeping, 29
TIA/EIA, 40-42
PBX management, 87-88, 102
PCs
Defining, 176-179
Labeling, 180
Naming, 179-180
Phones
Defining, 172-174
Labeling, 180
Naming, 179-180
Physical star wiring, 12
Planning, 335-340
Charts, 110-116
Timeline, 162-167
Platform options, 127, 129-130, 234
Premises cabling, 16-17
Printers
Defining, 172-173, 176
Labeling, 180
Naming, 179-180
Procedures
Administrative, 81-86
Data Entry, 373-376
Inventory, 341-360, 371-374
Service Order, 93-96, 329-330
Trouble Ticket, 97-100, 329-330
Project team, 105-108
Workflows (roles/responsibilities), 106-107
Project scope, 109-116
Protocols, 129-130, 234
Purchase order, preparation, 285-287
Purpose statement, 217

Q

Quantity, recordkeeping for, 28-30
Queries, 75, 255-256
Custom, 323-324
SQL tools, 323-324
Quotes, see Request For Proposal (RFP)

R

Recordkeeping, 21-32
  Aspects, 22-26
  Availability, for, 30
  Connections, for, 30-32
  Methods, 21
  Tools, 21-22
  Uses, 26-27
Records structure, 117-126
Relational database, 70-74, 129-130, 234
Reports,
  Defining requirements, 141-143, 212
  RFP, 253-254
  Samples, 145-148
  Service Order, 146, 238-239
  Testing, 318
  Trouble Ticket, 146, 240
Request For Proposal (RFP)
  Benefits, 8-9
  Evaluating, 261-278
  Preparing, 209-212
  Sample, 213-260
Required information, 118-120
Ring topology, 12
Roles & responsibilities, 106-107

S

Savings
  Inventory avoidance, 61-62, 161
  MACs, 59-60, 159
  Troubles, 60-61, 159-160
  Unused circuits, 61, 160-161
Scheduling, 335-340
Scope, project, 109-116
Screened twisted pair, 16-17
Security, levels, 127-129, 235, 291
Service orders, 93-96, 114
  Functionality, 93
  RFP, 211, 238-239, 245
  Testing, 315-317
  Using, 95-96, 329-330
Shielded twisted pair, 16-17, 30
Software customization, 231, 279-282
Source code, in escrow, 286-287
Space, 15, 39-40, 42, 47-48, 51-52
Splice, 37, 40, 42, 44-45, 51-52
Staffing, inventory, 64-67, 166, 334-338
Star topology, 12-13
Station cable, see Horizontal cable
Structured Query Language (SQL), 323-324
  Changes, 324
  Queries, 323-324
  Reports, 323-324
  Working with, 323
Structures, CMS, 77-80
Support tables, 169-173, 291-293, 295-298, 313-314
System selection, 283-284

T

Tables, 169-173, 212, 255-256, 291-292, 295-299, 301, 314
  Look-up, 255-256, 291-293
  Reference, 325-329
  Support, 169-173, 291-293, 295-298, 313-314
  Validation, 212, 255-256, 291-293, 295-299, 301
Team, project, 105-108
  Members, 105-106
  Responsibilities, 106-107
Terminals
  Defining, 172-175
  Labeling, 180
  Naming, 179-180
Termination hardware, 29, 38, 121-122, 195-206, 298, 362-363
  Blocks, 199-202, 206, 247
  Dataset Table, 121-122
  Defining, 195-206
  Labeling, 38, 202-203, 206, 362-363
  Naming, 198, 202-203, 205
  Outlets (jacks), 192-199
  Patch panels, 203-206
  Recordkeeping, 29
  TIA/EIA, 43
Termination position, 44
Testing, 289-294, 313-318
  Basic software, 289-294
  Catalog, 301-314
  Connection rules, 313-314
  File locking, 291, 294
  Multi-user, 291-294
  Reports, 318
  Service Order, 315-317
  Trouble Ticket, 317-318

TIA/EIA Standards, 3, 11,13-15, 39-52
568A, 11, 15, 39
569, 11, 14, 39
606, 3, 11, 13, 22-23, 39-52, 117-137
607, 11, 13-14, 39
Timeline, planning, 162-167
Topologies, wiring, 11-13
Tracing circuits, 311-312
Training, 331-334, 338-340, 371-376
Data Entry, 373-376
Inventory, 338-340, 371-372
System Users, 230, 331-334
CMS Administrator, 332-333
Customer Service, 333
Data Technician, 334
Help Desk, 333-334
Project Manager, 332-333
Voice Technician, 333-334
Transferring data, 368-369
Trouble tickets, 97-100, 329-330
RFP, 212, 240-241, 245
Savings, 60-61, 159-160
Testing, 317-318
Using, 329-330

U

Unshielded twisted pair, 16-17, 30

V

Validation Tables, 212, 255-256, 291, 294-298
Values (benefits), 53-62, 158-162
Vendors
Demonstrations, 262-263, 274-276
Evaluating, 262-264
References, 230, 263-264
Verifying data, 369-370

W

Warning messages, 249, 251, 291, 293, 375
Wiring closets, 336-358, 362-364
Inventory, 336-358
Labeling, 362-364
Workflows, 83-84, 149-156, 330
Administration, 83-84
Applying new, 151-152
Current, 149-150
Defining new, 149-152
Samples, 153-156
Workstations, see offices

Z

Zoom feature, 75, 140